THE STRUGGLE FOR POWER
AND INFLUENCE
IN CITIES AND STATES

THE STRUGGLE FOR POWER AND INFLUENCE IN CITIES AND STATES

DICK SIMPSON
University of Illinois at Chicago

JAMES D. NOWLAN
University of Illinois at Urbana-Champaign

BETTY O'SHAUGHNESSY
Loyola Academy, Wilmette

Longman

Boston Columbus Indianapolis New York San Francisco Upper Saddle River
Amsterdam Cape Town Dubai London Madrid Milan Munich Paris Montreal Toronto
Delhi Mexico City Sao Paulo Sydney Hong Kong Seoul Singapore Taipei Tokyo

Editor-in-Chief: Eric Stano
Supplements Editor: Donna Garnier
Marketing Manager: Lindsey Prudhomme
Production Manager: Eric Jorgensen
Project Coordination, Text Design, and Electronic Page Makeup: Electronic Publishing Services Inc., NYC
Cover Design Manager: John Callahan
Cover Designer: Laura Shaw
Cover Illustration/Photo: Chris Carroll/Corbis
Photo Researcher: Linda Sykes
Senior Manufacturing Buyer: Dennis J. Para
Printer and Binder: Courier Kendallville
Cover Printer: Lehigh-Phoenix Color/Hagerstown

For permission to use copyrighted material, grateful acknowledgment is made to the copyright holders on pp. 343–345, which are hereby made part of this copyright page.

Library of Congress Cataloging-in-Publication Data
Simpson, Dick W.
   The struggle for power and influence in cities and states / Dick Simpson, James D. Nowlan, Betty O'Shaughnessy.
      p.   cm.
   ISBN 978-0-321-10518-9
   1. Local government—United States.   2. Power (Social sciences)—United States.   I. Nowlan, James Dunlap, 1941-   II. O'Shaughnessy, Betty.   III. Title.
   JS331.S497   2010
   320.801—dc22   2009027479

1 2 3 4 5 6 7 8 9 10—CRK—13 12 11 10

Longman
is an imprint of

www.pearsonhighered.com

ISBN-13: 978-0-321-10518-9
ISBN-10:   0-321-10518-4

FOR

Sarajane Avidon (1941–2006)
—Dick Simpson

Samuel K. Gove, mentor and friend
—James D. Nowlan

My mother, Ruth Quinn Bremer
—Betty O'Shaughnessy

CONTENTS

3 STRUGGLES IN STATE LEGISLATURES AND CITY COUNCILS   34

4   EXECUTIVE POWER IN STATE AND LOCAL GOVERNMENT   73

8 CITIZEN PARTICIPATION IN STATE AND LOCAL GOVERNMENT 186

10   METROPOLITANIZATION AND GLOBALIZATION 247

PREFACE

The *Struggle for Power and Influence in Cities and States* has been written primarily for use in college courses on state and local government. But it is also useful for the general reader who seeks to understand state and local politics and for those who lead these governments closest to the people.

There are a number of unique features to this book. First of all, while we discuss institutional structures and public policies, our primary focus is on the struggle for power within those institutions which result in particular policies. Second, we seek to provide you not only with knowledge that can help you be a better citizen, but also with understanding and insights that can help you become a political leader. Third, we demystify how power and influence are obtained in a democratic society, and we examine the ethical limits on their use.

In Part I of our book, we discuss the importance of state and local government. Part II looks at the institutions of government—the constitutions and charters that provide the foundation, the three branches of state and local government, and the federal system that divides power among national, state, and local governments. You cannot be a successful political actor if you do not understand the rules of the game and the structures of government. We also look at our various court and criminal justice systems. In Part III, we turn to politics in the election process and to different types of political participation by citizens, political parties, interest groups, lobbyists, civic groups, social movements, and community organizations. Part IV focuses on the policies and problems of local government with chapters that examine budgets and taxes, metropolitan governance, and education. Part V provides reflections on democracy and ethics in politics and government.

Each of the chapters wraps up with a summary of the material covered, a list of key terms highlighted in the text, review questions to help you revisit what you have read and to test what you have learned, discussion questions so that in your class and with your friends you can deepen your understanding of the chapter material, and a set of practical exercises that guide you in discovering those who currently have power and influence in your own community and state and how your local and state governments actually work. We also provide a list of useful Web sites for further research on topics that particularly intrigue you, as well as a suggested reading list for more in-depth coverage of some topics. Throughout the text, you will find real-life stories and examples in special features that illustrate the themes discussed in the chapters.

All of this is meant to introduce you to the struggle for power and influence that is the lifeblood of politics. We find the study and practice of politics exciting and important. Our goal is to provide you with the tools and knowledge necessary to hold your state and community leaders accountable. Hopefully, one day you will join in the democratic process as knowledgeable citizen participants and effective civic leaders.

ACKNOWLEDGMENTS

Grateful acknowledgement goes to the reviewers who provided valuable feedback through-out the writing of this text.

Craig Emmert, *University of Texas of the Permian Basin*
Richard Fording, *University of Kentucky*
Joe Gaziano, *Lewis University*
Russell Harrison, *Rutgers University*
Ngozi Kamalu, *Fayetteville State University*
Kenneth Palmer, *University of Maine*
Christopher Reaves, *University of Alabama at Birmingham*
Carmine Scavo, *East Carolina University*
Maurice Sheppard, *Madison Area Technical College*
Brent Steel, *Oregon State University*
Nicholas Swartz, *Campbell University*
Jack Treadway, *Kutztown University*
Nelson Wikstrom, *Virginia Commonwealth University*

DICK SIMPSON
JAMES D. NOWLAN
BETTY O'SHAUGHNESSY

SUPPLEMENTS

Pearson is pleased to offer several resources to adopters of *The Struggle for Power and Influence in Cities and States* and their students that will make teaching and learning from this book even more effective.

Instructor's Manual/Test Bank

Written by text author Betty O'Shaughnessy of Loyola Academy, Wilmette, this resource includes Learning Objectives, Chapter Summaries, Key Terms, Lecture Outlines, and a Practice Test including Multiple Choice, Identification, Short Answer, and Essay Questions. Questions address all levels of Bloom's taxonomy and have been reviewed for accuracy and effectiveness. To download, please visit the Instructor Resource Center at www.pearsonhighered.com/irc.

Pearson MyTest

This flexible, online test generating software includes all Practice Test Questions found in the Test Bank. Questions and tests can be easily created, customized, and saved online and printed, allowing instructors ultimate flexibility to manage assessment. To start using, please visit www.pearsonmytest.com.

State & Local Government and Politics Study Site

This open-access, online resource includes practice tests, class activities, and Web links organized by major topics and arranged according to the table of contents of *The Struggle for Power and Influence in Cities and States*. To access, please visit www.pearsonhighered .com/stateandlocalgov.

Georgia Speaker of the House gets standing ovation.

STATE AND LOCAL POLITICS

State and local politics is animated by the struggle for power to control state and local governments. These governments pass laws, spend tax money, and undertake projects that directly affect citizens, businesses, and the communities in which we live. Because any society is made up of many competing interests, there is a struggle over who should exercise the power of government and influence policy decisions.

Our national government has the ultimate powers of war and peace, directs foreign affairs and trade, and levies the largest component of our taxes. The Supreme Court of the

United States provides the ultimate arbitration of our laws. The national government redistributes wealth through our welfare system and with federal grants. Most domestic tasks, however, are carried out by thousands of state and local governments that pass the laws that affect our daily lives, including traffic rules, business regulations, and punishments for crimes like murder and theft. Moreover, state and local governments provide most of the basic government-to-people services related to education, health care, child protection, police, fire fighters, garbage collection, and water and sewer systems.

Many of you reading this textbook spend hours each day in classrooms—perhaps in public universities or community colleges—built and paid for by state or local governments. Your tuition is set by a board of elected or appointed trustees. Even if you attend a private college, state and local governments provide the legal framework and often the financial aid that make college education in the United States possible.

Besides passing laws and regulations, state and local governments determine numerous taxes and fees, and they decide how that money will be spent. With some-times billions of dollars at stake, there is naturally a struggle for the power to determine what laws will be passed, what the tax policy will be, and which projects will be approved. We hope these matters will be decided by reasonable men and women based on consensus. But even individuals of reason and good will do not always agree on what should be done. Should marijuana use be lawful? Should certain books be banned in high school courses or from public libraries? Should the wealthy pay more than the poor for government services? These questions are matters of fierce debate decided in your hometown and state capital.

In addition to differences in ideology and policy views, there is a very human element to the struggle for power and influence in government. Politics offers symbolic and psychological rewards. For instance, it can be a matter of personal ambition and pride—as well as an honor for one's family, friends, ethnic group, or race—to become a public official with the title of governor, senator, or mayor and being addressed as the Honorable Mr. or Ms. Gonzales. The power to make things happen is an important motivation for engaging in the political struggle, and so is being given respect and honor. Having thousands of people applaud what you say, appearing on the nightly news broadcast, seeing your photograph and words on the front page of your hometown newspaper, being on a first-name basis with presidents and the heads of international corporations—this is heady stuff. These ambitions rule the struggle for power at least as much as policy preferences.

THE IMPORTANCE OF STATE AND LOCAL POLITICS

Politics is a serious game that can be great fun—exhilarating and exasperating for its participants. But the stakes are often high and the impact on people's lives can be dramatic. Local governments have the primary responsibility in society for protecting lives. On the other hand, some state governments take lives: the State of Texas, for example, has in recent years executed more than 100 of its citizens who have been convicted of murder.

State governments are primarily responsible for caring for the mentally ill. State and local government hospitals and health clinics are the primary providers of health care for many children, the poor, and the elderly. Local governments impose most of the rules on citizens, telling them whether they can build a house and even how they must build it. And, of course, states and local school districts are responsible for the education of the nation's children as well as providing services for adults who need remedial education.

States and urban regions are often big. Really big. If it were a nation, the state of California would have the world's seventh largest economy. The metropolitan regions of New York, Los Angeles, Chicago, Miami, and Houston are bigger in population and economic activity than many of the nations of the world.

LEARNING ABOUT LOCAL POLITICS AND GOVERNMENT

As experienced politicians and public officials as well as college professors and high school educators, we think a good way to teach state and local government is to follow the struggle for power, just as a detective trying to work out a complicated crime will often "follow the money" to find the guilty parties and unravel the story behind the events.

In following the struggle for power, we often look especially to the legislative and executive branches of government—the governor and state legislature, the mayor and the city council, the county board president and the county board, the school superintendent and the school board. We do not neglect the judiciary that interprets the laws nor the bureaucracy that administers the laws, but we have discovered that the greatest power struggles lie in the legislative and executive branches. We look to the relationships between these branches, such as strong mayor/weak council city governments; to the elections which determine who holds office; to the staff members who advise elected officials; and to the interest groups, lobbyists, and citizens who influence their decisions. In this text, we turn a bright spotlight on these aspects of real politics as we help you understand your own state and local governments.

Government is neither a bloodless abstraction nor just a state constitution or city charter. Nor is it just the more dramatic events reported on the nightly news. Government is about real people engaged in real battles that create tangible outcomes for us all. While outside forces—war and peace, economic prosperity and recession, national political tides and decisions—set the parameters of local politics, local political actors have a direct effect.

Many textbooks treat state and local government as if state politics is primary and governing towns, cities, and counties is secondary. They imply by their emphasis that the state is the big dog and local government is just its tail. A key state court decision of the 19th century, called "Dillon's Rule," declared: "local governments are the creatures of the state." That is, the state created local governments by statute and they exist only under the conditions and terms set by these state laws. For instance, in earlier centuries, local governments could not pass new taxes without permission in enabling legislation from the state. Unlike local governments, states are recognized as sovereign by the U.S. Constitution.

Yet while counties, cities, and towns may have historically been creatures of the state, most municipalities now have "home rule" powers and considerable freedom from state interference. Some metropolitan regions in the United States have a larger population and greater wealth than over half of the independent nations of the globe, so local governments are not completely dominated by the states. In fact, 30 percent of Americans live in central cities, 50 percent live in suburbs, and 20 percent live in smaller towns and rural areas. Because of their importance, town, city, county, and special districts receive equal treatment with state government and politics in this book.

State and local governance is also in the midst of a profound revolution. Today, most Americans don't just live in towns and cities, they live in metropolitan regions. These geographical locations do not have a single general government ruling them but a plethora of governments. For instance, in the Chicago metropolitan region, there are more than 1,200 local governments with the power to tax, and the Chicago urban sprawl spans three states: Illinois, Indiana, and Wisconsin. The Los Angeles urban region reaches all the way from the city of Los Angeles down the coast of California to cities in Mexico.[1] The New York urban region includes the suburbs of New Jersey and the city of Newark, and some scholars claim it stretches north all the way to Boston. The sizable geographic area, large and diverse populations, and multiple governments make the task of providing government services to these urban regions difficult. Coordinated planning and control of housing and economic development are virtually impossible. Thus, metropolitan regions create profound challenges for local and state governments.

At the same time, cities and states now exist in a global economy. They have sister cities all over the globe. American states and major cities compete and trade with countries such as China and India, which they may have ignored 10 years ago. They aspire to defined roles in the global economy, which they hope will bring them vast wealth. However, the global economy also brings with it economic instability and a widening gap between rich and poor citizens. Being part of the global society may also increase our fears of terrorism, and the efforts to combat terrorism may define the first half of the 21st century much as the "cold war" against communism defined the last half of the 20th century. We are not just dealing with routine politics and government as usual but with a challenging politics and government for which new techniques must be developed to compete successfully in the global struggle for power.

The study of "global cities" assumes that while the pressures of globalization and modern technologies may be much the same on all cities, each city responds to these forces by creating its own niche in the system. Each city's response depends on the history, special economic advantages, and decisions of local elites and government officials. These global forces and local responses are important if we are to understand cities and states in the 21st century.

THE SCOPE OF THIS BOOK

Case studies, discussion questions, and exercises in this book let you complete your own account of the struggle for power. We guide you step by step in learning about the social, economic, political, and governmental conditions in your own community and state. We will also teach you how you can join with others to affect the future of your town, city, and state. Having an impact requires that you first understand who has power, how they got it, and how they use it. Second, it requires that you formulate a clear political philosophy or set of political goals. Finally, it requires that you take political action. We will prepare you to do all three.

The Struggle for Power in Cities and States provides a clear look at state and local politics from the perspective of those inside the process. The authors have observed politics and government firsthand and also draw on the testimony of others in local and state politics and government. In the following chapters, we will explore four specific themes:

1. The role of *power and influence* in state and local government and politics.
2. The tension between *representative and participatory* politics.

3. The ways in which the *structures of government and the political processes* interact.

4. Whether the *outcomes* produced by the governmental institutions and processes are good or bad.

Power and Influence

Politics is about the *struggle for power and influence* in the real world of state and local government. It is a fight over who gets what, why, and how. We want to know *who* has power and influence; *what* rewards they, their allies, or supporters receive from the use of that power and influence; *why* they seek power, which requires time, effort, and money to achieve; and *how* they obtain their particular type of power and influence.

We introduce the essentials of state and local politics by projecting you into positions of power and influence. You can visualize yourself acting in positions of power to keep what is valuable in state and local government and to change what is wrong. Your goals can be achieved, rather than those of someone else—at least in your imagination. We want you to visualize what it is like to be a school board member, county board president, state legislator, judge, or high-level government employee in an agency, like one that issues drivers licenses or zoning permits. We want to convey to you the excitement of participating in an election campaign or of getting a bill passed that changes the future of your town or state.

Because we want you to think about politics and government analytically and effectively, we have to make certain distinctions. For instance, what is this power we write about? We think of power as the capacity to induce someone to do something he or she would not have done otherwise. Power can also be the ability to get people to join together to accomplish some important task, but often political goals have to be accomplished over the opposition of others. In the 1970s, Chicago's Mayor Richard J. Daley had the power to force aldermen, state legislators, and members of Congress to cast votes with which they personally disagreed, sometimes strongly. If they didn't vote as he wanted, as chair of the Democratic Party of Cook County, Richard J. Daley could deny them renomination and take their family members off the city payroll.

Consider also the power of George E. Pataki, the 53rd governor of New York State. First elected in 1994, he was reelected in 1998 by the largest landslide vote for a Republican governor in New York history. He won reelection to a third term in 2002 by a margin of over 725,000 votes, becoming one of the longest serving governors in the United States. Given his political popularity, he had the power to veto any bill, killing it instantly, because the state legislature usually did not have the votes to overcome his veto.

There are many other examples of power. Organized crime has exercised power among certain lawmakers by the threat of terror and blackmail on the one hand and monetary bribes on the other. Students exercised power in the 1960s by taking over the offices of university presidents, forcing them to hear demands and make changes in university policies.

In contrast to the power to *force* others to do as you want, influence is the capacity to *persuade* another person to do something he or she had not been planning to do. Intelligent, highly respected state lawmakers or city council members, for instance, can persuade their colleagues to change their votes by the persuasiveness of their arguments. Governors exercise influence when they suggest that a project desirable to a legislator will be approved before the next election, which will greatly help their reelection efforts—that is, *if* the legislator will support the governor's proposed increase in the sales tax. Lobbyists influence mayors and city council members to approve their client's zoning proposal or city contract bid by providing campaign contributions for their next election. Graduate students exercise influence and power when, in their effort to gain recognition and benefits from a university, they affiliate with a powerful union and withhold their teaching services in a strike. Undergraduates exercise power and influence when they organize a boycott of the school cafeteria until it provides better food and cheaper prices. To the degree students use arguments to persuade the university administration to grant their wishes and demands, they are exercising influence. When they organize an actual boycott, they are exercising power, not just persuasion.

Power and influence are not static qualities. They ebb and flow depending on how effectively they are used. Every player in politics and government has some latent power and influence. It can be their vote, information, organizing skill, contacts in the media, money, ideas, public speaking ability. It can be formal powers like a governor's veto or a city council leader's authority to assign proposed ordinances to committees where they will be buried or advanced.

In a game of checkers, each player's pieces have the same potential to move, so winning requires skill. The struggle for political power and influence is similar. To succeed requires skillful use of your particular potential power and influence, because other players also have potential power and influence. The most effective politicians have honed their skills in the use of their formal and informal powers. They expend their power and influence judiciously. Politics is a game of debts and credits. The object is to build as many credits and incur as few debts as possible while pursuing policy changes and sometimes material objectives. Effective political leaders spend their "political capital" to get projects built, for example, or to get laws passed.

Like money, power and influence can be used for good or evil. Throughout this book, we illustrate with real-world examples how players, both important politicians like governors and mayors and seemingly weak players like students, have gathered and applied power and influence in their quest for what they view as a better society. In this text, we may not always use the terms power and influence separately. Sometimes we may use the terms interchangeably, because power and influence are alike in many ways, but you need to keep in mind the difference. Usually, power is much more difficult to obtain. We have the ability to *influence* a decision more often than we have the *power* to control an outcome.

When Hillary Rodham Clinton was first lady of the State of Arkansas during her husband's terms as governor, the position was without any formal power. However, as a strong-willed, intelligent person, she transformed her role into that of a major advisor, becoming highly influential in the areas of mental health policy, children's issues, and education—just as she did later as first lady of the United States. Because of her abilities, fame, and political resources, she was then elected to a position of power in her own right as U.S. senator from New York. In 2008, she ran for president of the United States so that she could gain presidential powers herself. Although Barack Obama was the eventual winner, Clinton was his choice for secretary of state, one of the highest cabinet positions. During the course of her career, she moved from influence to power.

To take a different example, Robert Moses was for several decades arguably the most important official in New York State, yet he was never elected to office nor did he have much personal wealth. Instead, Moses was the executive director of several independent government authorities that built expressways, housing developments, and parks. He used funding from public bonds to mold the gratitude of hundreds of contractors, architectural firms, and bond houses into the power that created or killed massive infrastructure projects. Even President Franklin Roosevelt, who came to detest Moses, was unable to unseat him from his administrative positions of power. Mere governors of the state and mayors of New York City gave up trying. As these examples illustrate, power and influence reside in different political positions and among different personalities.

Representative Versus Participatory Democracy

The concept of democracy seems at first glance simple enough, especially at the level of local and state government. Democracy is defined by its Greek origin: *demos*, or "the people," plus *kratein*, "to rule." So democracy is the rule of the people. But how are "the people" to rule in a city or state of several million people, with a budget of several billion dollars, hundreds of pages of existing technical laws, and tens of thousands of government employees?

If we as citizens were willing to devote most of our time and effort to governing, we could do so by gathering every week in an open square and debate for long hours on the laws that were to be enacted by our decisions. This could be made somewhat easier by the use of modern technology such as cable television and computers, but then we would have to provide these to every citizen, and ensure that citizens spent hours each week studying the decisions to be made. This would be participatory democracy at its most basic level. Perhaps in the 21st century we can find a way of making more opportunities for participatory democracy, but this will only be so if citizens demand it and are willing to sacrifice the time, effort, and attention needed to make complicated government decisions.

The Athenians 2,500 years ago practiced a version of this direct democracy, but there were only 10,000 citizens in their city. The many slaves and women did not have a voice in the decision-making. Even today in smaller communities, participatory democracy can be maintained, as in the New England town meetings where the citizens elect town officials, approve the town budget, and change laws. In smaller towns and suburbs, attendance at the city council meetings can be high when there are controversies, and citizens can make their will known. Public officials will then often vote as the outspoken citizens direct. There are other instruments of citizen participation, such as initiative, referendum, and recall elections which allow us to pass laws directly or to remove from office officials who displease us by their actions. There have been successful citizen efforts to limit the terms of office of public officials or to restrict campaign funding for candidates running for public office in many cities and states today. But in general, we have mostly opted for representative democracy. We have been content to elect public officials such as governors, mayors, and legislators and then leave them to make the necessary laws, budgets, and appointments.

Throughout *The Struggle for Power in Cities and States*, we explore tensions between representative and participatory (or pure and direct) democracy. We want especially to know how the mechanisms of politics and government are used to create a government that is at least in some sense *of, by, and for the people*. We also want to explore how state and local governments sometimes become autocratic or tyrannical, or how they may fail to adequately represent the interests of citizens.

Institutional Structures and Political Processes

The struggle for power takes place within carefully defined institutional structures and processes. For instance, the legislative process by which a bill becomes a law and the legal and practical role of committees within that process are discussed in the chapter on legislative bodies. The powers of the veto and budget initiation are often lodged with governors, mayors, and other chief executives, and those official powers often determine the outcome of major decisions by state and local governments. The ability of the judicial branch to declare decisions of the legislative and executive branches illegal and unconstitutional provides a check and balance which members of the other branches must consider. The specific laws governing elections determine who may run for office, who may vote, and what may or may not be done in political campaigns.

You might think of the institutional structures and legal powers of various governmental officials as providing the rules (or checkerboard) on which the struggle for power takes place. Only certain moves can be made. And to win the game—that is, to pass a law or fund a program—requires that particular moves be made. But those moves are likely to be opposed and countered by others who want to prevent a law from passing or want to use government funds for their own causes.

To use another analogy, to become a good football or basketball or soccer player, first you have to learn the rules of the game and watch how experienced players move successfully within the rules. Similarly, students who would affect the political process have to understand the institutional structures, informal norms, and successful past strategies before they can have much impact on getting candidates elected, laws passed, or funds allocated (such as more state funds for higher education so students pay lower tuition). In future chapters, while we place the greatest stress on the political process and the struggle for power, we also discuss in some detail governmental and institutional structures that define the game of politics.

Political and Governmental Outcomes

The final test of state and local governments is not how the struggle occurred, or whether the players followed the rules to the letter of the law. Instead, the test lies in whether the laws, budgets, and policies enacted are effective. Are students in the cities and states better educated? Has crime been curbed? Are the people healthy, well housed, and well fed? Are the streets paved, is the garbage picked up, and is traffic well regulated? These are the final tests for state and local government.

How well governments are functioning is not always easy to determine. Certainly if there are no riots in the streets or political rebellions at the polls "to throw the rascals out," some minimal level of citizen satisfaction has been achieved. There are also objective standards for some governmental services. We can measure the cost to pick up a ton of garbage and the percentage of waste that has been recycled. These costs and environmental benefits can then be compared with those of other towns and cities. Likewise, we can measure the reading and math scores of students on national tests. We can determine for different localities the murder rates and the percentage of crimes that have been solved within the year. We can obtain expert opinions about government services. We can also make judgments for ourselves, based on what we read in newspapers or see in TV news coverage. For instance, most of us think that government at various levels failed to react adequately to the devastation caused in New Orleans by Hurricane Katrina.

It is useful to measure citizen satisfaction with different branches and departments of government, like the police department or school system, and particular services such as street sweeping. Of course, many governmental functions are difficult to measure directly. In the end, beyond all the fine theories of politics and government, we hope to help you make an assessment of whether your local and state governments are performing well or badly. From such assessments you can begin to determine what might be done to improve those governments closest to you.

THE EFFECT OF RECENT CHANGES ON STATE AND LOCAL GOVERNMENT

In 2008, Barack Obama was elected the first African American president of the United States. He ran on a platform and mantra of change. In the same election, Democrats won enough seats to gain majority control in both houses of Congress. With the support of only a few Republican members of Congress, President Obama was able to get his economic recovery package quickly passed, and many other major changes were put in place during the first hundred days of his administration. Less noticed were the political changes that occurred at the state and local government levels. For instance, after the 2008 elections, more state legislatures came under the control of one political party—usually the Democratic Party. The Democrats also did better in electing governors and city officials. Many of the state and local government officials were elected on the coattails of Barack Obama.

Such political changes may not be permanent. However, in 2008, more young people identified themselves as Democrats than Republicans, and the youth vote went overwhelmingly for Obama, which indicates that the recent political changes are likely to continue for some time. In 2000, voters between the ages of 18 and 30 supported Democratic and Republican candidates and the two major political parties equally. By 2006, 60 percent of these young adults supported the Democrats, and in 2008, two-thirds of this age group voted for Obama. But such political changes are not written in stone—and whatever the fate of the national political parties, Republicans still have a very strong hold on many states and cities. Several factors will determine whether the 2008 election was a permanently realigning election—that is, a watershed election that affects national and local politics for an entire generation of voters, as happened with the New Deal elections in the 1930s. Among the current determining factors are (1) the outcome of the wars in Iraq and Afghanistan, (2) the economic recession in the United States, and (3) the Republican Party's potential transformation. Politically, we are at a critical point in the struggle for power in the United States.

The world is a complicated and sometimes dangerous place in which major transformations can occur quite unexpectedly. One example is the collapse of the Soviet Union in the 1990s and the end of the Cold War. Another is the terrorist attacks on American soil in 2001. Foreign affairs often affect our cities and states and do so even more now in our globalized world. How successfully the United States deals with situations in Iraq, Afghanistan, Pakistan, and the Middle East and Asia in general will have profound impacts. The results of the Obama administration's changes in foreign policy regarding Iran, Cuba, and North Korea will have long-term effects and will directly impact the fates of the Democratic and Republican parties as they struggle for political ascendancy in each city and state.

Even more clearly, widespread economic recession is affecting all of us. Our nation has been experiencing the highest home foreclosure rate since the Great Depression. The unemployment rate is the highest in over a decade, and businesses have been failing in record numbers. The administration's success with its recovery plans will either cement support for the Democrats or cause voters to turn back to the Republican Party, which held presidential power and political advantage from 2000 to 2008 and had control of both houses of Congress for even longer. While the Obama administration will be held accountable for the national response to the recession, state and local leaders of both parties will be held accountable for how well each state and city responds to the more local economic crises. It is relatively easy to govern in times of affluence, when there is plenty of money for government services and when the private sector is providing plenty of jobs and economic rewards. It is much more difficult to govern in a time of scarcity, when hard choices have to be made among different priorities.

The outcome of the current political transition will depend on the ability of the Republican Party to reposition itself to gain support among young voters. Because of the perceived failure of the policies of President George W. Bush's administration, unpopular actions like the Iraq War, the economic downturn, and the platform of the Republican candidates in 2008, which particularly failed to appeal to young people, more youth voters were drawn to the Democrats and especially became supporters of Barack Obama, who seemed younger, more hip, and more like them. The Republican Party needs to find a message and candidates that will resonate with younger voters if it is to be effective in the future, as older supporters of both parties die off and today's youth become the dominant electoral force. This has happened many times before, to Democrats as well as Republicans, as the minority party has reinvented itself. But such transformation is always a challenge for the party out of power.

The greatest change since 2008—especially at the level of cities and states—has been economic, not political. The current economic recession has meant higher unemployment, and the great losses in the stock market have affected not just the very wealthy but also the working class, middle class, and retirees. The economic setback has made paying for a college education more difficult. And it has been a disaster for state and local government finances, even with the infusion of funds by the federal stimulus and recovery programs of 2009.

States like California in 2009 faced a nearly uncontrollable deficit of $42 billion. Most states faced multibillion-dollar deficits, and many cities had a shortfall of hundreds of millions of dollars. This was at the very time of an ever greater demand for state and local government services such as medical care, job training, education, and infrastructure repair. Any "rainy day funds" that had been set aside were soon expended. State and local governments had to consider significant service cuts and greatly increased taxes to continue to provide even the most basic level of services. While the federal government recovery plans helped, they did not provide all the resources needed.

It will take a number of years for our economy to recover. It will take much longer for states and cities to gain more taxes from a better economy, which allows high collections of income, sales, and property taxes. It will be several years after the economy recovers that each city and state will have sufficient revenues to meet all the legitimate demands for public services.

In the meantime, the struggle for power and influence in cities and states will intensify because the stakes are so much higher. Some citizens will argue that states and cities must raise taxes to meet service demands. Others will argue that in a recession, taxes need to be lowered to help businesses recover and to lessen the financial pressures on individuals and business taxpayers. Some groups will argue for massive expenditures on infrastructure improvements, public works, and job creation programs. Others will argue that subsidies to businesses and hard-pressed homeowners are the most important function for state and local governments in a recession. Some will agree that the budget needs to be cut but will argue that the allocation for education, roads, public safety, health, or prisons can't be cut because that government function is crucial. While most people might agree that, especially in a time of scarcity, public corruption should be curbed, there will be arguments over how that is to be accomplished. And each government will be in a struggle with other governments to get the funding necessary for the most basic functions.

With such critical governmental decisions to be made, it is more important than ever to understand the struggle for power and influence in our cities and states. Decisions taken now will influence our lives for decades to come.

THE STRUGGLE

One of the authors of this text, Dick Simpson, recalls his days as a reformer in the Chicago City Council during the reign of its most famous boss, Mayor Richard J. Daley. There were more than 40 pro-Daley Democrats on the 50-member council and as few as three independent, reform-minded aldermen by the time the mayor died. Nonetheless, despite losing immediate votes, reformers were able to bring about change. Simpson had as his objective bringing more democracy and justice to Chicago, which was dominated by what was known as a "political machine," an entrenched power structure of the local major party. For instance, Simpson opposed the appointment of sons and other relatives of powerful machine politicians to boards of city government and the giving of city contracts only to those with political clout or large contributors to the mayor's reelection campaigns.

When Simpson opposed such nepotism with a brief speech on the floor of the council, he was interrupted by the longest tirade of Mayor Daley's long and colorful career. In the end, Simpson lost the vote on a particular political appointment by 48–2. However, the

ALDERMAN DICK SIMPSON opposing Mayor Richard J. Daley's policies.

11

power and voter support of Daley's machine was eroded because of the strong stands that the few reformers took in the city council against patronage, nepotism, segregationist policies, and waste in government.

In later years, when African American reform mayor Harold Washington was elected in Chicago, he instituted new policies of affirmative action, governmental ethics, and more efficient government, all of which Simpson and other reformers had advocated a decade earlier. Washington passed the first ethics ordinance in Chicago's history. He hired more blacks, women, Latinos, and gays. He opened up city contracts to minority firms and created the largest neighborhood redevelopment and infrastructure repair program Chicago had ever experienced. Reformers like Simpson had earlier lost votes and city council battles, but they set the stage for later reforms in Chicago. Even with the election of Richard J. Daley's son, Mayor Richard M. Daley, who governs Chicago today, these reforms have remained.

Another of the authors, Jim Nowlan, served in the Illinois House of Representatives shortly after graduating from college. He set as his objective a new law to require coal strip miners to reclaim (or restore) land once they had mined it. The big multinational coal companies were against his bill, which they easily defeated in Nowlan's first legislative session. Fortunately, the American public was becoming sensitive about the degrading of our environment, and television reporters were looking for stories.

Nowlan convinced one of the major national television networks to fly to his rural western Illinois district to see firsthand the wastelands left in the wake of strip mining. The TV crew interviewed the head of the coal association and State Legislator Nowlan from the floor of a mine, where a 16-story-high mechanized shovel growled behind them. The story was on national news that evening, and the next morning Nowlan's bill had newfound respect among his colleagues. Two years later, his strip-mining bill was enacted into law. Nowlan didn't have personal power or influence as yet in the Illinois legislature. He did, however, have behind him the power of the

JIM NOWLAN campaigning for state representative.

media, intense public opinion, and a governor who came to see the importance of the environment as an issue.

Another of your authors, Betty O'Shaughnessy, was a high school social studies teacher who believed that her students would learn about civic engagement best if they had some hands-on experience. So she required her political science students to either work in a campaign or serve as election judges, and she encouraged her U.S. history students to do the same. To no one's surprise, her students became very interested in politics, even though most of them could not yet vote. Whenever there is an election, O'Shaughnessy gets e-mail from former students in colleges all over the country who are working in campaigns.

BETTY O'SHAUGHNESSY serving as trustee at township meeting.

O'Shaughnessy's classroom assignment had other unintended consequences. Believing that she could not ask her students to do what she was not doing herself, she became involved in the 2008 campaigns of Barack Obama and Dan Seals, the Democratic candidate for the Illinois Tenth Congressional District seat. While doing this, and scoping out candidates for her students, she became friendly with many local political workers and members of civic organizations, and she was eventually slated to run for the office of township trustee. O'Shaughnessy was elected in April 2009 and now has the responsibility of helping to guide this suburban township government through the recent economic downturn.

Power and influence come in many forms. Throughout this book, we will explore how you can gain power and influence at the local level. One of the lessons we teach is that you don't gain significant power without struggle. Most people want more power. Some seek power for noble reasons, others for personal ambition. Most seek power for a combination of reasons. But to gain power, you must enter the fray of politics.

In politics and government, you don't start by running for president of the United States. Even if you get your start by working in campaigns for presidential candidates, you start building your own power and competence at the level of state and local government. We focus this book on the struggle for power and influence at these levels.

POLITICS

While some of the struggles for power take place within state and local government, and all are governed by laws and institutional structures and norms, the struggles are mostly political. Politics is sometimes thought of as some dirty, unpleasant, near-illicit activity. Often, good people such as yourselves are advised to stay away from it. Yet all democracies, whether representative or participatory, depend on citizens practicing politics—electing officials, influencing laws and regulations, and determining the actions of their government. In the end, politics cannot be avoided in a democracy, and politics can only be learned by participation. You can't become a good golfer by just passively watching others play golf on TV. You can't become a good citizen without participating in the political process.

So we encourage you to vote in elections, participate in debates about public policy, and observe your state and local government closely. Books like this one can help you become more knowledgeable and more skillful. But it can't substitute for the actual practice of politics.

This is a very important time to engage in politics because state and local governments are engaged in critical experiments that will determine the course of democracy, justice, security, and prosperity for perhaps decades to come. State and local governments lead the way in finding methods to curb pollution and reduce energy expenditures. They provide safety and security for their citizens. They provide education and guarantee the quality of health care. They respond to crises like hurricanes, tornadoes, and terrorist attacks. They cushion the effects of economic recession. They are literally laboratories for democracy and successful public policy.

National politics is exciting and important, but state and local politics is much easier to enter. There you can have an immediate and significant impact while you learn politics close to home. In 1968, many students and others thought they would simply elect a new president and changes would work their way down to affect state and local politics. Whether or not you help elect a president you support, permanent change begins at community meetings, city hall, the county building, and the state capitol. If solutions to problems are to be found, as well as the paths of hope, they will be uncovered in the experiments in these arenas.

We do not expect that reading this book will catapult you automatically to positions of power or influence even in state and local government. Instead, we hope that it will help you to become better citizens and grassroots leaders who form the bedrock of democratic government. How you decide to participate and whether you advocate liberal, conservative, or even radical causes are choices you will make. But we hope this text will help you make wise choices and participate more effectively in politics. Benjamin Franklin was asked, walking out of the Constitutional Convention, what kind of government the Founding Fathers had just created. He is said to have answered, "A republic, if you can keep it."[2]

In the 21st century, you have been bequeathed by your forbearers an imperfect government grounded in democracy. The system you see today was achieved by over 200 years of political struggle and personal sacrifice. It is your task to keep it and to perfect it.

QUESTIONS FOR REVIEW

At the end of each chapter of this book, you will find questions that will help you review the main ideas of what you just read. In this first chapter, we ask questions that deal with how state and local politics fits into the study of political science and government in general.

1. Why is the study of state and local politics important?

2. What do the authors consider the best way to learn about politics and government, especially on the state and local level?

3. What is the meaning of the word *power*? What is the meaning of the word *influence*? How would you

distinguish between these terms in thinking about politics and government?

4. Using examples from state or local government and politics, how would you explain and distinguish between the following?
   a. Representative and participatory democracy
   b. Institutional structures and political processes
   c. Political and governmental outcomes

5. According to the authors, where is the best place to begin working toward gaining power and influence in politics and government?

6. Why does politics matter? Can it be learned just by reading this book? Why or why not?

DISCUSSION QUESTIONS

1. How are power and influence defined, and what is the difference between them? What is one example from the morning newspaper about the use of power and a different example of the use of influence in politics or government?

2. Why are state and local governments important in our lives? Why should we bother to study them?

3. In what ways are local governments "creatures of the state" and in what ways do they have "home rule" powers? What difference does this make?

4. Why is there a struggle for power?

5. Are power and influence good or evil?

6. What is the difference between representative and participatory (or pure) democracy? Why don't state and local governments all operate on the principles of participatory democracy?

7. If we are particularly interested in the political process, why do we have to pay attention to institutional structures?

PRACTICAL EXERCISES

Using a single sheet of paper, answer the first set of questions on one side and the second set on the back. Your answers will be sketchy but important.

1. Suppose in a few years you wanted to run for a seat on the city council or in the state legislature. How would you begin your campaign for the office? What platform would you run on? How would you win enough votes to get elected? Why might you decide to run for this office?

2. Suppose you were elected to the city council or the state legislature. What would be your goals? How would you go about accomplishing those goals?

At the end of the course, repeat the exercise and see whether your answers have changed due to what you have read in the text and what you have discussed in class.

The Virginia Constitutional Convention, 1830, by George Catlin

CONSTITUTIONS AND CHARTERS
THE FOUNDATIONS OF STATE AND LOCAL GOVERNMENT

Players in the struggle for power in state and local politics are embedded in a web of agencies, groups, and individuals that can help or hinder their objectives. This chapter introduces you to state constitutions and city charters. These frame the politics of state and local governance and provide a context for the four themes of this book: *power and influence, democracy*, *structures and processes*, and *outcomes*. For instance, a governor has more power and influence than a legislative staffer by virtue of his or her formal powers. But a governor is also more powerful because of the informal perquisites of the office, such as use of the governor's mansion to host a lawmaker and the lawmaker's constituents.

Aristotle thought of the *polis* (city-state or body of citizens) as a way of life, and the *politeia* (how a polis is run, or a constitution) as the plan for a way of life.[1] Constitutions may be seen as the rule of law rather than the rule of certain individuals. In the United States, we are all expected to operate within the lawful boundaries of the U.S. Constitution. Analogously, state constitutions and city charters provide the ballpark and the rules by which the game of politics is played. These documents establish the dimensions of the playing field, the numbers of players on each team, the length of the game, how to settle disputed calls, and other critical rules of the game that often determine who wins and who loses

State constitutions and city charters provide the formal institutions and the legal framework in which the struggle for power and influence in state and local politics takes place. Too much power in the hands of one or of the few threatens the democratic polity. Constitutions and charters set the boundaries and limit power. The constitutional tools of limitation and sharing of power include checks and balances such as both legislative and executive approval of legislation, advice and consent by the state senate of appointments by the governor, requirements that extraordinary majorities approve legislation in certain cases, and bills of rights that protect citizens from abuses of power.

Constitutions and charters make it possible for democracy to run smoothly. But they do not differentiate between the wishes of regular citizens and those of organized interests. Thus, groups like the Chambers of Commerce or labor unions wield much more power and influence at the state capitol or city hall than a lone individual. Moreover, while constitutions and charters create the structures of government, very often the practical processes that allow the government to run smoothly have evolved from political interpretations of these documents over time. Constitutions and charters do not guarantee that there will be favorable political or governmental outcomes for one group or another. Rather, they determine how these outcomes will be decided (such as in the courtroom or through legislation). They set forth the proper legal procedures to be followed.

Legislators and city council members are elected to represent their constituents. Interest groups and ad hoc citizen protest groups participate, often loudly, in efforts to influence the policy outcomes. All of this activity takes place within the institutions established by state constitutions and municipal charters. These provide the framework and shape the processes of government. They form the starting points for our study of the struggle for power.

THE BEGINNINGS OF STATE AND LOCAL GOVERNMENTS IN THE UNITED STATES

In order to better to understand why our constitutions and charters serve their purposes so well, we need to look at how and why local and state governments were first formed. In general, local governments were formed to impose order, facilitate commerce, and encourage harmony within the community. The roots of American local government reach back at least to ninth-century Anglo-Saxon England. The country was newly unified, and the king needed effective governance at the local level. England was divided into shires, hundreds, and townships, and each shire (later county) was headed by a powerful shire reeve, or sheriff.[2] Because sheriffs wielded enormous powers of taxation and justice, they were often unpopular with the peasantry. They were also distrusted by the king, to whom they reported. This provides the backdrop for the popular tales of Robin Hood and his guerrilla-style conflict with the Sheriff of Nottingham. England's Norman rulers also created the office of justice of the peace. Over time, these justices became powerful local executives as well as judicial officers, at the expense of some of the sheriff's powers.

In the North American English colonies, the *county* became the primary unit of local government in the middle colonies, whereas the *town* dominated in New England. County officers included the sheriff, whose duties generally included those of tax collector and treasurer; justices of the peace; coroners; and the county lieutenant, who was responsible for the local militia contingent. In 1691, the colony of New York created elective county boards composed of town supervisors. This became a model for many other states. Over time, the county boards took over the executive powers that had resided with the justices of the peace.

In New England, settlers generally formed into compact communities for protection against Indian attacks. As a result, local government formed around the settlements, or towns. Town government was originally responsible for highways, education, care for the poor, regulation of business, assessment and collection of taxes, land records, and judicial functions. Still active in many New England communities to this day, town government offers direct democracy in smaller towns.

State charters also had their origins in the colonial era. Each colony had its own charter suited to the character and customs of its residents. The Continental Congress called on the new states to draft **state constitutions** in 1776 to set out the framework of laws that would govern within the state. Although these documents all proclaimed American republicanism, they reflected the needs of their respective states.

In one of its last acts under the Articles of Confederation, the Continental Congress enacted the Land Ordinance of 1785. This act opened up vast areas of the United States for settlement from Ohio to the Mississippi River through an ingenious system of surveying and platting the uncharted lands into uniform 6-mile squares called townships. The townships were further divided into 1-mile square "sections" of 640 acres each. Townships became important administrative units of local governments responsible for roads, education, and care of the poor. This was accomplished with elected trustees, a supervisor, a clerk, and a road commissioner. Equally significant, the Northwest Ordinance of 1787 established the method by which unincorporated territories eventually became states. As a result, new landowners from the east could find the exact location of their property when they arrived in the newly platted Northwest Territory, later to become the midwestern states of Ohio, Indiana, Illinois, Michigan, and Wisconsin.

By 1870, almost the entire area of the United States outside tribal areas had been divided into counties. Whereas local government officials had been appointees of the British crown during the American colonial period, in the 19th century in America, most of the county and other local government offices became elective.

In the United States today, we have thousands of separate government units other than those of states, counties, and townships. As villages and cities emerged, needs developed for fire control, water, streets, and other services not provided by townships and counties. This led to creation of municipal corporations for villages and cities. Separate school districts also developed across most of the country to address the need for education. In recent years, school district numbers have been reduced dramatically as the one-room schools of the first half of the 20th century were consolidated into larger and larger districts. In the 20th century, there also developed a proliferation of special-purpose districts to provide services that could be administered and financed more satisfactorily than through counties, townships, and municipalities. These include sanitation, water, health, park, library, and fire districts, to name but a few of the types of special-district governments.

Often state constitutions severely limited the level of debt that counties, cities, and towns could have, and new taxes often had to be approved by the state legislature. As a result, when voters needed a new local government service such as schools or a library, the only way to fund it was to create a new unit of government with additional tax powers and the ability to borrow money for public projects. Many property owners today pay taxes to support numerous separate local governments. As of 2007, there were 3,033 counties, 16,519 townships and towns, 19,492 municipalities, 13,051 school districts, and 37,381 special districts. While some consolidation of local governments occurs from time to time, the total number of separate governments has remained approximately the same since 1962.[3]

CONSTITUTIONS AND CHARTERS: THE BALLPARK OF POLITICS

Americans are experienced in constitution-making. According to one scholar on the topic, "As of 1995, American states had held over 230 constitutional conventions, adopted 146 constitutions and amended their current constitutions (on average) over 120 times."[4] Nineteen states have worked with only one constitution, while nine states have enacted five or more constitutions over the years.[5] State constitutions may be revised by proposals from the legislature, citizen initiative, constitutional conventions, and in Florida by a constitutional commission. In all states except Delaware, proposals to revise or amend the state constitution must be approved by the voters.

Over our nation's history, state constitutions have reflected various attitudes toward government. With the British king's often harsh colonial governors fresh in their minds, early state constitution drafters reduced state governors to what one historian called "mere ciphers."[6] Early governors tended to be limited to short terms (1–3 years), possessed limited if any veto powers, and had limits as well on appointments and reelection. During the 1800s, however, legislatures often failed to lead, and constitution makers and voters reacted by strengthening the powers of governors and limiting the powers of elected officials to borrow.[7] The Progressive Era of the late 1800s and early 1900s influenced constitution drafters toward giving more power to the people. Between 1900 and 1950, 21 states provided citizens the constitutional or legislative initiative, the right to a voter referendum on important matters, and the right to recall elected officials from office. Some observers speculate that the "one person, one vote" decisions of the 1960s that eliminated badly apportioned legislatures may have improved the climate for constitutional modernization; from 1950 to the 1980s, about a quarter of the states adopted new or revised constitutions.[8]

Since 1780, the Massachusetts state constitution has stood the test of time, though it has been amended 120 times. Drafted by John Adams, the Massachusetts document served as a template for the U.S. Constitution.[9] Other state constitutions have not been as long lasting, and they all vary dramatically in length, from 8,295 words in Vermont to 310,296 (at last count) in Alabama.[10] Differences in length can be partly explained by the degree of difficulty each state requires for the enactment of amendments. For example, Vermont is one of a dozen states that requires proposed amendments to the state constitution to be adopted by two sessions of the state legislature—that is, in two separate biennial (two-year) periods—while sixteen states require only a simple majority vote from both houses for a proposed amendment to be adopted and sent to the voters for ultimate approval.[11]

Just as all the states create their own constitutions, half the states authorize cities to draft their own **city charters**. The states create local governments, and all powers of the local units emanate from their respective states. The structures and powers of local governments are embodied, for the most part, in the state legislative statute books as, for example, in the "School Code" and the "Cities and Villages Act." However, half of the states provide that some or all of their cities may frame and adopt their own unique city charters, in effect their municipal constitutions, subject to the higher law of the state statutes and constitution. Some states also provide that counties may create their own charters.[12] This charter-making authority confers on the cities and on some counties a significant amount of freedom in determining the structure of their government and the manner in which they will regulate local matters.

Since the late 20th century, many states have also granted "home rule" power to at least larger cities and counties. This has given many cities greater power to tax and to enact important laws (called ordinances in cities) without having to seek state approval. Today, some city and county charters or key ordinances adopted since being granted home rule powers provide the same type of framework as our national and state constitutions.

FINDING A MODEL STATE CONSTITUTION

The writing of some American state constitutions has provoked bloodshed. As the American republic continued its westward development in the 1840s, "slave state" or "free state" provisions in the constitutions of territories seeking statehood became a major divisive issue before the Civil War. For instance, in 1849, California ratified a state constitution that barred slavery. This hotly debated question of whether California would be a slave or free state was only solved with the Compromise of 1850, which admitted California as a free state yet provided for strong fugitive slave laws. After the 1854 passage of the Kansas-Nebraska Act, pro-slavery and free-state settlers fought violently in what became known as "Bleeding Kansas." Supporters of slavery packed a constitutional convention at Lecompton, Kansas, and enacted a state charter that declared the rights of slaveholders inviolate and prohibited any future amendments that might abridge those rights.[13] With support from President Buchanan, the U.S. Senate approved the pro-slavery Lecompton Constitution, but the House rejected it by a close vote of 120–112. In 1859, free-state Kansans passed an antislavery constitution that was approved by the U.S. Congress, which granted statehood to Kansas in 1861. By that time, however, the American Civil War had begun, precipitated in large part by such battles over state constitutions.

Most state constitutions are patterned after the U.S. Constitution, which in turn was heavily influenced by several colonial charters. The "model state constitution" preferred in 1963 by the respected National Municipal League—now the National Civic League—differs significantly from those in effect in most states.[14]

Scholars like neat, clean, brief state constitutions that give nearly all authority to governors and legislators as well as "home rule" provisions granting significant independent decision-making for local governments.[15] In contrast, framers attending constitutional conventions during the 19th century feared what their elected officials might do and thus often limited the powers of state governments to tax and enter into debt. These documents generally limited local governments' taxing and borrowing powers as well, requiring them to go to their state legislatures for authority to make even minor changes. Nearly every 19th-century state constitution required state permission for any new local taxes. After much revision and amendment to state constitutions over the years, few resemble the model constitution devised by the National Civic League, although rare is the constitutional revision study group that has not consulted that model as a starting point.

What follows is a snapshot of several of the primary provisions of the 1963 model state constitution and a brief discussion about what tends to exist in today's state constitutions. You will get the most out of this discussion if you compare these provisions with those in your own state constitution.

> **Article IV. The Legislature. Section 4.02.** The legislature shall be composed of a single chamber consisting of one member to represent each legislative district.

The model constitution proposes a **unicameral legislature**, a single legislative chamber where deliberations can be focused and decisive. However, only the State of Nebraska has a unicameral legislature. All other state constitutions today provide for **bicameral legislature**, a legislative body composed of two chambers, like the U.S. Congress. This is a model that Americans understand and support.

> **Article V. The Executive. Section 5.01.** The executive power of a state shall be vested in a governor.

The idea behind this provision is to have a single elected statewide executive official and to vest all executive powers in that person. Other state officials, like a state attorney general, state treasurer, or secretary of state, would be appointed by the governor, as the president appoints cabinet officials in the national government.

However, most states have a divided executive branch, with numerous additional officials elected statewide with powers independent of the governor. Typically, these include the lieutenant governor, secretary of state, attorney general, and auditor, and in some states even a superintendent of public instruction and a commissioner of agriculture. Other positions, such as state treasurer, have been added in many states. This has the effect of dividing the power of the executive branch of government.

> **Article VI. The Judiciary. Section 6.04.** The governor, with the advice and consent of the legislature, shall appoint the chief judges and associate judges of the supreme, appellate, and general courts.

The model state constitution envisions a judiciary appointed by the chief executive. It is thought that such a judiciary would be knowledgeable, wise, fair, and freed from partisanship in deciding court cases. Although the model state charter indicates that judges should be appointed, 31 states elect at least some of their judges. Of these, 13 states elect judges in partisan elections, and 18 elect judges in nonpartisan elections. For instance, Michigan elects state supreme court justices in partisan elections, the rest in nonpartisan elections. In Indiana, selection of judges varies by both county and level; some judges are elected in partisan elections, some in nonpartisan elections, and some are appointed.[16]

> **Article VIII. Local Government. Section 8.02.** A county or city may exercise any legislative power or perform any function which is not denied to it by its charter . . . and is within such limitations as the legislature may establish by general law.

The original fear codified in many state constitutions was that unrestrained local governments might make bad laws and go heavily into debt. Therefore, the states gave them only limited powers that today are often insufficient to govern global cities and sprawling metropolitan regions.

This power of "home rule" for cities and counties is still not provided in several state constitutions that have been amended in recent decades. Even those state constitutions with home rule provisions for cities often include prohibitions on imposing an income tax and often impose limitations on other kinds of taxation and indebtedness. Many state constitutions grant to local governments only those powers specifically enumerated in their respective constitutions. Local governments in those states fall under **Dillon's Rule**, enshrined in a 19th-century court decision, which holds that all local governments are creatures of the state and have only the powers explicitly granted by the state government.

> **Article IX. Public Education. Section 9.01.** The legislature shall provide for the maintenance and support of a system of free public schools open to all children . . . including public institutions of higher learning.

The U.S. Constitution does not mention education, since this was considered a function of state government. Most state constitutions do explicitly provide for state control of public schools. All states have a Department of Education and a chief state school officer. Of these, 12 state school officials are elected posts established in the state constitution.[17] However, there are major fights in most states about the level of funding which the state must provide and the amount that must be raised by local property taxes to fund schools.

> **Article XII. Constitutional Revision. Section 12.01.** (a) Amendments to this constitution may be proposed by the legislature or by initiative.

Most state constitutions make it much harder to amend the constitution than by a simple majority vote of those elected to the legislature. Approval, often by extraordinary majorities (two-thirds of the vote, for example), is required in both houses of the legislature. Often proposed amendments must then also be approved by a majority of the voters in a referendum.[18]

Most states provide the legislature with the authority to call a constitutional convention or create a revision commission in order to consider comprehensive change to a state's constitution. Fourteen states provide for periodically submitting to the voters the question of whether to call a new constitutional convention. A number of states provide that citizens may use the initiative process to amend the constitution. Generally, the way in which most state constitutions are amended in practice is for the amendment to be offered and passed in both houses of the state legislature (often by special majorities such as a two-thirds majority vote) and then, perhaps, submitted to the voters for final approval. When a constitution becomes hopelessly out of date, a state constitutional convention is called to rewrite the document, but voters still have to ratify the change.

The differences between the model state constitution preferred by scholars and the wide variations among state constitutions are rooted, of course, in politics. Conservatives tend to favor limiting the power of governments to tax. Liberals prefer to enshrine principles in a constitution that protect the rights of minorities. These philosophical contrasts—and myriad others—are played out in efforts to revise state constitutions and city charters. Change comes, but generally not easily. According to G. Alan Tarr, each American state has had an average of three constitutions since the nation's founding, although Louisiana has adopted 11 constitutions in its history. Individual amendments to state constitutions, on the other hand, have been frequent: 689 were adopted between 1994 and 2001 alone, and many others have been proposed and adopted every year since.[19]

Major reform of a highly restrictive, century-old constitution can be difficult when groups and citizens are deeply suspicious of their elected officials. The 1901 constitution of Alabama was enacted to deprive African Americans of rights, especially the right to vote. It also limited state and local governments from taxing and making changes without coming to the voters in a referendum to amend the constitution itself.[20] Governor James E. "Big Jim" Folsom made serious efforts at constitutional reform in Alabama during his two terms (1947–1950 and 1955–1958). A moderate on race, Folsom repeatedly called the legislature into special session to consider constitutional reform. Folsom decried the malapportionment of legislative districts, which favored the mostly white owners of plantations, and the inability of local governments to self-regulate. "Instead," as one historian notes, "legislators passed local laws for counties, often swapping favors among themselves to promote pet legislation."[21]

During Folsom's second term, desegregation and Alabama's general opposition to it overshadowed his reform program, including constitutional revision. According to H. Bailey Thomson, Alabama voters humiliated Folsom for his lack of enthusiasm for resisting civil rights by denying his bid for a place on the Democratic National Committee in 1956. (There was no effective Republican Party in the state at the time.) "His crusades over, Folsom limped through the rest of his term in an alcoholic haze."[22]

Two lessons can be learned from Folsom's story. First, constitutional revision can be affected or even overwhelmed by other issues about which voters feel more strongly. Second, politics is a tough business. In the struggle for power and influence and in the effort to affect public policy, no one always wins.

More recently, constitutional reform efforts in Alabama seemed to gain momentum. In 1994, the *Mobile Register* newspaper gained serious public attention with its major series on the deficiencies in the Alabama constitution. By 2000, an organization of statewide leaders called the Alabama Citizens for Constitutional Reform (ACCR) put money and prestige behind their effort. In 2003, newly elected reform-minded Republican governor Bob Riley put his career on the line by pushing for constitutional reform, especially tax reform and tax increases. He garnered the support of business leaders, who put millions of dollars into a campaign to make Alabama a regional leader. Riley's reform efforts passed the legislature and went before the voters in a required referendum in 2003, when voters, by a two-to-one margin, sent his package to ignominious defeat. According to H. Bailey Thomson, "The election confirmed a deep, almost pathological, distrust among Alabama's electorate for government at all levels, while putting even more pressure on an antiquated constitutional system to produce at least some temporary fix for long-festering problems."[23]

Nevertheless, constitutional change can be achieved, as proved in 1998 in Florida. It followed strong voter rejection two decades earlier of a more ambitious set of proposals. The successes in the more recent Florida effort, as well as those in other states in the past 30 years, suggest that the keys to adoption of constitutional change include the following:

- Lay good, thorough groundwork by a respected study commission.
- Avoid allowing the process to fall into the trap of partisanship.
- Avoid making unnecessary enemies by steering clear of emotional issues that may not be central to fundamental objectives.

- Put issues on the ballot separately rather than in a take it or leave it omnibus proposal that cumulates opposition.
- Mount a strong, well-coordinated, and well-funded campaign for enactment.[24]

Beyond wholesale rewrites of constitutions and city charters, 18 states authorize constitutional and statutory amendment by the **initiative process**, which is direct democratic action by the voters.[25] An individual or group of voters, for example, drafts a proposal to amend the state constitution and circulates petitions in support of putting the proposal on the statewide ballot. If enough valid signatures are gathered—the minimum is generally some percentage of those who voted in the most recent election—the proposition goes to all the voters of a state. The vote required to enact the proposition is generally a majority of those voting on the issue. Sometimes a majority of all those voting in the election is required, even if they don't vote on the constitutional amendment.

Direct democratic action is not always the smoothest method of changing a constitution. Between 1964 and 2002, for example, legislative action in Colorado proposed 62 constitutional amendments, of which 49 (79 percent) were adopted by the voters in the required referendums. In the same period, citizen groups used the initiative process to put 61 proposed amendments on the ballot by successful petition drives, yet only 23 (31 percent) of these were approved by the voters. These measures were more controversial than those proposed by the legislature.

In an analysis of proposals to change campaign finance laws in Colorado by statutory initiative (1994 and 1996) and constitutional initiative (2002), political scientist Anne Campbell notes that proponents for change saw their successful statutory initiative of 1996 (passed by 66 percent of those voting on the issue) "gutted" by a hostile legislature and later by the courts. As a result, the groups seeking change put their 2002 initiative (passed by the same 66 percent favorable vote) into the constitution, where the legislature and state courts could not change it.[26] The Colorado case demonstrates how citizens can propose constitutional amendments as a defensive mechanism against government officials who have proven to be particularly hostile to their policy proposals.[27]

Even in the citizen initiative process, compromise is often necessary in order to win voter enactment of a proposition. In Colorado, since the "proponents believed that their optimal solution (public financing of campaigns) would not be approved by the voters," notes Campbell, "they compromised and proposed an initiative that they thought would improve the political process *and* would pass on election day."[28]

In some states, direct democratic action is not used to initiate amendments. Instead, these states hold periodic referendums on the state constitution. For instance, Illinois requires that its voters decide every 20 years whether or not a new constitutional convention shall be called. The referendum to hold a new convention was voted down in 1988 and again in 2008. While specific interest groups have issues they want to be considered, voter approval for a call for a convention—which cannot be restricted as to subject matter—could open a Pandora's box of unexpected and possibly unwanted proposals. Controversial issues such as prohibiting same-sex marriage, outlawing abortions, or barring new taxes could be written into the state constitution if a new convention were held. So usually a proposal for a state constitutional convention is voted down. But this mechanism offers an opportunity to consider whether a constitution still provides the framework necessary to cope with real modern-day problems.

A major problem in governing states today is that many of their constitutions are old and difficult to amend. However, a major strength is that the basic institutions of representative government are enshrined in these documents and are not easily challenged by current political passions.

CITY (MUNICIPAL) CHARTERS

In states that provide for city charters, a charter commission of three to 25 or more local citizens is convened to draft the charter. In every instance, the charter must be referred to the local voters for their approval before becoming effective. In Ohio, for example, an elected charter commission develops a charter which is then submitted to the voters at the next election, a fairly standard procedure.[29] Big cities like New York and Los Angeles have revised their city charters in the last decade or so in an attempt to bring their governmental structures in line with modern needs—such as conforming to "one person, one vote" laws, or creating stronger mayoral powers, which are now important to governing large cities. Generally, making major revisions to city charters or adopting new ones is not lightly undertaken. Efforts, for instance, to merge city and county governments are often defeated by the voters even if such changes might seem sensible to political scientists.

During the actual playing of the game of politics, candidates and political operatives may only lightly regard constitutions and city charters, but the rules in these documents guide the political process.[30] For example, there is the case of the New York City charter and New York mayor Rudy Giuliani. In the wake of the September 11, 2001, terrorist attacks on New York City, then-Mayor Giuliani provided stirring leadership that rallied his 8 million fellow New Yorkers. Following his bravura performance that fall, Giuliani could have been reelected by a landslide—except he was barred from seeking a third term by New York's city charter, which limited elected city officials to two successive terms.[31] Charters can be changed, however. Giuliani's successor, Mayor Michael Bloomberg, managed to avoid the two-term limit himself by getting the city council to pass legislation in 2008

MAYOR GIULIANI speaking after the 2001 attack on the World Trade Center.

extending the number of terms a city officeholder could serve; instead of two 4-year terms, the limit is now three.

In an elegant article in the *National Civic Review*, Lawrence F. Keller contends that municipal charters, while they are generally overlooked in the study of politics, "are of major significance for the quality of public life for most Americans."[32] Keller traces the roots of municipal charters back to the Roman republic and empire. The word *municipal* refers to the internal affairs of a government. As Keller explains, "Roman leaders realized that local policies could vary without affecting their ultimate control over the empire. They conceived of the city as a separate legal entity, called a municipality, and granted it the authority to direct its own local affairs."[33] Today, many cities and suburbs have a charter that provides the framework of their local government.

As with state charters, the National Civic League has put together a model charter for cities. Some of those provisions are discussed below. Once again, you will get the most out of this discussion if you compare these model provisions to those in your town or city's charter.

> **Article I. Powers of the City. Section 1.01.** The city shall have all powers possible for a city with self-governing powers to have under the constitution and laws of this state as fully and completely as though they were specifically enumerated in this charter.

These powers include the ability for cities and towns to make their own laws, called municipal and town ordinances; to tax and create their own government budgets; to appoint city officials such as the police chief and the entire city bureaucracy, which can range from a few hundred employees in the average large suburb to tens of thousands in larger cities. These self-governing powers are used to create public safety and a multitude of government services, such as streets, sidewalks, traffic lights, stop signs, fire protection, garbage pickup, zoning and building regulations, economic development, emergency shelters for the homeless, and public housing for the poor and senior citizens. While cities, towns, and counties complain that they do not have all the powers they might like, for the most part they have the powers they need to govern effectively.

> **Article II. City Council. Section 2.01.** All powers of the city shall be vested in the city commission [or city council]; except as otherwise provided by law or this charter, and the council shall provide for the exercise thereof and for the performance of all duties and obligations imposed on the city by law.

The model city charter by the National Civic League differs from the model state constitution by placing the principal powers in the hands of a city commission or city council, not in the hands of a strong mayor with powers similar to a president or state governor. Most often, the mayor is chosen by the council from among its own members. Even if the mayor is elected directly by the voters, the intent is to create a weak mayor/strong council (or commission) form of government. A few city councils parcel out responsibility for different departments to different city commissioners.[34]

In any case, the city council or city commission is meant to be a strong legislative branch of government with all the critical powers of government. This includes budgets and taxes, appointments, and legislative oversight of city employees and activities. Particularly in larger cities, city councils have often been unable to carry out these responsibilities effectively. But many suburbs carefully guard their prerogatives, much to the chagrin of mayors and city managers.

> **Article II. Section 2.03. Mayor.** The mayor shall be a voting member of the city council and shall attend and preside at meetings of the council, represent the city in intergovernmental relationships . . . and perform other duties specified by the council. The mayor shall be recognized as head of the city government for all ceremonial purposes and by the governor for purposes of military law but shall have no administrative duties.

Just as the city charter determines the structure of the city council, it also determines the council's official relationship with the mayor's office. Nevertheless, while this institutional structure of the city council can help or hinder a mayor's ability to make policy, the power of any individual mayor is often determined by her or his political clout.

In smaller suburban communities and towns, mayors have fewer powers, often not even veto power. In larger cities and counties, mayors tend to be strong executives similar to governors. However, their powers are often not granted directly in the city charter but by state law, city ordinance, or tradition.

> **Article III. City Manager. Section 3.04.** The city manager shall be the chief executive officer of the city, responsible to the council for the management of all city affairs placed in the manager's charge by or under this charter.

The model city charter envisions for cities and towns a city manager, usually with a master of public administration (MPA) degree and years of professional experience "managing" city government. Most large cities and counties in the United States have an elected mayor or county board president with broad executive powers. Most of these do not have a city or county manager. But middle-sized cities, counties, and smaller towns have tended to follow the model city charter and appoint rather than elect city, town, or county managers to run the local government bureaucracy and to ensure that services are delivered professionally, effectively, and impartially. Even big city and big county mayors usually employ a chief of staff or chief administrator who functions more or less like a city manager. As the "Struggle for Power" feature shows, such officials are often given considerable power in the interest of more responsive government.

The reform model of local government assumed that if politics were taken out of the delivering of city services, local government would run like a business—efficiently and inexpensively so that citizens would get good services for low taxes. While this view has proven in some ways to be naive, city managers do tend to increase the professionalism and effectiveness of the governments they direct. They have not, however, succeeded in taking the politics out of local government. Their presence does affect the struggle for power and influence locally, and the accountability of local government to its citizens. We will explore how this occurs in more detail in Chapter 4.

The history of New York City is, in some important ways, the history of its municipal charters. The many charters under which the city has operated since its founding in 1656 have largely represented a struggle over the formal powers of the mayor.[35] Originally, the governor of New York State appointed the mayor of the city. In 1821, the state finally granted the city the power to elect its own mayor. By 1897, when an even greater city of New York emerged with the annexation of Brooklyn, Western Queens, and Staten Island, the mayor was accorded a veto power that could only be overridden by a vote of five-sixths of both houses of the city legislature. In addition, the mayor dominated the budget process during that era. Soon jealous borough leaders began chipping away at mayoral power

STRUGGLE FOR POWER

New Clout in a Big County

L.A. County supervisors have done something unusual—given up power voluntarily. Nearly all local governments struggle to find the right balance of power between the elected officials who set policy and the manager and staff who far outnumber them and do most of the work. In order to make the lead decision-maker more responsive to the public, several large cities have moved away from the manager form of government toward a strong-mayor system. In Los Angeles County, the board of supervisors has gone in the opposite direction, giving up much of its power to the unelected county administrator.

Los Angeles is the nation's largest county, with 100,000 employees and a $20 billion budget. Running it would be a big task for anybody, but it's been particularly tough because so much managerial authority rests with the five elected supervisors. After veteran administrator David Janssen announced plans to step down, two candidates quite publicly passed on the job, while others demurred from seeking it. They felt they just wouldn't have the clout to do it right.

"It's the problem that's pervasive with counties everywhere," says Jackie Byers, director of research for the National Association of Counties. "Administrators don't have the same level of authority that a city manager would have."

Stung by the job-candidate turn-downs, the L.A. County board recently voted to cede some of its power to the administrator, letting him hire and fire most appointed department heads. Not all the supervisors were happy doing that, but most of them admitted it was preferable to the hydra-headed system the county has had in place. "It's become clear to the L.A. board that they've got to formalize a lot of the authority to make sure that they get a viable replacement for David," says Jim Keene, of the International City-County Management Association.

One veteran supervisor, Mike Antonovich, took exception. He compared a county government dominated by an administrator to colonial government under King George III. "What it does," a spokesman for Antonovich argued, "is create an additional barrier between the public and the vital public services it pays for."

It is tough to strike the right balance between ballot-box accountability and the fact that most government officials are unelected. Whoever the new county administrator is, he or she will still have to answer to the board of supervisors. That should ease some of the concerns. The other option—electing a countywide mayor—has been rejected by voters several times.

Source: Alan Greenblatt, "New Clout in a Big County," *Governing*, May 2007. Copyright 2007, Congressional Quarterly, Inc.

---

through charter revisions. In 2004, New York City amended its city charter to grant strong powers to the mayor and create a unicameral city council made up of the public advocate and 51 council members elected by district.[36] So far, this seems to be the most effective structure of local government that New York City has ever achieved.

FORMS OF LOCAL GOVERNMENT

Mayors and other elected officials must achieve their objectives within the rules established by the municipal charter and state laws. Generally, three basic organizational frameworks have evolved in American cities: commission, mayor-council, and council-manager.[37] Each has its strengths and deficiencies.

Early in the 20th century, in the wake of a devastating tidal wave, Galveston, Texas, instituted the **commission form of government** to replace a council and mayor who had governed badly. It was an attempt to provide a more responsive form of city government. In the commission form, which is now rare, three or five commissioners are elected citywide. Generally in the order of highest number of votes received in the election, each commissioner selects an executive portfolio, such as commissioner of police, fire, streets, public works, or administration. The commissioners then serve as executives and as a legislative body. Some suburban government may call their city council a city commission, but few local governments follow the commission form today.

The **mayor-council form** of local government provides for a mayor with all executive powers and an elected city council, from separate districts or wards or city-wide. The city or town council serves as the legislative body. In a **strong mayor government**, the mayor is the sole elected executive, prepares the budget, and has veto power over city ordinances passed by the council. The city council must muster a supermajority of at least 60 percent to override the mayor's veto. In a **weak mayor government**, the city council performs the legislative function, and the mayor lacks veto power and shares executive powers with, for example, an elected city clerk and treasurer. Nonetheless, politically powerful mayors use their influence to get their policies adopted regardless of the form the government takes. For instance, the weak mayor form is dominant in America's largest cities, including New York, Los Angeles, and Chicago, but these cities still have strong mayors who wield considerable political power.

Throughout the 20th century, the **council-manager form of government** became ever more popular, particularly in towns and suburbs. In this approach, the city council and mayor appoint a professional manager who is the chief executive of the city. The city manager reports to the mayor and council and serves at their pleasure. The council-manager form was enthusiastically recommended throughout the 20th century by the National Municipal (Civic) League in the organization's "model city charters," as we saw above, and city managers have become a well-trained and highly professional corps of public servants. The master of public administration degree (MPA) is almost a prerequisite for city managers, and a number of universities across the country offer this practical, professional degree. The International City Managers' Association provides strong support, training, and continuing professional development for its members.

The smallest cities and villages often believe they cannot afford a full-time professional manager, and mayors and councils in the biggest cities are often loath to give up their power to an appointed official. But the council-manager form of government dominates most local governments. Many specialized districts of local government also appoint an executive. Although they frequently have an elected school board, school districts appoint a school superintendent to recommend policies, provide educational leadership, and manage the school bureaucracy.

Such structures do not control all government outcomes. Legendary Chicago Mayor Richard J. Daley operated in a weak mayor form of government from 1955 to 1976. Mayor Daley shared executive duties with several other elected city officials, including Chicago's city clerk, city treasurer, city council, and elected judges. The power he lacked in his official elected role, he made

MAYOR RICHARD J. DALEY holding a press conference.

up for in his role as chair of the Cook County Democratic Party Central Committee. Daley ran the party operation with an iron fist. He determined who would be nominated and therefore elected—as Chicago was a one-party city—to the other government offices, and he personally doled out thousands of patronage jobs.

CHAPTER SUMMARY

State constitutions and city charters provide the framework of the four themes of this book: *power and influence, democracy, structures and processes*, and *outcomes*. Constitutions and charters make it possible for democracy to run smoothly and for political players and ordinary citizens alike to move through the web of politics. These charters have been around since the beginning of our nation and have been instrumental in determining who participates, what the rules are, how disputes are settled, and other important aspects of governance.

Local charters and state constitutions have their roots in colonial times. Local charters set out who was responsible for roads, education, and care for the poor. The various colonies established government charters keeping in mind the character and customs of their respective citizens. Since there were limits in the constitutions and charters as to the amount of debt that could be incurred by a local government, over time new types of governments such as school or library boards were created to borrow money for such projects. At present, there are over 86,000 different local governments in the United States.

State constitutions are patterned after the U.S. Constitution, which in turn was heavily influenced by colonial charters. The National Civic League has issued a "model constitution," which differs from the actual state constitutions in effect. Today, most state constitutions set the rules for three branches of government, with a governor, bicameral legislature, and a judiciary. Most states also provide for education and for amending the constitution. Although it is difficult to completely rewrite a state constitution, hundreds of amendments are regularly added by the state legislatures or the people. The differences among the states in some issues—such as local home rule, election of judges, and the existence of the initiative, referendum, and recall processes—are rooted in politics.

City charters are often drafted by a commission of three to 25 or more local citizens, but the charters must be ratified by voters before taking effect. Although at times these documents may be taken lightly by politicians, they nevertheless provide the framework in which local government can operate. Generally, three basic types of local governments have evolved: the commission, mayor-council (with a strong or weak mayor), and council-manager forms. In the now rarely used commission form, three to five elected officials are the executives and also the legislature. In the mayor-council form, a mayor has executive powers, and an elected council serves as the legislature. In the strong mayor system, the mayor is the lone elected official and has budget and veto powers over city ordinances passed by the council. In a weak mayor system, the city council is elected and acts as the legislature. In this system, the mayor lacks veto power and shares executive power with other elected officials. In the council-manager form, the mayor and council appoint a professional city manager. Regardless of the type of city government, politically powerful mayors can get their policies adopted.

The formal structure of a government will not completely define how it functions in practice. State constitutions and city charters provide a framework for the struggle for political power and influence, but they are not the sole determinants of the outcome.

KEY TERMS

bicameral legislature A legislative body composed of two chambers, as found in most states, p. 21.

city charter A document that provides the framework for the laws of a given town or city, p. 20.

commission form of government Local government in which three to five elected commissioners serve as executives and as the legislative body of city, p. 28.

council-manager form of government Form of city government in which the mayor and city council appoint a professional city manager, who reports to them, p. 29.

Dillon's Rule All local governments are creature of the state and have only the powers explicitly granted by the state government, p. 22.

initiative process Direct democratic action by voters through a petition to place a measure on the ballot for the next election, p. 24.

mayor-council form of government Form of city government in which an elected mayor serves as the chief executive, and an elected city council serves as the legislature, p. 29.

state constitution A document that provides the framework for the laws of a given state, p. 18.

strong mayor government Mayor-council government in which the mayor serves as the sole executive, prepares the budget, and has veto power over the city council, p. 29.

unicameral legislature A legislative body composed of one chamber, as found in Nebraska, p. 21.

weak mayor government Mayor-council government in which the mayor shares executive power with other government officials and lacks veto power, p. 29.

QUESTIONS FOR REVIEW

1. What legislation passed under the Articles of Confederation set up a blueprint for the formation of local governments? For states?

2. What is a state constitution, and why is it difficult to find a "model" constitution that can be used by any state?

3. What are the three basic forms of local government, and who controls most political and governmental power in each?

4. What is one way that state constitutions and city charters can limit, shape, or control the struggle for political power and influence?

5. In what ways can the limits imposed by constitutions and charters be frustrating for governmental officials like governors, mayors, or legislators?

3. Do you favor being able to amend the state constitution or city charter by citizen initiative? Why or why not?

4. What do you believe is the best form of city government? How much power should the mayor and city council (or commission) have?

5. Should all cities be required to have a city manager? Why or why not?

6. Do you and your friends believe that "you have to know someone" in order to get a job in your state and local governments? If you have to know a politician or someone who works inside government, is that a good or bad thing? Why?

7. Would you support or oppose a requirement in your state constitution that declared free tuition in public colleges to be a right of all citizens of your state? Why?

8. Do you think that a state constitutional convention or city charter convention should be called to revise these important documents in your state or city? Why or why not?

DISCUSSION QUESTIONS

1. Do you think it ought to be easy or hard to amend a state constitution or city charter? Why?

2. Imagine that you are a delegate to your state's constitutional convention or your city's charter convention. What is one aspect of the current constitution or charter that you would want to change? (Find a copy of your state constitution or city charter at the library or on the Internet if you are uncertain of the provisions.)

PRACTICAL EXERCISES

Throughout this text, we will be suggesting practical exercises that require you to do research on your own state and local governments. Much of this research can be done on the Internet, but some may require that you research library documents or ask public officials. It is important that you discover how your own state and local governments actually work if you are to understand the struggle for power as it actually occurs.

1. Obtain a copy of your state constitution from the library or your state government's Web site. Use it to answer the following questions:
   a. What is the approximate length of your state constitution? Do you think it is too long or too short? Explain.
   b. Does your state provide for citizen initiatives in elections? What are the restrictions on this power of amendment?
   c. List 10 key amendments that have been adopted since this constitution was first approved.
   d. How are judges selected in your state?

2. List the officials that are elected statewide in your state. Does it make sense to have each of the officials elected? (State governments have an official state Web site that usually lists all elected state officials, and many officials have their own sites in which they explain the duties of their office.)

3. What types of license applications can be filled out online in your state and city? Go to the appropriate Web site and list at least one state or one city license that can be applied for online.

4. Can you register to vote online in your state? If you are eligible to vote but are not registered, see if you can register online. If not, go to your city or township hall or to a public library and register to vote.

5. Find out what form of government your own municipality has. (Most cities and suburbs have an official governmental Web site that lists elected public officials and may have a copy of the city charter.) Based on the information you have obtained, describe your local government in a paragraph. Who has the executive and the legislative powers?

6. How many aldermen, city council members, or city commissioners does the largest city in your state have?

7. Look up when the city charter (or state law governing your city) was last rewritten or amended. Should it be amended now to keep up with new trends or "best practices" in local government?

8. Draft an amendment for your city charter. As a class, discuss the proposed amendments that students have drafted, then vote on them.

WEB SITES FOR FURTHER RESEARCH

www.nlc.org The National League of Cities (NLC) is a nonprofit association of cities and state municipals leagues.

www.ncl.org The National Civic League (NCL) is a nonprofit, nonpartisan, organization that fosters innovative community building and political reform, assists local governments, and recognizes collaborative community achievement through technical assistance, training, publishing, research, and the All-America City Awards.

www.icma.org The International City/County Management Association (ICMA) is the professional and educational association for appointed local government administrators throughout the world.

Both small towns and large cities have Web sites. For instance, www.nyc.gov provides access to New York City's agencies, programs, and services. Follow the links at the site to find a copy of New York City's charter.

www.bozeman.net is an informational site for the small town of Bozeman, Montana. Its public documents links shows how a town can adapt the National Civic League's model cities charter to its own needs.

All states have their own government Web sites, such as www.vermont.gov for Vermont and www.ca.gov for California. You can find the constitutions of most states by following the links on their Web pages.

Most counties in the United States have Web sites, such as www.kingcounty.gov for King County, Washington (the Seattle area). Following the links will bring you to the county's charter.

For a copy of a model state constitution that you can compare with the constitution of your own state, go to www.laits.utexas.edu/txp_media/html/cons/features/0301_02/modelcons.pdf.

SUGGESTED READING

For History of State and Local Government
Snider, Clyde F. *Local Government in Rural America*. New York: Appleton-Century-Crofts, 1957, chapter 1.

For State Constitutions and Charters
The Book of the States. Council of State Governments, various years.

Tarr, G. Alan, Robert F. Williams, and Frank P. Gad, eds. *State Constitutions in the Twenty-first Century*. State University of New York Press, 2007. Provides informative articles, many historical, on individual state constitutions.

Wilentz, Sean. *The Rise of American Democracy: Jefferson to Lincoln*. W. W. Norton, 2005.

For City Charters

Frederickson, H. George, Curtis Wood, and Brett Logan. "How American City Governments Have Changed: The Evolution of the Model City Charter." *National Civic Review* 90 (Spring 2001): 3–18.

Siegel, Fred. *The Prince of the City: Giuliani, New York, and the Genius of American Life.* Encounter Books, 2005. A riveting biography of a hugely effective but controversial mayor. Siegel discusses how Giuliani took advantage of new authority granted the mayor by revisions in the New York City Charter.

Viteritti, Joseph P. "The Tradition of Municipal Reform: Charter Revision in Historical Context." *Proceedings of the Academy of Political Science*, 37, no. 3, *Restructuring the New York City Government: The Reemergence of Municipal Reform* (1989): 16–30. A comprehensive analysis of the history of charters for New York City.

Opening
session of the Florida
House of
Representatives

STRUGGLES IN STATE
LEGISLATURES
AND CITY COUNCILS

Why would anyone want to run for the state legislature or city council? The pay is
generally bad, and it's hard for a state legislator to keep a good job and also serve in
the state capital 50 or more days a year. On the local level, it is hard for city council members
to work up to 80 hours a week, attend hundreds of meetings a year, and receive calls for city
services even at home on weekends. Worse, the public seems to think that many elected

officials are lower than pond scum.[1] Constituents verbally beat up on their elected lawmakers without mercy, and even if you are honest and hard working, some constituents will believe you are a crook. Conservatives who don't like how you vote might call you a communist, while disgruntled liberals could label you a fascist.

Sometimes nobody runs. Officials in small cities and villages have been known to plead with citizens to run for vacancies on their councils. In northern Arizona in 1998, nobody ran for two open seats in the state house of representatives. The district in which the two seats were located encompassed 147,000 constituents and sprawled across 300 miles of northern Arizona, a long way from the capital in Phoenix. Serving in the legislature that year was even more unattractive than usual. That year the legislature met for 121 days in faraway Phoenix. The post paid only $15,000 annually. Being gone so much of the time made it difficult for a legislator to hold a good job back home in the district.[2] Write-in candidates finally came forward to fill the two seats. (In cities and states, it is not unusual for at least 20 percent of the incumbent members of city councils and state legislatures to run unopposed, and more than half to have only token opposition.)

While aldermen, county board members, and other local officials may not have to travel to the state capital to serve, they must forego time with their families and progress in their profession if they are to play this role in local government. Many able citizens are unwilling to serve in even local government legislative bodies. The sacrifices of time and career are too great.

SERVING IN THE CITY COUNCIL OR STATE LEGISLATURE

Despite the disadvantages in running for and serving in a legislative branch of government, in the 1970s author Dick Simpson ran for the Chicago city council as an anti–Mayor Richard J. Daley candidate in a city that loved its mayor. After winning his seat in the council, Simpson led a small band of independent liberal Democrats in traditionally losing causes, but he felt that he and his independent colleagues made a difference as watchdogs and constructive provocateurs. They initiated new legislation and acted as spokespersons for their constituents and the voters who opposed the powerful Daley machine. Simpson loved the battles on the floor of the city council over the future of the city. He didn't always relish dealing with endless constituent complaints, but he recognized that such service is also part of the job of a legislator.

About the same time, downstate Illinois resident and University of Illinois graduate student Jim Nowlan was elected to a seat in the Illinois house, representing a rural district that was 100 miles long by 50 miles wide. As a legislator for two 2-year terms, Nowlan championed environmental causes and was a voice for students opposed to the Vietnam War. Although he voted in favor of a new state income tax, he was still reelected. Nowlan hobnobbed in Springfield and Chicago with lobbyists, and he mingled under the gilded capitol dome with 235 often colorful, even outrageous house and senate members from the inner city of Chicago to the state's farmlands 400 miles away. In exciting times with so much at stake, why wouldn't a citizen want to run for office?

Serving in a state legislature or a big-city council or on a county board puts the lawmaker in the eye of a policy storm. Swirling about each lawmaker are the governor or mayor's people, agency staffers, reporters, and interest group lobbyists—who outnumber lawmakers.

STRUGGLING FOR POWER AND INFLUENCE IN LEGISLATIVE BODIES

Legislative bodies—from local school boards and city councils to state legislatures—illustrate the central themes of this book. There is a constant struggle for power and influence stemming from both noble and base motives. This is an arena in which efforts to represent the interests of the people in a democracy are demonstrated most starkly.

No **representative democracy** can exist without a legislature composed of elected members who represent their district constituents and the city or state as a whole. In cities of a million or more residents and states with perhaps tens of millions of residents, **participatory democracy** (or **pure democracy**) as practiced 2,500 years ago in Athens with only 10,000 citizens is not possible. Thus, in modern democracies, regular elections are held to choose the officials who will make the actual decisions of government. Participatory democracy, however, is still seen in places like the small towns of New England, where the citizens hold town hall meetings to choose officials, pass budgets, and determine governmental policies.

The legislative bodies of a representative democracy are constrained by their institutional structure and rules of procedure. The committee process, for example, ensures that proposed legislation, budgets, and key executive branch appointments get a fair hearing and that proposed bills and budgets are improved before being taken up by the entire legislature, which will also debate and possibly amend the proposed legislation. Committee hearings also provide the opportunity for citizens and interest groups to express their opinions and sway the votes of legislators.

These critical legislative processes depend on institutional structures to set the framework within which the legislature operates. The division of powers between the legislative, executive, and judicial branches of state governments is established in the state constitution. The separate branches and their powers within city and county governments are set forth in state laws and, in some cases, in city charters. The constraints on the legislative process itself are set forth not only in the constitutions, charters, and laws but in much greater detail in rules of procedure adopted in the first session of the legislative body.

Beyond the straightforward struggle for power, the processes of representative democracy, and the institutional restraints, we must determine whether the laws, budgets, and appointments approved by the state legislatures, county boards, city councils, and innumerable independent governmental boards and commissions are good for citizens. Do they promote or detract from effective governance? Do the individual legislators provide good representation for their constituents? Do they follow open, democratic procedures that allow citizens to participate in and observe the process? Are the resultant laws, budgets, and executive appointments promoting efficient and honest government?

LEGISLATIVE LIFE

Because of the wide range of differences among the states and among local communities, the various legislatures meet on very different schedules. Some state legislatures meet only part of the year. While some state legislatures might hold 30-day sessions, others are in session through most of the year. Sessions are generally **biennial**, that is, for two-year periods. In some states, the legislature meets only during the first year of the biennium, though lawmakers find themselves drawn to the capitol for special sessions as well as committee meetings throughout each year. City councils, school boards, and other local legislatures like county boards generally meet year-round at least once a month and usually several times a month. Local government committee meetings usually occur every week or at least monthly.

While the chief duty of a legislator is to pass laws, legislators find that in order to do so effectively, they must meet with many different people and perform myriad tasks. A typical day in the state capitol starts with breakfast with an interest group, committee hearings, a caucus among one's own party members and leaders, and a floor session to debate proposed legislation and cast **voice votes** (members answer in groups, saying yea or nay) and **roll call votes** (each member votes when called on by name). With roll call votes, the way each legislator voted is entered in the legislative documents that become public record.

If possible, a lawmaker squeezes in some time with office staff to dictate letters and work on constituents' problems. A mother might want her son transferred from one prison to another; a father might be irate that his daughter has been rejected by a state university; a woman who is struggling to get off welfare may have been denied a job at a nursing home because she has a felony record.

Later in the afternoon or evening, the lawmaker may drop by receptions or events sponsored by retail merchants, the Chamber of Commerce, a nursing home association, or community bankers. Afterward, the legislator might have a dinner meeting with like-minded members from a particular caucus or from both parties to discuss stratagems for passing or defeating a particular bill. The day may end near midnight with a gathering of fellow members, staffers, lobbyists, reporters, or visiting influential voters from the district.

At the local government level, members of city councils, school boards, and county boards meet regularly throughout the year: almost every week in big cities and as little as once a month in village halls. Local lawmakers also must attend several hundred community meetings and political gatherings each year. City council members must answer phone calls from constituents and may meet with a thousand or more individual citizens each year to discuss government service complaints and community problems. Council members are **ombudsmen**, officials who look into individual concerns or complaints and seek a solution. Ombudsmen serve as a check on government officials to ensure that government services are delivered honestly and properly. Your local council members might assist you to overcome an injustice or a delay in delivery of a government service. They deal with problems from parking fines to immigration issues, from barking dogs to crimes, from health and building inspections to federal programs like Social Security and state programs like welfare. The list is endless because local government officials are truly the closest and most easily reached of all elected government officials. They can solve myriad problems for their constituents.

Some of the most basic concerns of citizens are shown in Table 3.1, on constituent services in a Chicago ward. As the table demonstrates, most of the top service complaints or requests that a Chicago alderman receives were solved more than half of the time; all of the complaints about street cleaning, snow removal, sewers, and fire hydrants were solved. The only problems showing less than a 50 percent solution rate were chronic gang problems and major building complaints that required court action of 18 months or longer to resolve. Ending neighborhood crime problems such as street gangs requires organizing block clubs and police department surveillance, a solution that takes longer than constituents want.

TABLE 3.1 Constituent Services in Chicago's 44th Ward for One Year

Ten Most Requested Services	No. of Requests Received	Percentage Solved
Abandoned cars; avg. number of days for removal: 30	350	95
Garbage and bulk pickup	143	97
Building complaints	123	
Major complaints requiring court action; avg. length of time: 18 months		12
Minor complaints with no court action required		73
Broken streets, sidewalks, curbs[a]	94	54
Parking and traffic	77	75
Tree trimming removal[b]	77	60
Street cleaning and snow removal	70	100
Sewers and fire hydrants	50	100
Zoning, licenses, and permits	44	53
Gangs and police problems[c]	33	33
All others	425	72
Total	**1,486**	

[a]Some curb requests are 2 years old.
[b]Alderman's office also provides trees for planting each spring.
[c]Chronic problems require organizing citizen block groups.
Source: Dick Simpson, ed., *Chicago's Future in a Time of Change* (Champaign, IL: Stipes, 1993).

Imagine being a city legislator who, with your staff, must deal with nearly 1,500 service complaints and requests a year. In fact, the number of complaints handled by a single Chicago alderman has grown such that some aldermen now handle thousands of complaints a year.

The ombudsman role places a severe limit on the time needed to write general citywide legislation. On the other hand, service requests and direct experience with constituent problems provide fertile ground for writing new laws that address the actual problems that citizens are experiencing.

In addition to dealing with service complaints and attending legislative meetings, local lawmakers attend receptions, breakfasts, and lunches just like their state legislative counterparts. And they are always "on call." Every constituent and every organization feels free to bend their ear with what ought to be done in the city, the schools, and the county. But solving a constituent's problem with her landlord, seeing that a senior gets "meals on wheels" delivered to his door, or getting funding for a new school or park are all satisfying achievements for a local lawmaker. There is pride and a sense of accomplishment in making a real difference in the lives of neighbors.

Stress affects many lawmakers in both state and local government. Promises made to an interest group are hard to keep when a powerful party or a government leader demands you vote otherwise. If your government service keeps you away from home more than expected, your full-time job and family relationships could suffer. Too many rich dinners and long, hectic days can affect your health. And much as you may try, you can't please everyone, not even all of your own backers.

WHO ARE THE LAWMAKERS?

More than 7,000 individuals are elected to our 50 state legislatures. There are more than 500,000 elected local government officials in the country. These state and local elected offices are unparalleled opportunities to become a part of the political process and government.

State legislatures and city councils fulfill the representative democracy function in government.[3] Each lawmaker represents a particular district with its unique mix of interests. Inside city hall or the state capitol, a tugging and pulling between elected officials and those who wait in the lobby outside the legislative chamber ultimately results in budgets, state laws, and city ordinances. Legislators also have the final say on the appointment of department and agency heads such as police chiefs, school board superintendents, and cabinet officials who actually handle the nuts and bolts of state and local governments.

Reflecting Constituents and Their Wishes

In the **descriptive representation theory**, a state legislator, alderman, or county board member best represents constituents by reflecting them like a mirror, or like a photograph represents its subject. The representatives are of the same race, ethnicity, gender, and socioeconomic level as their constituents.[4] In practice, there is a bias in favor of electing individuals who have the money and time to participate or are willing to sacrifice much of these precious resources. For example, professionals such as lawyers with high incomes and schedules flexible enough to fit the legislative timetable can partially finance their own campaigns and can take the time to serve in the state legislature. City council, school board, and county board members usually represent a wider range of professions than state legislators. They include former city department employees, homemakers, and teachers, but these categories may not reflect the majority of citizens in a district.[5]

Other theories of representative democracy downplay the degree to which elected officials share descriptive racial, ethnic, and gender characteristics. These theories focus on how well the votes and actions of representatives reflect the wishes and the economic interests of their constituents. In other words, it doesn't matter if you are black if you vote to continue racial segregation or discrimination. It doesn't matter if you are a woman if you vote against women's rights or better day care services for children. Ultimately, how a legislator votes and advocates programs for constituents may matter more than skin color, gender, or other descriptive characteristics.

Demographic Changes

Once the preserve of white men who had wealth, leisure, and social prestige, legislative bodies now more closely reflect the demographic makeup of their states and cities. Blacks, Hispanics, Asians, and other minorities have increased their percentages in state and local government. Once rarities in legislative bodies, women now comprise from one-tenth to one-half the membership of legislatures and councils. Similar percentages of women are in leadership roles in these bodies. As the Struggle for Power feature shows, women have assumed leadership roles in many state legislatures.

These demographic changes in legislative bodies have come about for several reasons: The U.S. Supreme Court declared in *Baker v. Carr*[7] and subsequent decisions that all legislative districts must have the same population as others in a state or city, and that minorities

STRUGGLE FOR POWER

Women in Government

Female state lawmakers are moving into leadership roles in unprecedented numbers, overseeing their legislatures' daily business, shaping states' political agendas, and laying the groundwork to get more women elected. In 2007, 58 women lawmakers were chosen as legislative leaders—senate presidents, house speakers, presidents pro tem—a 20 percent gain over the 48 in 2006 and more than double the female leaders in 2000.

The gains come at a pivotal moment for female politicians, with Hillary Rodham Clinton having run for president and Nancy Pelosi becoming the first female speaker of the U.S. House of Representatives. Having female legislative leaders will influence the public and fellow lawmakers to change their attitudes so more women seek public office and more voters support them.

"If you're not at the table, you don't get heard," said Massachusetts Senator Therese Murray, a Democrat who made history in March 2007 when she was chosen by fellow lawmakers to serve as senate president—the first woman in her state to do so. She holds one of the most powerful positions in state government.

There were only four women legislative leaders in the nearly all-male political world of the late 1970s. That figure rose in the 1990s to 28—still 8 percent or less of all legislative leaders. But after 2000, the numbers began to climb: to 30 in 2001, 42 in 2003, 48 in 2006, and 58 in 2007. Credit goes to the women who broke ground and paid their dues, became committee chairs, and built coalitions. To quote Debbie Walsh, director of the Center for American Women and Politics at New Jersey's Rutgers University: "Part of it is they've been in there and they've earned their spots."

But that's only part of the equation, she and others say. "It isn't just about individuals," said Marie Wilson, president of the White House Project, a nonprofit group that encourages women to lead in business and politics. "There has been a change in the country and the culture. The culture matters. . . . You start to see women as leaders. You get more comfortable with women, whether you like their policies or not."

Taking on leadership responsibilities helps erase the outdated, misguided perception that there are "women's issues," said Kentucky Senate President Pro Tem Katie Stine. She not only wields the gavel but oversees debate and decides what gets heard and when. "I didn't get the impression that my gender was even an issue," Stine said when she was elected. "When you're in the forest, you don't notice the trees."[6]

Source: Adapted from Robert Tanner, "Female State Leaders Double Since 2000," *Chicago Sun-Times*, April 3, 2007.

---

must have the same opportunities as whites to be elected as members of representative bodies. Redistricting occurs at least every decade after the census, and because minority populations have been growing at a rate much greater than that of the white population, their numbers have been increasing in state and local government.

The Los Angeles City Council elected in 2005, for example, shows this diversity. Of its 15 council members, 8 were white, 4 were Hispanic or Latino, and 3 were African American; no Asians were on the city council in 2005. Women held only 3 of the 15 seats, or 20 percent of the council.

In the Chicago City Council of 2009, 19 of the 50 members were African American, 9 were Hispanic or Latino, and 22 were Caucasian. By comparison, in the 1970s only 8 of the 50 aldermen were African American and no Hispanics or Latinos had served on the council. Likewise, while the council is still a male-dominated legislative body, the gender split has improved since 1971 when only 2 women served. There are now 18 women, or 36 percent of the current Chicago City Council.[8]

Racial and gender representation is not an exact match to the population in city councils like those of Los Angeles and Chicago, or in the various state legislatures, but women and minorities (including openly gay men and women) are now serving and leading state and local legislative bodies throughout the country. Nonetheless, although lawmakers increasingly represent the diversity of their states and cities, they are different from their constituencies in that, overall, they are above average in income and education and more ambitious for approval and achievement.

LAWMAKING ROLES

Whether serving in a city council or a state legislature, a lawmaker plays many different roles, often simultaneously. Legislatively, you vote on issues, and you try to resolve any legislative conflicts that might occur. You also perform many services for your constituents, and you act as a watchdog against governmental or political abuses.

Voting on Issues

When it comes to voting on legislation, all legislators—whether city council or school board members, state legislators or county board members—have a dilemma on how best to represent constituents. You must represent the particular district that elected you, but at the same time, you are supposed to represent the public good of the entire town, city, county, or state. These are sometimes conflicting demands. Should you act like a delegate representing a nation at the United Nations, or should you act like a trustee on the board of a corporation? Legislators who take on the role of a **delegate** will vote based on the opinions of the majority of their constituents. But legislators in the role of **trustee** will use their own judgment about what is in the public's interest, even if many or most constituents disagree. Sometimes with noncontroversial issues like the placement of a stop sign or the request for grants from the federal government, there is no conflict and the appropriate vote is the same for both delegate and trustee.[9] Legislators acting as delegates will vote according to their best understanding of how their constituents would want them to vote. A trustee will sometimes act as a lawyer doing what is in a client's best interest even if the client doesn't know enough about the law to ask, or as a bank trustee would invest funds for a minor who doesn't know enough about stocks, bonds, and real estate to know the best investment.

Added to these delegate and trustee cross-pressures is the moral and ethical requirement for legislators to represent their own conscience, especially on complicated issues that their constituents don't fully understand, or on issues of human rights that override parochial concerns. Issues involving immigrant rights, abortion, and same-sex marriage often must be decided based on conscience, even if a vote may be unpopular in the lawmaker's district.

Beyond their roles as delegates or trustees in representing their constituents, lawmakers tend to play out one or more distinct roles in the legislative process. David Barber helped capture these roles by observing the Connecticut legislature in action half a century ago.[10] Barber said that lawmakers tended to become experts, brokers, lawmakers, and ritualists. **Experts** focus on topics like education or welfare and become respected authorities to whom other lawmakers turn for cues in voting. **Brokers** enjoy the game of brokering deals among varied interests and across party lines on controversial issues—and getting the deals enacted into law. **Lawmakers** enjoy sculpting proposals into the best shape possible and then advocating their adoption. Both experts and lawmakers enjoy going to national conferences of legislatures or national conferences of city and county officials to learn about the latest policy ideas and management innovations. At the other end of the spectrum, Barber found **ritualists**, those who enjoy the status of serving but leave the heavy lifting to others. Yet these ritualists provide the support their leaders and parties need to enact and implement programs.

With each separate vote, a legislator must reconcile these different roles and demands. And this is not the end of the conflict. Most officials are elected with the support of a political party or faction to whom they owe allegiance. The position of their party is well known on each important issue before the legislative body. Legislators are generally expected to vote with their party on at least the most critical issues. Yet sometimes the demands of conscience, or the needs and views of their constituents, mean they must vote against the party's position. These are difficult decisions. They must be made in the face of legitimate but conflicting demands.

Cynics observe that all elected officials are so ambitious that their votes can be predicted by knowing which vote is most likely to get them reelected. Idealistic or naive observers believe that public officials always act in accordance with their conscience for the public interest. Neither view adequately explains all legislative votes. Legislators try to balance, as best they can, what will serve their political ambitions *and* what is best for the citizens they are elected to serve.

Resolving Conflicts

If men and women were angels, then legislative bodies would not be needed. Since we are not angels, representative institutions have developed in democracies to resolve conflict among us. Legislative bodies formally enact bills and budgets, consent to appointments made by executives to major posts, and oversee the management of government by the executive branch. In doing so, state legislatures and city councils are responding to pressures from interest groups, the executive branch, and citizens in their own districts. After assaying proposals and weighing pressures from all external and internal forces, these legislative bodies enact some bills, reject most, and pass budgets that regulate government spending.

To accomplish these tasks, power and influence by those who have it are imposed on those who have less of it. Enactment of laws that affect us all—whether they are local no smoking laws or state tax increases—requires compromise, sometimes under duress. Legislative leaders, governors, mayors, and interest groups often exercise their power and influence on lawmakers to force or persuade them to vote in a way they would not have done otherwise.

Conflict and duress are not fun. To relieve tensions that build in a setting where people hold strong, often opposing views, informal "rules of the game," or **norms**, have developed in legislative bodies. One important norm is that deliberations in a legislative body—whether a state legislature, county board, or local city council—are *impersonal*. Views of a citizen or a legislator may be attacked legitimately, but not the person advocating them. This allows a legislator to bitterly oppose another lawmaker's bill on the floor yet meet her for dinner later when the day's work is done.

Civility, or civil conduct, is another important norm of most legislative bodies. Legislators are expected to treat each other with politeness, thoughtfulness, and courtesy. A visitor in the legislative gallery may be surprised to hear an alderman or state legislator say, "I have the greatest respect for my esteemed colleague from across the aisle, but this bill is absolutely the most ill-conceived piece of gibberish ever to come before this body."

Reciprocity, the return of a favor or a vote in kind, is another legislative norm: do me a good turn in committee or on the floor of the legislature and I'll return the favor when I can. For instance, a state legislator from an urban district may vote to pass another legislator's farm policy proposal out of committee even though the urban legislator may not care about agricultural issues. The expectation is that the legislator from the rural district will return the favor when inner-city school lunch programs are up for a vote. In city councils, decisions about zoning changes or stop signs in a single ward are a matter of "aldermanic privilege." In such a case, other members of a city council will vote for or against minor decisions affecting a single neighborhood as long as that alderman will support their decisions on matters affecting their wards.

These norms do not always benefit the legislative process. For instance, one may hear a legislator say in committee: "I think this is a bad bill, but out of respect for the sponsor I will vote aye in committee to move the bill to the floor so the whole chamber can discuss the issue." This reciprocity can clog the floor with bad bills. Yet this norm also softens the blow of conflict when a member must oppose the same colleague on another matter.

Serving Constituents

Legislators do more than vote on issues, and they do more than play particular roles on the floor of the legislative body. They also provide services to constituents. Good legislators serve as ombudsmen who ensure their constituents get the government services they deserve. When the bureaucracy bogs down and fails to issue a welfare check, pick up the garbage, or approve a necessary building permit, the state or local legislator intervenes to get the service delivered. With better pay, staff support, and district offices, many state and big-city lawmakers have expanded their constituent services role. A visible district office operation attracts residents who have trouble navigating the many agencies—local, state, and federal—that exist to regulate and assist citizens and businesses. When a license to operate a business becomes snarled in the state capitol or city hall, when a small business owner has a dispute with the tax department, or when a nonprofit organization needs help identifying grant opportunities, elected legislators and their staffs step in to help solve these problems.

Following the lead of the U.S. Congress, many state legislators, city council aldermen, and county board members have become more aggressive at winning "pork barrel" funding for projects such as new parks, swimming pools, paved streets, sidewalks, and community buildings for their districts. The premise is that local lawmakers know more about needs in their communities than bureaucrats in the state capitol or in large departments at city hall. This is good local politics for incumbents. Problems arise, however, when expensive projects viewed as important to one town or city ward look like a waste of taxpayer money to other legislators or to the general public.

Watchdog Functions

Some lawmakers want to create conflict, or at least controversy, rather than resolve it. This happens when they find abuses of power in an entrenched majority. Chicago Alderman Dick Simpson and his small band of independent Democrats—in opposition to more than 45 regular Democrats loyal to Mayor Richard J. Daley and later Mayor Michael Bilandic— aroused the press and public to matters of racial injustice, budgetary waste, and corruption in city contracts. This opposition bloc offered alternative ordinances or amendments to those of the Chicago administration. These aldermen also offered an alternative city budget each year. The council's opposition aldermen helped find candidates to run against those selected by Chicago's entrenched Democratic political machine. They laid the groundwork for the later election of African American reform Mayor Harold Washington, who enacted many ordinances once brushed aside under Mayor Richard J. Daley. Lesson: A small minority in a legislative body can prepare knowledgeable people to take over and operate as a majority at a later time. They can arouse the public to change the politics and government of a city or state.

Before lawmakers can fulfill their many functions, however, representative bodies must organize themselves at the beginning of each new session. That process is described in the next section.

ORGANIZING AND OPERATING A LEGISLATIVE BODY

Legislatures meet in diverse settings: ornate rococo chambers in the East and Midwest, clean-lined settings in the Southwest, computerized city halls in suburbs, and makeshift settings in tiny villages. They all serve the same purpose. In their different shapes, they are the ballparks in which the legislative game is played out.

As we discussed in the last chapter, state constitutions and city charters set the basic rules for the game. Most state legislatures are bicameral with two chambers: a house of representatives or assembly and a senate. Nebraska is the only state with a unicameral legislature, a single legislative chamber. In contrast to state legislatures, most U.S. cities, towns, and counties operate with a unicameral legislative body.[11] "Special districts" of local government such as school, park, and waste management districts are governed by a small board of directors, commissioners, or trustees.

Membership in state legislatures ranges from California's 30 state senators and 50 assembly seats for more than 30 million residents to New Hampshire's 500 house members for its 1 million citizens. City councils range from 5 members in small communities to 51 city council members in New York City.

At the beginning of each legislative term, legislators organize their legislative body. The elected members cast votes for the person to lead the representative body, whether speaker of the house of representatives, president of the senate, or floor leader of the city council. The **speaker of the house of representatives** and the **president of the senate** have the power to appoint committees and guide proposals through the legislative process as determined by the specific rules of their respective chambers. They are the most powerful officials in the house and senate. The **floor leader** is the legislative member who determines the agenda and sometimes the procedural rules. The floor leader is usually from the majority party and chosen by the majority party caucus.

The selection of legislative leaders illuminates the four themes of this book. First, it is a *struggle for power* because the leaders have much more power than the average legislator. Second, the leadership selection is constrained by the *institutional restraints*—both laws of the state and rules of the legislative body about how the election of leaders is to be held and decided. Third, *representative democracy* is practiced as the members of the legislative body struggle to choose their leaders. Lastly, the choice of leaders directly determines how *effectively the legislative branch will play its proper governmental role.* If good leaders are chosen it is likely, but not guaranteed, that good laws will be made.

Most states and major cities operate on a partisan, political party basis. Smaller cities, towns, and suburbs usually are officially nonpartisan, although city council and county board members' party loyalties are usually well known to politicians and voters. Legislatures may also include several caucuses. A **caucus** is a group of legislators of the same political party or faction, or an informal, closed meeting of the group. The women members of a legislature, for example, might form a women's caucus, or there might be a Latino caucus. In partisan legislatures, each of the major parties holds a party caucus to select the person they want to be their leader for the legislative term.

Because of the powers that come with leadership of a legislative chamber, members of a party caucus are expected to vote formally on the floor of their chamber just as they did in the earlier caucus. The leader selected by the party caucus with a majority of members in the chamber is almost always elected a few days later on the floor as the leader of the chamber, becoming the speaker of the house or majority leader of the senate. The leader selected by the party with the minority of members becomes the minority leader of the chamber.

Party caucuses represent the first stage in the allocation of power and influence within most legislative bodies. Often the same division in the vote on legislative leadership will recur in later partisan votes on the most important legislation. These first key votes not only organize the legislative body but also create a template for party-line voting over the next several years.

Sometimes there are glitches. In 1979 in Illinois, the African American caucus in the Illinois state senate withheld its votes for senate president from its own majority Democratic Party. As a result, the Republican governor presided over the senate for three months. The power struggle was finally resolved when these Democrats agreed to vote for a particular senate president in return for more powerful committee and leadership positions for African Americans and their white liberal reform allies. When the Indiana house of representatives was once evenly divided between the two parties, it elected co-speakers who presided on alternate days. The Florida senate for a time agreed to switch senate presidents between different parties on alternate years.[12]

One of the most interesting modern leadership fights occurred in Tennessee. In the November 2004 election, the Republican Party won a majority of seats in the Tennessee senate for the first time since Reconstruction. So Republicans expected to elect the leader of the chamber, who in Tennessee is called the speaker of the senate. But when the votes were counted, Democrat John Wilder won by a vote of 18–15. Wilder had been speaker since 1971, and two Republicans voted for him "out of loyalty for the years of power-sharing he had instituted" in the years of Democratic Party control. In previous years the Senate Democratic caucus had tried to dump Wilder as speaker, but he had built a coalition of support between some Democrats and Republicans to keep his office. In return for their support, he had appointed a number of Republicans to lead committees.[13]

If there is competition among two or more candidates for a party's leadership, then supporters of the winning candidate tend to receive the best committee chairs and committee membership assignments from their new leader. Legislators are careful about whom they support, since it will affect their own power and prestige.

Like the leaders of the state legislative chambers, the heads of city councils and county boards generally have the power to appoint committee chairs and members. The leaders shape the rules by which the legislative body operates. They control the flow of legislation, assign offices, and hire and direct the staff of the party in the legislature. Because of these substantial powers, the head of the legislature is also able to raise significant campaign contributions, which can be bestowed upon party members in their reelection campaigns. This makes individual legislators even more beholden to their legislative leaders. As a result of these powers and their enhanced ability to raise large amounts of campaign funds, legislative leaders have great leadership authority.

Leadership power has its limits, however. In California, Arnold Schwarzenegger, a strong Republican governor facing powerful Democratic legislative leaders, initially issued a challenge to the legislators: "The train has left the station and there are three things [legislators and their leaders] can do. One is they can join and then jump on the train. Number two, they can go and stand behind and just wave and be left behind. Or number three, they can get in front of the train—and you know what happens then." But the governor's own train jumped the tracks. His ballot initiatives in November 2005 failed miserably and his approval rating dropped from 54 percent to 34 percent. In order to win reelection in 2006, he changed tactics and negotiated his programs with the very Democratic legislative leaders he once mocked as "girlie men."[14]

Although the majority party leader is most often the presiding officer of a legislative chamber, elected executive officers sometimes preside. In Texas, for example, the powerful lieutenant governor presides over the state senate and manages the business of that body. In most cities, the mayor presides over city council meetings. In counties, the county board president or a county mayor chairs the county board meetings. But most state legislatures and many local government bodies are presided over by the majority party leader. This provides the majority party with control over the floor debate and procedures.

Since the leadership in the legislature is very powerful in most states, a changeover in parties is momentous and often results in a major change in the bills that are proposed and passed. For instance, in the November 2008 elections, Democrats made a net gain of 3 new governorships from 2006 to 2008 and controlled 27 statehouses (up from 23), as Figure 3.1 shows. As a result of Democrats gaining greater control, a number of state legislatures began enacting social and economic policies that had been tabled during Republican control. For example, New Hampshire lawmakers approved a civil union bill, Iowa and Colorado banned discrimination based on sexual orientation, Colorado made gay adoption legal, Maryland passed a smoking ban, Iowa raised the minimum wage, and Maryland's governor signed into law the country's only state "living wage," requiring state contractors to pay their employees a minimum of $8.50 per hour.[15] Since the 2008 Democratic victory, Massachusetts, Connecticut, Iowa, Vermont, and Maine have adopted laws allowing for same-sex marriage.

In recent decades, a number of state elections have resulted in equal numbers of Republican and Democratic members in legislative chambers. In order to prevent stalemate over election of legislative leaders, accommodations must be made. In Arizona in 2001, when 15 members from each party were elected to the Senate, the lawmakers agreed that one party would select the presiding officer while the other party would select the chairs of powerful committees.[16] In Montana, state law says that in a tied house, the speakership goes to the party of the governor. In 2004, this rule decided a 50–50 split between Democrats and Republicans in the Montana house of representatives, giving the speakership to the Democrats because the newly elected governor was a Democrat.

It needs to be stressed that legislative leaders are not just arbitrarily given power. Legislative leaders are expected to make the legislative system manageable "by presenting issues clearly, narrowing the alternatives, organizing public hearings, and promoting the party or administration's point of view on bills."[17] In most legislative bodies, whether they are city

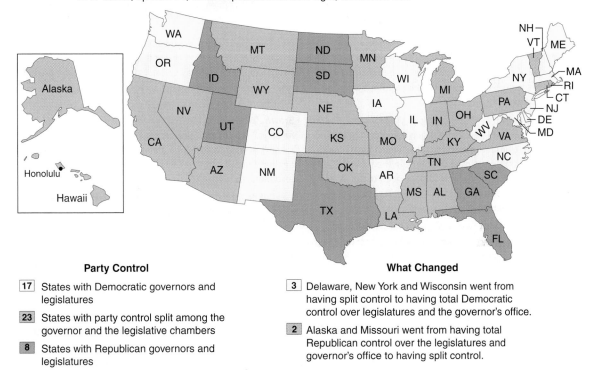

THE STATE OF PARTISAN POLITICS

The 2008 elections gave Democrats control of the governor's office and both legislative chambers in 17 states, up from 14, while Republicans control eight, down from ten.

Party Control

17 States with Democratic governors and legislatures

23 States with party control split among the governor and the legislative chambers

8 States with Republican governors and legislatures

What Changed

3 Delaware, New York and Wisconsin went from having split control to having total Democratic control over legislatures and the governor's office.

2 Alaska and Missouri went from having total Republican control over the legislatures and governor's office to having split control.

FIGURE 3.1 The State of Partisan Politics: 2008 Elections

councils, county boards, or state legislatures, the majority leader is the most powerful figure next to the mayor, county board president, or governor. The leaders use their powers to "run" the legislature. They control committee appointments (and therefore legislative perks and rewards for individual legislators) and the flow of legislation (which makes all the difference as to whether a legislator gets a proposed bill or ordinance passed or defeated). They are key figures in the legislative body's organization and operations, ensuring that laws get passed, budgets get adopted, and appointments to the executive branch get confirmed.

LEGISLATING EFFECTIVELY

Throughout most of American history, state legislatures, city councils, and county boards were part-time, poorly paid, and lacking offices and staff. Consequently, lawmakers generally conceded the operation, budgeting, and management of government to governors, mayors, county board presidents, and school superintendents. On the big issues such as tax changes and education funding, lawmakers made the formal decisions. But even on these issues, they were largely dependent on governors, mayors, and lobbyists for the information on which to base their decisions. States in earlier years only provided a general framework

Montana's Easygoing Roy Brown

For a few weeks in the fall of 2004, Roy Brown was the speaker-elect of the Montana house of representatives, chosen by fellow Republicans in November to lead their 50–49 majority in the 2005 legislature. Yet a last-minute court decision changed everything.

Six days before the session began, the Montana supreme court invalidated a half-dozen votes cast in one house district, transferring victory from a third-party candidate to a Democrat, creating a 50–50 tie in the House. State law says the speakership in a tied house goes to the party of the governor, so Brown lost his new post before he ever took it, because the governor was a Democrat.

House Republicans were outraged over the reversal, threatening on the eve of the legislature to boycott the opening day or stage a mass walk-out on the floor. Yet Brown, an oilman turned real estate investor from Billings, Montana's largest city, counseled against it, saying it would make Republicans look like "a bunch of kids" who threw a tantrum after not getting their way.

The advice is right in character for the easygoing Brown. "Whatever happens, I just try to make the best of it," he says. "I try to persuade people with facts and logic and reason, and if they don't go along with that, I don't hold it against them. People have to vote the way they feel comfortable with."

While term limits ended Brown's career in the Montana house, in 2006 he was elected to serve in the Montana state senate.

Source: Adapted from Mike Dennison, "Big Skies, Big Leaders," *State Legislatures*, July/August 2005.

---

of laws, and they had much smaller budgets before World War II. Cities back then were much smaller and easier to run. Most aldermen and city council members were part-time.

In the 1960s and 1970s, an unusual combination of a powerful assembly speaker in California, Jesse (Big Daddy) Unruh, and charitable "good government" foundations such as Ford and Rockefeller began to modernize the California legislature. Most other big cities and states followed suit. By the end of the twentieth century, most large cities like New York, Chicago, and Los Angeles had modernized their governing bodies. Today, many smaller communities have significant staff support, along with up-to-date computer and cable television systems, to assist local council members in keeping the public informed. By the 1960s, internship programs in the legislatures were created to provide information-gathering staff for lawmakers, briefing them on important issues and showing that independent information had value. Legislative leaders began to provide offices for members and to publicly support higher salaries. Permanent legislative staff eventually augmented the interns. Members began to ask more informed questions in committee meetings.

Author Jim Nowlan had no office and no staff when he was first seated in the Illinois house in 1969. The job paid $6,000 a year. He returned calls to constituents from a tiny pay phone in the house corridor as there were no cell phones back then. Before he left the legislature four years later to run for lieutenant governor, Nowlan had a small allowance for staff and office and his salary was $18,000 a year. By 2006, Illinois state legislators earned $57,619, plus expense accounts to cover the costs of attending legislative sessions. They now have offices in both the capitol building in Springfield and in their home districts. Legislative leaders, including committee chairs, receive supplemental salaries of $9,000–$23,000 a year in addition to their regular legislative salaries. As Table 3.2 indicates, state legislative salaries are increasing. In 2009 they ran from a high of $116,208 in California to a low of $200 or

TABLE 3.2 Salaries of State Legislators in 2009

State	Annual Salary
California	$116,208
Michigan	79,650
New York	79,500
Pennsylvania	76,163
Illinois	65,353
Ohio	60,584
Massachusetts	58,237
New Jersey	49,000
Wisconsin	47,413
Maryland	43,500
Delaware	42,750
Washington	42,106
Oklahoma	38,400
Hawaii	36,700
Florida	31,932
Missouri	31,351
Minnesota	31,140
Colorado	30,000
Connecticut	28,000
Iowa	25,000
Alaska	24,012
Arizona	24,000
Maine	23,713
Oregon	19,884
Tennessee	18,123
Virginia	18,000
Georgia	17,342
Louisiana	16,800
Idaho	16,116
Arkansas	15,060
West Virginia	15,000
North Carolina	13,951
Rhode Island	13,089
South Dakota	12,000 2 yr term
Nebraska	12,000
Indiana	11,600

(continued on next page)

TABLE 3.2 Salaries of State Legislators in 2009 *(continued)*

State	Annual Salary
South Carolina	10,400
Mississippi	10,000
Texas	7,200
Vermont	614 per week
New Hampshire	200 2 yr term
Kentucky	185 per day
Wyoming	150 per day
Nevada	137 per day
Utah	130 per day
North Dakota	120 per day
Alabama	101 per day
Kansas	86 per day
Montana	83 per day
New Mexico	0

Source: Salary data from National Conference of State Legislatures (NCSL).
Notes: 1. Source of data: National Conference of State Legislators, 2008 spring survey.
2. Salary for Washington Legislators is as of September 1, 2008.
3. In Virginia, the annual salary for Senators is $18,000 and $17,640 for House members.

less in New Hampshire. A number of large states, such as Florida ($30,336) and Texas ($7,200), don't pay their legislators much for their services. But most states are increasing the salaries and the support staff to make public service more attractive and to create a more professional legislative body.

Author Dick Simpson, who was first seated in the Chicago city council in 1971, had an office tucked away in the back of city hall, a small stipend for an office in the community, and a single staff member paid by the city. The job of alderman paid $8,000 a year. To provide the services his constituents needed, he donated his entire salary to run his aldermanic office. In addition, he raised money in the community so that he could have a staff of three people, phones, and a photocopy machine. When he left office in 1979 to return to university teaching full-time, Simpson's aldermanic salary had been raised to $17,500. As shown in Table 3.3, Chicago aldermen in 2007 were paid $98,000, with generous expense and staff accounts to provide the administrative support. These salaries are indexed to inflation. As of 2009, Chicago aldermen were paid $104,000 for what some of them still regard as a part-time job.

Table 3.3 also shows that Los Angeles, with only 15 council members, pays the highest city council salary in the United States at $139,476. At the other end of the scale, the 10 city council members in San Antonio earn the least of any large city at $1,040 a year. City council members in smaller cities naturally earn less than those in bigger, wealthier cities. In East Lansing, Michigan, an alderman earns $6,600 a year while Hialeah, Florida, council members get a salary of $3,254 but an expense account of $38,000. Today, most state legislators and big-city council members have substantial personal and committee staff support, salaries from less than $20,000 to more than $100,000 annually, good pension and health benefits,

TABLE 3.3 City Council Salaries

	No. of Council Members	Salary
Los Angeles	15	$139,476
Chicago	50	98,000
New York	51	90,000
Philadelphia	17	85,000
Detroit	9	81,700
San Diego	8	71,796
Houston	14	40,000
Dallas	14	35,000
San Antonio	10	1,040

Source: *Chicago Tribune*, November 7, 2002, Metro Section, 14.

and the confidence that the legislative branch is independent of and co-equal with the executive branch of government. Those legislative bodies that haven't raised legislative salaries and provided support staff are, for the most part, in the process of doing so. The current economic recession has not caused any cuts in salaries of legislators themselves and almost no staff have been cut.

Members of local government bodies in suburban villages and towns often have even fewer resources; many township clerks, for example, serve part-time. Most local government trustees have no additional staff of their own and are paid small stipends or serve on a voluntary basis. As a trustee of the township of West Deerfield, Illinois, author Betty O'Shaughnessy relies on the township supervisor and clerk and their staff to provide her with the information and help she needs to carry out her duties, which include approving the budget, setting general policy, and making decisions on township responsibilities such as road maintenance and general assistance. With a small budget, the amount of services local governments can provide on their own is minimal, and West Deerfield Township officials search for ways to serve residents without increasing taxes, such as directing constituents to appropriate government or community service offices, and finding alternative ways to finance and address local needs. For instance, West Deerfield houses the local food pantry, which is supported through donations and run by volunteers, and the township received a grant to purchase a van for taking mobility-challenged residents to medical appointments.

Information is indeed power. Almost all lawmakers want to be remembered well. They want to make sound, rational decisions. Good information is necessary to assess the severity of a problem and to develop responsible options to correct it. In most state legislatures and big cities, lawmakers have committee staff experts and often their own personal staffs to do analysis and develop policy options. Many of these staff members begin right out of college serving as interns and "fellows." They often stay in government to become respected experts, relied on by elected lawmakers for legislative recommendations. You may want to consider a legislative or government internship as a part of your own education. At a minimum, every student and every citizen should attend at least one city council or state legislative session to better understand what representatives do.

COMMITTEES AS THE FOUNTAIN AND DEATHTRAP OF LEGISLATION

The eight-member Arizona Senate Education Committee opened its hearing on bills one sunny March day in 2001. Thirty or so lobbyists for groups like the state teachers' union, the school board association, the state universities, and the state civil liberties union sat in the audience of the hearing room, facing the committee. Behind the senators, two young committee staff sat, poised to provide information as requested. Earlier the staff had prepared one-page synopses, without recommendations, of the dozen bills to be heard by the committee that morning. One state senator, not on the committee, went to the witness table to present her bill to provide for college hazing penalties. "Parents in my district asked for this bill," the senator explained, not sounding enthusiastic about her legislation.

Lobbyists for the Arizona Civil Liberties Union and one of the state universities had registered their opposition to the bill. Each testified briefly (the senate had a five-minute time limit) that the bill was unnecessary. "The campus administrations have authority to deal with hazing infractions," said one. Members found other weaknesses in the bill, yet the committee decided to vote it out favorably so that the full senate could hear it. And so a typical legislative committee hearing goes.

Committees can be valuable in finding the weaknesses in proposed legislation and proposing amendatory language. Committees can sift the wheat from the chaff. When committees are strong, they can stop bad bills from clogging the chamber floor. At the same time, the committee process can seem undemocratic if a bill favored by a majority of the legislators has been prevented by a small minority of committee members from reaching the floor for a formal vote.

The fact is that few legislative bodies, especially the large ones, could effectively evaluate on the chamber floor the thousands of proposals introduced each biennium. Less than one-fourth of the bills or city ordinances introduced in a legislative session actually become law. In large states such as California and New York, 10,000 bills are introduced in a single legislative session, but only 2,000 or so will be enacted into law.[18] In Illinois, about 6,000 bills are introduced over the two-year state legislative session, but fewer than 1,000 pass.[19] In smaller states, a typical session may produce fewer than 300 pieces of legislation. In a large city like Chicago or New York, as many as 2,000 ordinances, orders, and resolutions may be introduced each year. Because most city council orders, ordinances, and resolutions are noncontroversial and are introduced by the city administration, most proposed city legislation becomes law. In smaller suburbs with city councils in which there are few conflicts, fewer than 100 ordinances may be introduced and passed each year.

The vast number of proposed laws (and the technical nature of some of them) makes it virtually impossible for a legislator, city council member, or county board commissioner to actually read and study every bill or ordinance. Zoning ordinances and antipollution laws, for instance, may run dozens and dozens of pages with technical descriptions of property or scientific descriptions of pollution levels.

A legislator's task of voting on all this legislation is simplified in three major ways. First of all, most of the proposed legislation is routine, governed by strict standards, and noncontroversial. Second, on important legislation, the political parties and the party leadership in the legislative body have a clear position, and good party members are expected to vote in accordance with the party decision. The danger is that the dominant party will have such strong control over the legislature that individual legislators play no meaningful role. Party control that is too strong can undermine representative democracy, but it does make the job of the individual legislator very easy.

Finally, to simplify their decisions, many legislators follow recommendations from the committee that has studied the legislation most thoroughly. The committee, and often its chair, will lead the discussion, answer questions, and provide a recommendation during the floor debate. Their less-informed colleagues often follow the committee recommendations because the committee members have most thoroughly studied and debated the legislation.

Types of Committees

Generally, legislative bodies are divided into standing committees and special committees. **Standing committees** are permanent legislative committees that write laws or statutes for a specific issue area, such as transportation, education, public safety, and budget or finance. In a small suburban town, there may be only a handful of standing committees covering finances, administration, and zoning. In larger cities or counties with larger legislatures, there may be more than two dozen standing committees. State legislatures tend to have a fairly large number of standing committees, usually 20–30. In addition, **special committees** may be appointed for a short period of time to study a particular issue such as bilingual education, a major corruption scandal, or an increase in crime. Each member of the legislature serves on a number of committees. The longer they serve in the legislature, the more likely they are to be appointed to the most powerful committees, such as budget or finance.

Committees are meant to allow for debate and discussion of proposed legislation and to receive testimony from experts as to the wisdom of a course of action. Moreover, committee members generally have acquired a certain amount of expertise in their issue areas. In **committee-dominated legislatures**, the committees do the major work on legislation, and the committee recommendations are often used as guidelines for voting by both parties. Recently, we have seen the development of **party-dominated legislatures**, in which the majority and minority party leaders determine how their members will vote, with the minority party usually voting in opposition.

Sometimes committees are used to kill legislation without a hearing if the administration or legislative leaders oppose it. For instance, in the Chicago city council, the committee chairs usually refuse to hold hearings on legislation the administration or leadership opposes or which has been introduced by opponents. They hold hearings only on legislation they plan to pass, based on the instructions from the city council floor leader or the mayor.

Routine Legislation

Some legislation is routine and may have little effect on a legislator's constituents. For example, an aldermen or state legislator may introduce a resolution to congratulate a local couple on their 50th wedding anniversary. If there is no objection, aldermanic or legislative privilege prevails (that is, other members of the council or state legislature defer to the representative to whose community the legislation pertains). The resolution passes unanimously and the couple are sent a copy of the official resolution to show their grandkids.

A slightly more complicated case of routine legislation is an ordinance to place a stop sign at a particular intersection. The ordinance is sent to the traffic committee of the city council. The intersection's traffic patterns and accident rates are studied by the city's street department, to ensure that the intersection meets state standards for stop sign placement.

The department then recommends either passage or defeat of the legislation based on its findings, and the committee recommends the same action to the council in compliance with state laws. This stop sign ordinance passes or fails on a unanimous vote.

More significant and controversial legislation is processed through legislative committees. They handle proposed legislation on important subjects. Committees break up the work of studying legislation among small groups that are supposedly more informed than the average legislator on the topic under consideration. Committee hearings are meant to find solutions to problems. After hearings are held, the committee may initiate new legislation or "perfect" proposed legislation by introducing amendments to it.

Killing a Bill

While committees can facilitate the passage of legislation, they can also be the deathtraps for proposals party leaders want buried. For example, Indiana has long been divided county by county into Eastern and Central time zones, but the federal government decided in January 2006 to leave St. Joseph and Marshall counties in the Eastern time zone year-round and put nearby Starke and Pulaski counties in the Central time zone. Democrats in the Indiana legislature wanted to force a statewide referendum to put the entire state in one time zone or the other, but the speaker of the house, Brian Busman, blocked it by assigning it to the rules committee, where it was killed. Such are the powers of a house speaker or senate president—and this is one use of committees: to kill rather than promote legislation.[20]

If the leadership opposes a bill that a committee is about to vote out onto the floor, the legislative leaders in some states have the power to substitute committee members for a single vote to prevent the committee's regular membership from exercising its will. Fortunately, this is the rare exception, not the rule, in most state and local governments, because it undermines the committee process in a legislative democracy.

When the committee process works well, it means that a small group in the legislature, one with the most expertise on an issue, can explore specific problems. Committees provide a place for citizens and concerned groups to testify and offer diverse perspectives on legislation. Unfortunately, in some legislative bodies, the committee process is a sham. The resulting legislation may be the worse because of it.

THE POWER OF REDISTRICTING

Fundamental to the organization of legislative bodies is the question of the makeup of the legislative districts, which are at the heart of representation itself. The U.S. Constitution provides that each state shall provide for representation of the members apportioned to it by the Congress, following each 10-year census. State constitutions also provide for periodic redistricting of the state legislature, and cities are usually required to redistrict at least every decade following the census.

District boundaries are decided by the legislature, and the task of redistricting provides the major political parties with ways to exercise or gain power. Legislative districts are primarily drawn for partisan advantage and to protect incumbents. Although a guiding principle is supposed to be equal representation, many state legislatures have resisted redrawing the lines of their districts to accommodate changes in population, even when their state constitutions required redistricting. In the 1960s in California, one state senator represented a county district of 8,000 people while the senator representing Los Angeles County covered

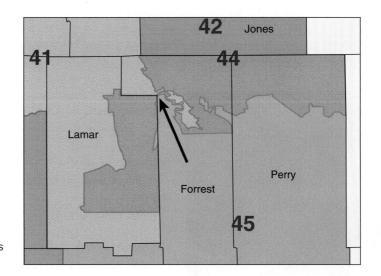

MAP OF MISSISSIPPI STATE SENATORIAL DISTRICTS, with gerrymandered districts 44 and 45.

3 million residents. These grossly unequal districts, however, are no longer allowed, since the U.S. Supreme Court declared in *Baker v. Carr* (1964) and subsequent decisions that "one person, one vote" shall be the law of the land. Today, each of the 30 senate districts in California has the same 1 million residents. Each of the 15 members of the Los Angeles city council represents about 254,000 residents.

Nonetheless, redistricting and drawing fair and equal districts remains contentious and very political, and the requirement for equal population among districts has not eliminated **gerrymandering**—drawing district lines to benefit a particular political party or ethnic, racial, or other group—which is still common in states, cities, and counties. For example, by drawing district boundaries that concentrate or dilute the minority party's voting strength, the majority party can create more districts likely to elect its party candidates than warranted by the overall voting strength of the party.

An extreme example of gerrymandering occurred in Texas in 2003, when the newly elected Republican majority in the Texas legislature adopted a redistricting plan that helped Republicans gain 21 of the state's 32 seats in the U.S. Congress. Although such redistricting between census counts is unusual, in 2006 the U.S. Supreme Court ruled 5–4 that it was constitutional.[21] This opinion, unless reversed, will allow state legislatures and city councils to continue to create districts that advantage one party or another more often than every 10 years.

Some cities and suburbs still elect part or all of their city council or county board at-large from the entire city or county, but most have switched to district or ward elections, and the district boundaries are redrawn at least after every federal census. Gerrymandering is thus a power struggle within local governments as well as state legislatures.

MAKING YOUR BILL A LAW

Passing any bill requires effort, but passing a controversial bill takes a lawmaker's perseverance as well as support from other public officials and interest groups. It requires hard work and proper timing. As the saying goes, "in politics, timing is everything." Martin Luther King Jr. preached that achieving civil rights for African Americans in the 1960s was "an idea whose

Enacting a Law

Effects of strip mining.

As a state legislator, Jim Nowlan pledged to his voters that he would pass a bill requiring coal-mining companies in Illinois to reclaim land stripped for coal, restoring the land to almost its original topography, rather than leaving it scarred and barren. First he had to draft the idea into language for a legislative proposal, or bill. Nowlan asked undergraduate student Larry Custer to do some volunteer research work. Custer found an agronomist at the University of Illinois at Urbana who knew about land reclamation. The three of them drafted Nowlan's idea into general language, which the Legislative Reference Bureau put into proper legal form.

House Bill 1626, an act to amend the Department of Mines and Minerals statutes, was set for a hearing at the regular Tuesday afternoon meeting of the Conservation Committee. Six other bills related to state parks and mining issues also appeared on "the call" for the hearing, but the small committee room was crowded only with individuals opposed to Nowlan's bill.

By tradition, the committee chair, a genial veteran lawmaker, first called sponsor Nowlan to the witness table to testify in behalf of HB1626.

"This bill, Mr. Chairman," Nowlan said, "simply seeks to require coal companies to return land stripped for coal mining to approximately the same topography that existed before mining. This will not restore the land to its original value, but will reclaim some of the value of the marvelous farmland of my district."

"What will it cost to reclaim the land?" asked one committee member.

"I don't know," Nowlan had to answer.

"How can you return huge, gaping gravel pit holes to their original topography?" asked another.

"This bill is not intended to apply to gravel pits," Nowlan responded.

"But it does," pointed out the committee chairman.

And so it went. Nowlan had no good answers, only an idea and an objective.

Then the chairman asked for testimony from the many persons who had registered their opposition to HB1626, including the chief lobbyists for the Illinois Coal and Illinois Aggregate (sand and gravel) Producers Associations. Opponents declared that the bill would bankrupt the coal companies, that it was vague and unenforceable, and that it would not achieve its purposes. HB1626 was defeated by an 8–3 vote, and Nowlan's idea was dead for the remainder of that two-year session of the legislature.

Following this disastrous initiation into lawmaking, Nowlan prepared for another try at a reclamation bill. He and his backers on county boards in central Illinois formed a Coalition to Reclaim Our Land, which raised money from individuals and from county boards of counties that were losing farmland valuation to the mining shovels' voracious bites.

Nowlan and a new bill co-sponsor, A. T. McMaster, led a small delegation of Illinois lawmakers from the Conservation Committee on chartered planes (paid for by the coalition) to visit sites of strip-mining reclamation in Pennsylvania and Ohio. The mining situations were different, but Illinois legislators saw that significant reclamation was possible without bankrupting the coal companies. The coalition also hired expert agronomists and others knowledgeable about the technology of reclamation to help draft a detailed, defensible bill.

Then Illinois's new governor, responding to rising public concerns, came out strongly in favor of environmental reform generally and strip-mining reform specifically. His agencies for conservation and mines and minerals, opponents in the previous legislative session, were now proponents of the reclamation legislation.

When a governor declares support for a lawmaker's bill, the situation changes markedly in favor of the legislation. It makes clear that the governor will sign the bill if it reaches his or her desk. Governors also bring the weight of their office to bear in behalf of a bill that is seen as part of "the administration's program." The governor's own lobbyists "work the bill," that is, seek support from key committee members and then other legislators when the bill reaches the house floor. Since legislators would like to gain the governor's support for their own legislative initiatives, many committee members and other legislators will rush to support the administration's position.

When Nowlan's bill came up for a committee hearing, the opponents returned. The bill would increase the cost of mining without increasing the market value of the coal mined. But this time, proponents, orchestrated by the Coalition to Reclaim Our Land and the governor's Institute for Environmental Quality, packed the hearing room and had answers for all the questions about the impact of the bill.

ABC Nightly News sent a two-helicopter crew to Nowlan's district to interview him and the president of the Illinois Coal Association. Filming from the floor of a mine, with walls of earth rising 80 feet on either side, and 10-story coal shovels and drag lines groaning in the background, ABC had compelling footage for the evening news. This coverage caused other lawmakers to have a new respect for Nowlan. The bill passed, but not without amendments from the coal association, adopted over Nowlan's opposition, which weakened the bill's requirements.

This is how an idea gets translated into a bill that becomes a law, and how draft legislation is improved over time but also subject to compromise.

time has come." In later decades, environmental protection was an idea whose time had come. As the Struggle for Power feature reveals, changes in public opinion over time helped pass a bill introduced by author Jim Nowlan in the Illinois legislature.

Today, state lawmakers generally have more staff and research support than in the past. Passing a controversial bill, however, is just as difficult as ever. It often requires years to perfect a bill, educate lawmakers and the public, build support, and get through the legislative gauntlet.

The legislative process favors inertia. Nothing happens until a lawmaker activates the process. It is much easier to beat a bill than to pass one, because at any of a dozen steps, a bill can be blocked. In the U.S. Congress, committees often take control of shaping important legislation. That is not generally true in state legislatures and city councils, where the primary sponsor of a bill is solely responsible for shepherding it through one or both chambers and seeing it enacted.

One of your authors, Chicago Alderman Dick Simpson, was one of a handful of reformers in the Chicago city council who discovered that an alderman in the minority can force passage of a city ordinance. When he was first elected, Simpson found major traffic problems in his ward at a certain intersection. The previous alderman had introduced legislation to put in a stoplight, but the proposal had been defeated because the intersection did not meet state standards for a light, according to the city's Department of Streets and Sanitation. Nevertheless, cars at the intersection had hit children, and parents were irate. Simpson reintroduced the stoplight ordinance.

Working with the local community organization, Simpson led a delegation of 50 mothers and children (including babies in mothers' arms) to a meeting with the commissioner of streets and sanitation. Although uncomfortable with a citizen delegation in his office, the commissioner listened as the parents and Simpson described why a traffic light was needed. The department once again surveyed the intersection—and this time it decided that a stoplight was needed. Clearly, the commissioner did not want parents and crying babies back in his office or a press conference by Alderman Simpson declaring that the department was unresponsive to community needs. Workmen installed the $100,000 stoplight quickly.

Several lessons can be learned from this account. First, while technical standards are important, the bureaucracy has latitude in how to interpret the standards. Second, humanizing the problem and applying political pressure within an agency can get results.

Third, by winning a battle like this with city hall, the parents and the community organization can feel empowered rather than victimized by a faceless bureaucracy.

Actions by legislatures and councils range from important life-and-death matters, like capital punishment and health care for low-income children, to harmless resolutions celebrating wedding anniversaries. Several years ago, the Texas house unanimously adopted a resolution commending a person who was a serial killer. A lawmaker had introduced it to make the point that few knew what was in such resolutions.

Thousands of bills are introduced in state legislatures each biennium and in city councils each year, but less than one-quarter become law. Those left buried in committee or defeated on the floor are found to be unnecessary, poorly drafted, bad ideas, or too politically hot to handle. Some, like strip-mining bills, may have to be introduced again in future sessions. Some, like city council reform ordinances and budget amendments, pass only years later.

Routine matters rather than controversy dominate many meetings of legislatures and city councils. As the Insider's View feature reveals, local councils deal with both major and minor issues. At many council meetings, the members and key city staff discuss issues publicly before a live audience as well as a cable television audience, and citizens have their chance to make comments on all major legislation. Such public decision-making and open legislative processes are the warp and woof of democratic government.

LIMITS ON LAWMAKERS

Elected officials are not free to act on their every whim. Among the ways of limiting the power of legislators are institutional limits, direct democracy, and term limits. Each is discussed below.

Institutional Limits

The state constitution, city charters, and state laws governing cities, counties, and independent units of local government place institutional limits on them. These founding documents from which these governments were created set forth the power of their offices, just as the U.S. Constitution does at the national level. There are often significant restrictions. For instance, home rule cities in some states can create their own taxes, levies, and fees *except* for income taxes. That power is reserved for the state government. In these states, no city council or mayor, no matter how powerful, can adopt an income tax without an explicit law from the state legislature granting the city a part of the state income tax.

Legislators must also convince the governor, mayor, or county board president to sign the legislation they pass. If the chief executive is unwilling to support a bill or ordinance, she or he can veto the legislation, which then cannot become law unless the legislative branch overrides the veto by a special majority—usually two-thirds of both houses of the state legislature, city council, or county board.

The courts frequently overturn laws that they judge do not conform to the U.S. Constitution, state constitution, or city charter. Thus, legislative bodies must pass legislation that the courts will uphold.

Direct Democracy

An alternative to deciding policy by legislative bodies is direct or participatory democracy. Three common methods by which voters may play a direct role in government are recall,

A Hollywood, Florida, City Commission Meeting

Hollywood, Florida, a suburban town north of Miami, has a population of 149,000. In attendance at the city commission meeting on February 24, 2004, were the mayor, the six elected commissioners, the city manager, 50 city employees, and 40 citizens. The meeting started at 1 p.m., followed a fixed agenda until 4 p.m., and then had public comments for a half-hour beginning at 5 p.m.

Following the Pledge of Allegiance, the city manager placed a "consent agenda" of 16 noncontroversial resolutions before the commissioners. Before taking a single roll call on the consent agenda, commissioners asked that 5 of the 16 items be taken off the consent agenda for discussion, which is the right of commissioners. The remaining 11 items were adopted unanimously without debate.

Following this vote, the city presented awards to several people: the woman who was Miss Florida in the 2004 USA pageant, who had participated in a Hollywood parade that week; the firefighter of the year; police employee of the month. Proclamations were issued to recognize Pet Adoption Day and to honor city employees who served as mentors in local schools.

The commissioners then discussed the more substantial items that had been removed from the consent agenda. For example, questions were asked about improving a golf driving range renovation in a city park at a cost of $86,000, and commissioners recommended a needs assessment for a children's summer program.

Commissioners then reviewed proposed blacktopping of traffic circles and the installation of speed tables as part of "traffic calming" experiments in Hollywood. Commissioners approved four traffic projects at a total cost of $183,000. Commissioners requested additional stop signs be installed at several intersections as part of the "traffic calming" effort. "Speeding is the number one issue for the city," declared a commissioner.

The finance department reported to the commission that a small surplus of $500,000 had been achieved for the fiscal (budget) year ended September 30, 2003, out of a total budget of $118 million. The public safety department spent $81 million, or two-thirds of the total city budget, for police, fire, and emergency services.

In final comments in the main commission meeting, Hollywood Mayor Mara Giuliante said the commission needed to revise the moratorium on issuing occupancy permits for businesses along Highway 441 until a new economic development and zoning plan was adopted and major highway repairs were completed. The city commission voted to do so at the next commission meeting after further debate.

The meeting ended with a half-hour of public comments on diverse problems and issues concerning the citizens who testified. These public concerns sessions often point up new problems to be taken up at future commission meetings.

---

initiative, and referendum. In the **recall** process, voters who are dissatisfied with an elected official sign a petition requesting a special election that will decide whether to remove the official from office. If the petition receives enough signatures, the recall election is held. The recall mechanism is a way of keeping legislators and chief executives accountable between elections.

The recall process was used in California in 2003 when voters recalled Democratic Governor Gray Davis, who was replaced in a subsequent election by Republican Arnold Schwarzenegger.[22] Illinois in 2008 considered holding a recall election to remove Governor Rod Blagojevich from office, but the effort failed and impeachment by the legislature was used instead. Although 16 states and their cities have adopted the recall process, few actually use it. The recall of Governor Davis, for example, was the first statewide recall election in California's history.

A device available to voters in 18 states is the direct **initiative**, a process in which voters petition to place proposed laws and constitutional amendments on the ballot for a vote of the electorate, rather than leaving lawmaking on the issue up to the state legislature or city council.[23] California is noted for its extensive use of the initiative; the ballot in that state is often scores of pages long because of numerous initiatives to be voted on. But initiatives do not always succeed. For instance, Governor Schwarzenegger in 2005 was hoping to bypass the California state legislature through a series of initiatives on the ballot, but the measures were voted down.

Many states have severe restrictions on the use of initiatives to pass laws. They require legal petition signatures of 10–15 percent of the voters to get an initiative on the ballot. Some states, like Illinois, restrict lawmaking by initiative to a narrow range of topics, such as the size and composition of the legislative branch. Unlike the states, few cities and counties use initiative procedures.

Almost all cities and states have a **referendum** process by which voters can petition to place a policy question on the ballot of the next regular election. However, a referendum is purely advisory, not binding law. Generally speaking, if the referendum passes by a wide margin, the legislative branch is quick to adopt the spirit if not the specific language of the referendum.

Advocates of more participatory democracy have advocated expansion of initiative and recall, but these proposals have generally not succeeded. Table 3.4 shows which states have established provisions for the use of initiative, recall, and referendum processes.

Term Limits

Another method of curbing the power of legislators has been term limits, one of the most successful movements in modern political history. **Term limits** are laws that restrict the number of times a public official can run for reelection. Usually, public officials at the state and local level can be reelected indefinitely, and the longer they serve in public office, the more powerful they become. In the 1990s, a movement was started to curb this power by enacting

TABLE 3.4 States with Initiative, Recall, and Referendum Provisions				
State	Initiative	Recall	Citizen Petition Referendum	Legislative Referendum
Alaska	D[a]	Yes	Yes	Yes
Arizona	D	Yes	Yes	Yes
Arkansas	D		Yes	Yes
California	D	Yes	Yes	Yes
Colorado	D	Yes	Yes	Yes
Florida	D			Yes
Georgia		Yes		
Idaho	D	Yes	Yes	Yes
Illinois	D			Yes
Kansas		Yes		
Kentucky			Yes	Yes
Louisiana		Yes		

TABLE 3.4 States with Initiative, Recall, and Referendum Provisions

State	Initiative	Recall	Citizen Petition Referendum	Legislative Referendum
Maine	I		Yes	Yes
Maryland			Yes	Yes
Massachusetts	I		Yes	Yes
Michigan	D & I	Yes	Yes	Yes
Minnesota		Yes		
Mississippi	I			Yes
Missouri	D		Yes	Yes
Montana	D	Yes	Yes	Yes
Nebraska	D		Yes	Yes
Nevada	D & I	Yes	Yes	Yes
New Jersey		Yes		
New Mexico			Yes	Yes
North Dakota	D	Yes	Yes	Yes
Ohio	D & I		Yes	Yes
Oklahoma	D		Yes	Yes
Oregon	D	Yes	Yes	Yes
Rhode Island		Yes		
South Dakota	D		Yes	Yes
Utah	D & I		Yes	Yes
Washington	D & I	Yes	Yes	Yes
Wisconsin		Yes		
Wyoming	D[a]		Yes	Yes
US Virgin Is.	I		Yes	Yes

D = *Direct initiative;* proposals that qualify go directly on the ballot.

I = *Indirect initiative;* proposals are submitted to the legislature, which has an opportunity to act on the proposed legislation. Depending on the state, the initiative question may go on the ballot if the legislature rejects it, submits a different proposal, or takes no action.

[a]Alaska and Wyoming's initiative processes exhibit characteristics of both the direct and indirect initiative. Instead of requiring that an initiative be submitted to the legislature for action (as in the indirect process), they require only that an initiative cannot be placed on the ballot until after a legislative session has convened and adjourned. The intent is to give the legislature an opportunity to address the issue in the proposed initiative, should it choose to do so. The initiative is not formally submitted to the legislature.

Source: National Council of State Legislatures, www.ncsl.org.

term limits, and 15 states adopted limits of 6–12 years as the time in office that legislators and other elected officials may serve (see Table 3.5).

Advocates of term limits say that they force beneficial turnover in elected office. They result in the election of more women and minorities and their promotion to leadership positions more quickly. While it was assumed that term limits would also result in the election and promotion of minorities, in the earliest studies of term limits in state legislatures this does not seem to be the case. Later studies seem to indicate that the higher turnover rate did speed underlying demographic trends. States with rising Latino population such

TABLE 3.5 Term Limits in State Legislatures

State	Year Enacted	House		Senate		% Voted Yes
		Limit	Year of Impact	Limit	Year of Impact	
Maine	1993	8	1996	8	1996	67.6
California	1990	6	1996	8	1998	52.2
Colorado	1990	8	1998	8	1998	71
Arkansas	1992	6	1998	8	2000	59.9
Michigan	1992	6	1998	8	2002	58.8
Florida	1992	8	2000	8	2000	76.8
Ohio	1992	8	2000	8	2000	68.4
South Dakota	1992	8	2000	8	2000	63.5
Montana	1992	8	2000	8	2000	67
Arizona	1992	8	2000	8	2000	74.2
Missouri[a]	1992	8	2002	8	2002	75
Oklahoma	1990	12	2004	12	2004	67.3
Nebraska	2000	n/a	n/a	8	2006	56
Louisiana	1995	12	2007	12	2007	76
Nevada[b]	1996	12	2010	12	2010	70.4

This list represents the 15 states that currently have term limits for legislators. They are ordered by the year of term limits' impact—the first year in which incumbents who were serving when the term limits measure was passed are no longer eligible to run for reelection.

[a]Because of special elections, term limits were effective in 2000 for eight current members of the house and one senator in 1998.

[b]The Nevada legislative council and attorney general have ruled that Nevada's term limits cannot be applied to those legislators elected in the same year term limits were passed (1996). They first apply to persons elected in 1998.

Source: National Council of State Legislators, "The Term Limited States," June 2009, www.ncsl.org.

as Arizona and California were able to increase Latino representation in the legislature more quickly.[24] Supporters of term limits maintain that democracy is better served by changing elected officials and by bringing in new blood and ideas. It is too soon to see if term limits will also improve legislative bodies by making them more responsive to their constituents.

Critics of term limits maintain that long service is beneficial because it allows officials to master the craft of governing and use their seniority and experience to serve their constituents well. Former Wisconsin state representative and speaker Tom Loftus (1977–1991) observed that term limits are "a crude way to dislodge the influence of political money" even if they fostered the founders' goals that power be restrained and balanced. Loftus argued that "by themselves they will not bring about much real change. They are based on the mistaken notion that if only the right individuals were elected, all would go well."[25] After following the progress of 2,000 bills introduced before and after term limits, California's Public Policy Institute found that legislators with term limits are less likely to act as gatekeepers and more likely to push through unpolished bills that are gutted later in the legislative process. Term-limited legislators also "made half as many changes to governors' budgets than their more experience predecessors and offered weaker executive branch oversight in general."[26] As Table 3.6 shows, term limits have even been repealed in six states, either by statute or court order.

But voters are still passing term limit referendums and defeating many legislative attempts to rescind existing term limits in cities and states. Florida State Representative

TABLE 3.6 Term Limit Repeals

State	Year Repealed	Year Enacted	Who Repealed?
Idaho	2002	1994	Legislature
Massachusetts	1997	1994	State Supreme Court
Oregon	2002	1992	State Supreme Court
Utah	2003	1994	Legislature
Washington	1998	1992	State Supreme Court
Wyoming	2004	1992	State Supreme Court

Source: National Council of State Legislators, "Term Limit Repeals, February 2006, www.ncsl.org.

TABLE 3.7 Legislators Forced Out by Term Limits in 2008

State	House	Senate
Arizona	7 of 60	2 of 30
Arkansas	28 of 100	4 of 35
California	24 of 80	10 of 40
Colorado	8 of 65	7 of 35
Florida	28 of 120	5 of 40
Maine	15 of 151	6 of 35
Michigan	44 of 110	0 of 38
Missouri	21 of 163	4 of 34
Montana	17 of 100	10 of 50
Nebraska	(one chamber)	13 of 49
Ohio	21 of 99	4 of 33
Oklahoma	7 of 101	5 of 48
South Dakota	13 of 70	6 of 35

Source: National Council of State Legislators, www.ncsl.org, 2008.

Baxter Troutman says the real argument now is about how long members need to serve before they can be effective when handed real power as leaders. "We've tried eight years," Troutman says, "and eight isn't enough. We need to take a hard look at 12."[27] There is no doubt, however, that term limits have had real effects on the composition of city councils and state legislatures. Each year, more legislators and city council members are forced to retire because of limits on the number of years that they can serve. Table 3.7 shows the number of legislators forced from office by term limits for one year, 2002.

THE LAWMAKERS' STRUGGLE TO DECIDE

State and local lawmakers make thousands of decisions a year in party caucus, in committees drafting and voting on amendments to legislation, and in votes on the floor of the legislature. The rules of a state legislative body generally require a bill to have two or three "readings" before it can be voted on for final passage. Often amendments come at the second reading and third readings. In local governments, legislation is normally referred to committees for hearings and for a committee recommendation before it comes to the entire body for a final vote. And each legislator must decide how to cast that vote—often a recorded vote.

Making Decisions Is Not Simple

A legislator in local or state government may vote on hundreds of bills and ordinances. How are lawmakers to make these decisions? Should elected officials represent themselves, just those who voted for them, or all their constituents? Should they vote according to their own political philosophy, follow the requests of their party leaders in the legislature, or vote parochially in the interest of their districts? Should lawmakers cast votes solely on the merit of a substantive issue before them, or should they vote to help a colleague who will then help them with their own legislation later?

The answers depend on the values, ambitions, and strategies of each legislator. In his famous "Address to the Electors at Bristol," English parliamentarian and political philosopher Edmund Burke explained that while his constituency wanted him to vote in favor of tariffs on trade from other countries, he felt a responsibility to vote in behalf of the larger interests of the nation. Burke believed that he was a "trustee" of the district, with authority to vote as he saw best for England, rather than a "delegate" or "agent" of the constituency who would vote only as they wished him to vote.[28]

Making decisions in a legislative body is not always easy. There are many demands on legislators which must be resolved with each vote they cast.

Ethics in Legislative Decision-Making

Lawmakers face numerous pressures to vote a certain way on an issue. If you are a public official asked to cast a vote that gives you a queasy feeling, the obvious solution is not to cast that vote. Unfortunately, big bribes to favor big interests seems to give some legislators little queasiness.

In 1909, 40 Illinois house members accepted up to $2,500 each to cast a vote for William A. Lorimer for the U.S. Senate, at a time when a new Model T Ford cost $850.[29] In that era in Illinois, U.S. senators were not yet directly elected by the general voting public. Another tradition was for powerful business figures to assemble a money "jackpot" that would be distributed at the end of a legislative session in accordance with the support individual lawmakers had provided to business interests. Lorimer, who was elected by bribes to the Illinois lawmakers, was expelled from the U.S. Senate in 1911 for having purchased the office. Unfortunately, bribery continues to exert power in Illinois, as it does elsewhere. Since the 1970s, for example, 30 Chicago aldermen have been convicted of corruption. A total of 1,500 public officials and businessmen have been convicted of public corruption since 1970 in Illinois.[30]

Scandals in cities like Miami, Las Vegas, and San Diego show the power of corruption. In San Diego, previously considered a reform city, Mayor Dick Murphy was forced to resign in the midst of a federal probe. The city was found to be several billion dollars in debt to its city workers' pension fund. In July 2005, San Diego city council members Michael Zucchet and Ralph Inzunza were convicted of conspiracy, extortion, and fraud along with a Las Vegas lobbyist and a city council aide. A lobbyist who worked for a strip club had bribed the council members with campaign contributions in exchange for changing city strip club regulations.

A third council member, Charles Lewis, had died after being indicted, before he could go to trial. Legislators at all levels of government and in all regions of the country are tempted to accept bribes in return for their votes.

What prompts lawmakers to accept bribes in return for their votes? A number of rationalizations have been offered. Some say, "That's the way it has always been done," or "I went along with the others." Others say, "I sacrificed financially to be an elected official, and I earn less than the lobbyists and businesspeople that come before me." Still others claim, "Nobody was hurt by my vote," or assert, "I had financial problems."

Some political cultures are more tolerant of corruption than others. And definitions of corruption change over time. A hundred years ago, New York City political boss George Washington Plunkitt distinguished between "honest graft" and "dishonest graft." Plunkitt said that if elected officials used their insider knowledge to buy land where they knew the city was going to build a park, and then sold the land to the city at a big profit, the officials were practicing "honest graft." Outright stealing from the city treasury was "dishonest graft." Today, both would be absolutely illegal. The basic standard is that a public official should not personally benefit in a material or financial way from government votes. To receive a benefit such as a free holiday in the Caribbean or an envelope of cash is public corruption and punishable under federal and state law, as council members in San Diego and too many other cities have learned.

Many states and municipalities attempt to legislate higher ethical behavior by passing laws that regulate campaign contributions and gifts, as well as laws that require public disclosure of income and potential conflicts of interest. Some states and cities now provide partial public financing of campaigns. However, writing laws that will succeed in preventing corruption is very difficult.

Florida passed an ethics law that requires state and local officials to report gifts they receive from anyone but relatives. Gifts from lobbyists cannot have a value of more than $100. Sounds simple enough. Yet application of the law has created confusion. For instance, Broward County officials went on a trip to Puerto Rico sponsored by the county's Puerto Rican Chamber of Commerce, which provided free airline and hotel accommodations. The objective of the trip was to promote tourism and business between county and commonwealth. Meetings among officials of the two entities occupied much of the time in Puerto Rico. "I was working," declared Miramar, Florida, Mayor Lori Moseley. "It definitely was not a pleasure trip."[31] None of the local officials reported the trip as a gift, yet there is no exception in the law for a trip such as this.

Did the Broward County officials violate the law? Were the sponsors "lobbyists"? The meaning of "lobbyist" in the Florida state law is broader than in most cities and states. According to one member of the Florida State Ethics Commission: "For purposes of the gift law, lobbyists may also include anyone who is paid to influence the government, such as a developer seeking city approval for a new housing development."[32] There is also confusion as to whether the expenses covered by the sponsors were gifts. Attorneys for some of the officials argued that the mayors and county officials had provided important information about economic development in exchange for the free travel and hotel rooms. Public officials are required to exercise ethical standards and to adhere carefully to laws meant to curb corruption. But temptations are many and the application of the laws uncertain.

Often major issues are decided in state and local government not on the ethical issues but on the practical and financial costs and benefits of a proposal. For instance, as the Insider's View feature demonstrates, the stance of the city of Hollywood, Florida, on increasing the number of slot machines at race tracks was decided on practical not moral grounds. Suburban governments often decide issues—ranging from permitting gambling to allowing adult

A Decision on Gambling

In 2005 in Florida, a binding referendum was to be held in Miami-Dade and Broward counties to determine whether slot machines could be installed at existing gambling facilities such as dog and horse racetracks. The city commission of Hollywood, Florida, faced a decision whether or not to endorse this expansion of gambling that promised to bring more money to schools throughout the state (not just in the two counties where the slot machines would be located).

The expansion of gambling is for many a moral issue about which there are strong feelings. The racetracks were lobbying heavily for the referendum, spending more than $3.5 million in support of the measure, while Christian organizations were strongly opposed. The Hollywood city commission decided to propose to the racing industry that they would support the measure if they were given a part of the profits. If they were not, they would oppose it.

Mayor Mara Giuliante, at the city commission meeting where the issue was debated, said: "It's a financial decision—what is in our residents' best interest financially." Vice Mayor Beam Furr and other commissioners argued that the increase of gamblers at the slot machine sites would strain traffic, police, ambulance, and other city services. "We are surrounded by gaming. We're going to have everything but the benefits," Vice Mayor Furr said. He argued that Hollywood should get the same 1 percent of the profits slots generated that the neighboring cities, which actually house the sites, would receive. The mayor concluded, "We will be against [the slot machine referendum] unless there's a benefit to the residents of Hollywood. We are not going to stand by and let our residents be kicked in the teeth."[a]

The racing industry supporting the slot machine referendum refused to offer Hollywood a part of the profits. Therefore the city commission voted 4–2 to oppose them and allocated $50,000 of city money to the anti-slot-machine campaign. Mayor Giuliante became a major spokesperson against the referendum.[b]

The referendum failed in Miami-Dade County by a vote of 52–47 percent. It passed in Broward County, where Hollywood is located, by a vote of 57–43 percent. The turnout was light in both cities, 14–18 percent of the voters.[c] The chief reason it was defeated in Miami-Dade County was that the governor, Jeb Bush, campaigned against it.

[a]Robert Nolin, "Hollywood Wants Share of the Profits If Slots Win O.K.," *Sun-Sentinel*, February 3, 2005, 3B.
[b]Jerry Berrios, "City Antes Up $50,000 to Fight Slots," *Miami Herald*, February 17, 2005, 1B.
[c]Jack Dolan and Erika Bolstad, "Slots: Yes and No," *Miami Herald*, March 9, 2005, 1A.

entertainment—on practical grounds, even though the issues may be emotionally charged. With questions ranging from religious freedoms to allowing the sex industry to flourish, legislators must often make difficult decisions.[33]

Even if ethics laws are written and passed, they don't work if they lack implementation and compliance. In Florida, as in most states and cities, lobbyists have generally failed to disclose small gifts to public officials. And the state legislature has refused to give the Florida Ethics Commission authority to investigate possible violations of the law. In Miami-Dade County between 1997 and 2002, four sitting county commissioners, one city commissioner, one city manager, a former Miami mayor, and one school board member were removed from office for abuse of the public trust.[34] Ethics laws are one tool to discourage corrupt behavior and curb the power of corruption in legislative decision-making. Aggressive prosecution of existing laws, a watchful press, and an outraged citizenry are other keys to more ethical behavior.

"HOW'RE WE DOIN'?" IN OUR LEGISLATIVE BODIES

When Edward Koch was the mayor of New York City, he greeted constituents with the question, "How're we doin'?"—meaning, how was he doing in leading his great city. Most constituents probably answered him on the basis of a vague sense of the city's civic health along with some knowledge of how the police, schools, garbage collection, and street services in their neighborhood were working.

Public Opinion

Citizens find it difficult to evaluate legislative bodies which pass laws governing their lives but don't deliver government services directly. Most citizens tend to have a low regard for state and city legislators overall, but high regard for their own local legislator. In some public opinion polls, only 13 percent of the respondents thought state legislators were doing a good or excellent job. In Chicago in the 1990s, 15–28 percent of Chicagoans thought the city council was doing a good job (with African Americans rating the council most poorly and whites rating the council more highly). Yet 30–50 percent thought their own individual alderman was doing a good job (again dividing along racial lines).[35]

Citizens of Providence, Rhode Island, in 2002 rated most of their services other than public schools as generally good or excellent. But here, too, there were differences between racial groups and neighborhoods in rating the services. For example, 70 percent of whites rated police protection as excellent or good, while only 49 percent of nonwhites gave them the same rating. Nonwhites rated public schools somewhat higher than whites.[36]

Scholars, journalists, and legislative observers generally believe that "one person, one vote" court rulings, modernization, improved salaries, term limits, and the increase in women in legislatures have had salutary effects on the legislative branches of government.[37] Criticisms continue, however. As a longtime observer of one state legislature put it: "Two generations ago, some lawmakers might take a little money here and there, but their word was good as gold. Today, legislators won't take money under the table but their word is not worth a damn."

Good and Bad Legislative Bodies

To determine whether any particular legislative body is effective at representing citizens, we must define an ideal legislature. Advocates of representative democracy have come up with many conflicting goals for legislatures: (1) representativeness, (2) democratic procedures, (3) effectiveness, (4) efficiency, and (5) achieving desired policy outcomes. If we are uncertain about what is a good legislature, we are confident that we know about bad city councils, county boards, school boards, and state legislatures.

According to writer Alan Ehrenhalt, "A bad legislature is unable to attract talented members, is lacking in staff and resources, succumbs to the pressures and personal favors of lobbyists, meets only infrequently and allows the governor [or mayor] to treat it like a doormat."[38] Most state legislatures, city councils, and local governing boards are now more professional than before. Reforms which began more than three decades ago in California under the strong speaker of the general assembly, Jesse Unruh, and spread by the Ford Foundation have taken hold. Observers now tend to agree that "there is no question that state legislatures [and many city councils] as an institution possess more resources, more information, and a higher degree of independence than virtually any of them did in the bad-old pre-reform days."[39]

As Alan Rosenthal argues in his book *Heavy Lifting: The Job of the American Legislature:* "there are three major functions of the modern legislature: representation, lawmaking, and dealing with the executive branch. The ideal legislature does all of them well, and keeps them in proper proportions." Currently, Rosenthal argues, "legislatures are performing the representative function extremely well. Members are listening to constituents and solving people's problems. The law-making function isn't quite as strong, and executive-relations function is the weakest of the three."[40]

City councils are also frequently criticized as poor legislative bodies. Rob Gurwitt in his article "Are City Councils a Relic of the Past?" lists a number of dysfunctions, including the following:

- The Baltimore City Council pursues tangents so frequently that the *Baltimore Sun* newspaper has called it "the hot-air council."

- The Philadelphia City Council was called by a former mayor "the worst legislative body in the free world."

- An observer of the Pittsburgh City Council says it "is a group of people who deal primarily with very mundane, housekeeping things in their district."

- In St. Louis the aldermen attend mostly to development issues in their own ward and not to the big decisions affecting the city as a whole.[41]

It is undeniable that many city councils fail to function as well as we would like. Still many individual aldermen and city council members work hard to do the legislative and ombudsmen functions with which they are charged. How well the city councils themselves are functioning is the issue.

Vera Vogelsang-Coombs in her studies of city councils has argued that a city council governs well if its members adopt: "(1) a democratic group dynamic; (2) active policy deliberations based upon democratic knowledge, and (3) constituent relations that reflect community building values."[42] She claims that some city councils like those in Norfolk, Charlotte, and San Jose function well for the most part. Others fail in their procedures and leadership. They fail to follow democratic procedures and become either "groupthink" councils or "adversarial" councils. They fail to be deliberative and become "rubber stamps," or they fail in their constituent relations and become "co-opted" or "disengaged."

COUNCILMAN ANTONIO VILLARIGOSA, now Los Angeles Mayor, speaking at a City Council Debate.

Generally speaking, state legislators today are better paid, have better informational support, and are in session longer each year than in the past. There have definitely been improvements in staffing and capacity. Likewise, some city council members now have computer software to track citizen complaints and larger staffs to assist them with the work of governance. Council meetings are often televised on cable television or webcast, and council voting is reported to interested voters and community organizations on official city hall Web sites. No analysis has been done, however, to show if these changes have resulted in a process that resolves conflict more effectively and produces laws that are better for our states and cities. The legislative branches which are the heart of state and local government still face significant challenges.

CHAPTER SUMMARY

Legislative bodies illustrate this book's central themes of *the struggle for power, representative vs. participatory democracy, institutional structures and processes,* and *political and governmental outcomes.* Legislative bodies, no matter how small, have to work out differences among factions who want to control the use of power. State and large city governments cannot function without elected officials representing their constituents, and even small village councils have to be responsive to the voters in getting local ordinances passed. State legislatures and township boards function according to the rules and structures set up by both custom and their charters. Regardless of procedure, however, all elected and appointed officials must determine whether their decisions are good for the people they represent, whether they allow citizens to be able to have a voice in decisions through open meetings and transparent, efficient, and honest government.

Elected officials may act as *trustees,* using their own best judgment, or as *delegates,* voting according to the wishes of their constituents, or they may change their roles depending on the issue at hand. David Barber found that officials acted as *experts* in one or more fields; *brokers,* who bargain to get laws passed; *lawmakers,* who work for the best policies; or *ritualists,* who support their party leaders but leave the hard work to others.

All lawmakers perform several roles. Besides voting on issues, they also try to resolve conflicts within the legislative body, relying on principles of civility and reciprocity. A very important function they perform is to provide services to constituents. Finally, all legislators must serve as watchdogs over the government agencies to ensure that they are providing services honestly and efficiently.

The organization of any legislative body is determined by its charter or constitution, but also by political considerations. For instance, all states but Nebraska have a bicameral legislature, but selection of the leadership is a power struggle among the politicians through a vote of the legislative body. The majority leader in a state legislature or large city council is generally the most powerful figure in the state, next to the governor, mayor, or county board president.

It is impossible for a legislator to know the details of every bill proposed, but most legislation is routine and noncontroversial. For the most important or complicated legislation, the party leadership has a clear position, making it easy for party members to vote that way. A legislator can also follow the recommendations of committees who have studied proposed legislation thoroughly. Committees are of two types: permanent standing committees which deal with routine issues such as education or transportation, and special committees appointed for a short period to deal with a particular issue.

Every 10 years, state legislative districts are to be redrawn to reflect the U.S. Census, and the redistricting function of the legislature sets up another power struggle, as district boundaries may be redrawn for partisan political advantage. Each legislator also has a struggle when trying to move a particular bill or ordinance through the legislative process. Blocking a bill is easier to do than passing it.

Lawmakers are limited by the institutional constraints of constitutions, charters, and custom. Another curb on lawmakers' power is the citizens' exercise of direct democracy through the recall, initiative, and referendum processes. Term limits are used in many states and cities to limit the time that an elected offical can remain in office.

Political change is incremental. In order to maintain effective democracy, voters must demand that their legislators follow guidelines encouraging representativeness, democratic procedures, effectiveness, efficiency, and those which are most likely to let them achieve desired policy outcomes.

KEY TERMS

biennial Continuing or lasting for two years. A two-year period is a biennium, p. 36

broker Legislator who makes deals among varied interests and across party lines on controversial issues, getting them enacted into laws, p. 41

caucus A group of legislators of the same political party or faction, or an informal, closed meeting of the group, p. 45

civility Civil conduct: acting with politeness, thoughtfulness, and courtesy, p. 42

committee-dominated legislature A legislature in which the committees do the major work on legislation, and committee recommendations are guides for voting by individual legislators of both parties, p. 53

delegate A legislator or other elected official who will vote and act based on the opinions of the majority of constituents, p. 41

descriptive representation theory The theory that constituents are best represented by legislators who are similar to them in characteristics such as race, ethnicity, and socioeconomic level, p. 39

expert Legislator who becomes a respected authority to whom other lawmakers turn for cues in voting, p. 41

floor leader The member of the legislature who determines the agenda and sometimes the procedural rules, p. 44

gerrymandering Drawing district lines to benefit a particular political party or ethnic, racial, or other group, p. 55

initiative A process in which voters petition to place a proposed law or constitutional amendment on the ballot for a direct vote by the electorate, p. 60

lawmaker Legislator who enjoys putting proposals into the best shape possible and advocating their adoption, p. 41

norm A standard developed to guide the behavior of members of a group, p. 42

ombudsman An official who is designated to look into individual concerns or complaints and seek a solution, p. 37

participatory democracy or pure democracy A system of government in which citizens make the governmental decisions directly in assembly together, p. 36

party-dominated legislature A legislature in which votes are determined by party affiliation, p. 53

president of the senate The official leader of the state senate, with the power to appoint committees and guide legislation through the legislative process as determined by the specific rules of the body, p. 44

recall A process in which voters petition to hold a special election that will decide whether to remove a public official from office, p. 59

reciprocity A return of a favor or a vote in kind, p. 43

referendum A process in which voters petition to place a referendum question on the ballot; a referendum is advisory to the government, not binding law, p. 60

representative democracy A system of government in which citizens elect government officials who make the actual governmental decisions, p. 36

ritualist Legislator who enjoys the status of holding office but does not work hard to develop legislation, p. 42

roll call vote Each legislator votes when called on by name, p. 37

speaker of the house of representatives The official leader of the state house of representatives, with the power to appoint committees and guide legislation through the legislative process as determined by the specific rules of the body, p. 44

special committees A legislative committee appointed for a short period of time to study or address a particular issue, p. 53

standing committee A permanent legislative committee that writes laws or statutes for a specific issue area, such as transportation or education, p. 53

term limits Laws that restrict how many times an elected official can run for reelection, p. 60

trustee A legislator or other elected official who will vote and act based on personal judgment about what is in the best interests of constituents, p. 41

voice vote Legislators vote by saying "Aye" as a group or "Nay" as a group when asked whether they approve or disapprove of the proposed legislation, p. 37

QUESTIONS FOR REVIEW

1. What is the difference between participatory and representative democracy? Which do we find in state legislatures? City or town councils?

2. What is the descriptive theory of democracy?

3. What are the differences when a legislator votes as a trustee or a delegate?

4. What are four important roles that legislators play?

5. Why is the choice of legislative leaders important in organizing and operating a state legislature or town council?

6. What are the functions of legislative staff in getting laws passed?

7. What are the basic types of committees? How can a committee member kill a bill?

8. How can redistricting affect the outcome of elections and future lawmaking?

9. How does a bill become a law in a state legislature? How does political power affect getting a bill passed?

10. What are three common processes by which voters may play a direct role in government?

11. What are some institutional limits on lawmakers?

12. How can ethics laws change how legislation is passed?

13. How can we judge if a state legislature, city council, or local governing board is doing a good job?

DISCUSSION QUESTIONS

1. Would you want to be an elected official? Why or why not? What are the advantages and disadvantages of running for and serving in a state and local government body?

2. How does a bill or ordinance become a law in your state or city? How many steps are in the process? Is this a good way to make laws? What changes would improve the process?

3. Imagine that you are a legislator facing a tough reelection battle, and you know you must have the vote of a representative whose ethics you question to get your key legislative initiative out of committee. Success of this initiative is important for economic development in your economically depressed district. Coincidentally, this legislator asks that you provide the key vote she needs to get a bill for expanded gambling out of a committee you sit on. You are opposed to expansion of gambling, but she will vote for your bill in committee if you vote for hers. What is the ethical thing to do? What would you recommend?

4. Are there term limits in your state and city? What are the arguments for and against them?

5. Should city council, county board, and state legislative meetings be televised? Why or why not?

6. Should legislatures be committee centered or party centered? What are the benefits and drawbacks of each type?

7. Is your state legislature and city council doing a good job representing constituents like you, making good laws for the state and city, and dealing with the executive branch on a more or less equal basis? Or do the members kowtow to lobbyists and influential groups and act as a "rubber stamp" for the chief executive?

PRACTICAL EXERCISES

It is important to discover key facts about your state legislature and local governing bodies. To do so, you will have to consult government Web sites and various newspaper articles. Organizations such as the League of Women Voters also provide booklets and Web site information on various government bodies. As a last resort, you can call the offices of elected representatives and various "watchdog" agencies which monitor the state legislature and city hall. These will vary in different cities and states. Yet everywhere these facts about your governments are available. They just take careful research.

1. Investigate to what extent the members of your state and local government legislatures are representative of the ethnic, racial, and gender divisions of your city and state. What percentages of your state legislature, city council, school board, and county board are white, black, Latino, Asian, and American Indian? What are the percentages for women and men? Usually, the descriptive information on state legislators and city council members can be found on city and state government Web sites.

2. Does your state legislature and city council provide staff for its members? How many staff members are provided? Does the staff work on a partisan basis (for one party or the other) or nonpartisan basis? You may want to invite a staff member of the state legislative or city council to your class to talk about their work.

3. How many standing committees exist in your city council, county board, and state legislature? Is the number the same in each body? Do they serve the same functions? Are any special committees or commissions currently investigating problems and holding hearings about them?

4. Using sources such as your state's Web site, local newspaper articles, and library resources such as Lexis-Nexis, find out the following:
   a. Determine the number of members in both houses of your state legislature; the number of Democrats and Republicans in both houses, and the names of the four top legislative leaders.
   b. Find at least one divided roll call vote in the state legislative session on a controversial issue such as same-sex marriages, abortion, or ethics. What is the issue? Did the proposed bill become law or did it fail? How many legislators voted for and against the issue? Why did they vote the way they did? How would you have voted on this issue?
   c. Determine the number of legislators on your county board and local city council or city commission.

What is the division between factions or parties in each body? This may be more difficult to determine because your local aldermen or city council members may run in nonpartisan elections.

5. Party or ideological factions may be more important than separate party labels at the local level. But the pattern of divided roll call votes will reveal the political divisions, particularly the votes to elect leaders, appoint committees, and adopt rules early in the legislative session. Using roll call vote data (available from Web sites or newspaper archives), find out what the usual vote divisions are in your local legislative bodies.

6. Find at least one divided roll call vote at the county and city levels. If there are no divided roll call votes (many times at the city level all votes are unanimous), what does this mean about city government? Using the local newspaper or the newspapers' Web site, follow at least one controversial issue voted on in the county board and city council. Who favored it? Who opposed it? Did it become law? How would you have voted on it?

7. Try your hand at redistricting. Below is a hypothetical matrix of 30 voting blocs of equal population in which all the voters are either Republicans or Democrats. The voting strength of the parties is evenly balanced at 15 blocs each. For this exercise, choose a party, either Republican or Democratic, and create 6 contiguous districts of equal population (5 blocs) in which you optimize the districts for your chosen party. How many districts can you create for your party? How easy is it to provide either party the advantage where voters are evenly divided between the two parties?

R	R	D	D	R
D	R	R	D	D
D	D	R	R	R
R	D	D	D	R
R	R	D	D	R
R	D	D	D	R

WEB SITES FOR FURTHER RESEARCH

Nearly every city and state government has an official Web site. Simply type the name of your city or state in your search engine to get a listing. Most of them end with ".gov."

To obtain general information on state legislatures go to the site of the National Conference of State Legislatures: www.ncsl.org.

To track the voting and background of most public officials, including state legislators, use the information at Project Vote Smart: www.votesmart.org.

To get further information on term limits, go to www.termlimits.org.

Information on cities can be obtained at the National League of Cities site, www.nlc.org, and the U.S. Conference of Mayors at www.usmayors.org.

Information on counties can be obtained at the National Association of Counties: www.naco.org.

Information on the nature of corruption in a single city and state can be found at www.uic.edu/depts/pols/ChicagoPolitics/anticorruption.htm.

SUGGESTED READING

Barber, James David. *Lawmakers: Recruitment and Adaptation to Legislative Life*. Westport, CT: Greenwood, 1980. Gives a detailed explanation of the various roles a legislator can assume during the lawmaking process.

Burke, Edmund. "Letters to Langriche (1792)." *Burke's Politics*, ed. Ross J. S. Hoffman and Paul Levack. New York: Alfred A. Knopf, 1949, 495. Burke's original discussion of descriptive representation. See also Burke's "Speech to the Electors at Bristol (1774)," 115, which presents his explanation of trustee voting.

Kurtz, Karl T., Bruce Coin, and Richard G. Niemi. *Institutitonal Change in American Politics: The Case of Term Limits*. Ann Arbor: University of Michigan Press, 2007.

Pitkin, Hanna Fenichel. *The Concept of Representation*. Berkeley: University of California Press, 1967. Gives excellent explanations of Edmund Burke's thoughts on representation, along with alternative theories of representative democracy.

Rosenthal, Alan. *Heavy Lifting: The Job of the American Legislator*. Washington, DC: CQ Press, 2004. One of the best books analyzing state legislatures and their performance.

Simpson, Dick. *Rogues, Rebels, and Rubber Stamps: The Politics of the Chicago City Council from 1863 to the Present*. Boulder, CO: Westview, 2001. A study of one city council and one city's political history over 150 years.

Smith, Kevin, ed. *State and Local Government, 2007*. Washington, DC: CQ Press, 2006. A comprehensive general work on state and local government.

EXECUTIVE POWER IN STATE AND LOCAL GOVERNMENT

Mayor Ray Nagin in the 9th Ward of New Orleans after Hurricane Katrina

There are chief executives in the various governments within the United States. Just as the president of the United States is the chief executive of the federal government, a governor plays that role in each of the 50 states. In addition, mayors serve as chief executive in America's cities and small towns. The **chief executive** is the highest elected or appointed official of a state, county, city, or separate local unit of government such as the local school system. The executive branch of any government, however, is not limited to one person. The executive branch also includes all of the chief executive's staff and all of the government agencies and offices that carry out—or execute—the functions of government. Collectively, we call these offices, agencies, and staff the **bureaucracy**.

Governors and big-city mayors manage multibillion-dollar budgets, protect lives by managing the police department as well as prisons and jails, save lives by directing fire and emergency services, restore lives through health care services that state and local government provide and regulate, and even take lives when governors in capital punishment states authorize executions. Governors are responsible for the care of children who have been abandoned or removed from the homes of their parents or guardians. Mayors in large cities often appoint school boards that operate public schools.

Governors often become presidents of the United States; George W. Bush, Bill Clinton, Ronald Reagan, and Jimmy Carter are among the most recent. Although it would be difficult for someone to move directly to the presidency from the office of mayor, big-city mayors have often been "president-makers." As heads of major political organizations in their states, they can deliver convention delegates and electoral votes to their preferred candidates. County board presidents and school board presidents most often do not go on to higher office.

Mayors often find themselves in charge during catastrophes. The September 11, 2001, terrorist attack on New York City's World Trade Center is a graphic illustration of the powers of a local chief executive, while also illustrating the limits on those powers. On the day of the attack, the city council was not in session and was unable to act. The courts were closed. Mayor Giuliani was accepted as the legitimate official responsible for dealing with this unprecedented crisis. He was soon aided by New York Governor George Pataki, who brought financial help and state employees to aid the city. Only later did funds come from the federal government as well.

Of course, chief executives do not always succeed in emergency situations. In September 2005, New Orleans and parts of the southeastern United States were devastated by Hurricane Katrina. The usual assumption is that state and local leaders will handle the first 72 hours of a disaster, but in the case of Katrina, state and local resources were insufficient and the federal government wasn't there in time to help. Mayor Ray Nagin of New Orleans and Louisiana Governor Kathleen Babineaux Blanco, as the local and state chief executives, were the prime actors on the scene. They had some major positive accomplishments. About 80 percent of the population with access to automobiles successfully evacuated before the storm hit. The day the storm hit, the governor asked President George W. Bush for "everything you've got." The governor had declared a state of emergency, put the National Guard on alert, and arranged to have the traffic patterns changed to aid evacuation. However, when Katrina hit land, the local officials were without adequate communications or sufficient resources to cope with a crisis of this magnitude. Most of New Orleans and much of the state and region suffered heavy damage. Without sufficient local resources or federal help from offices such as the Federal Emergency Management Agency (FEMA), Mayor Nagin and Governor Blanco were unable to cope with Hurricane Katrina.[1]

Nonetheless, state and local chief executives play a critical role during times of crisis as well as in the day-to-day tasks of governing. Despite criticism of how he handled the crisis, Ray Nagin was reelected mayor of New Orleans with 53 percent of the vote, having carried the majority of black precincts and a significant number of white votes.[2] Pollster John Zogby believes that Katrina may have been more of a defining moment than September 11, because it revealed structural failures of our system of government, with Washington getting most of the blame.[3]

STATE AND LOCAL CHIEF EXECUTIVES STRUGGLE FOR POWER

Governors, mayors, county board presidents, and chief executives of the many separate, independent units of local government are constantly engaged in the struggle for and the proper use of power, beginning with often fierce election campaigns. Once elected, these officials find themselves in a constant struggle with the legislative and judicial branches over the laws to be passed and how they should be interpreted and enforced.

We face two dilemmas in representative democracies. First, we want citizen participation and a strong legislative branch of government to provide informed debate, yet trying to respond to so many voices can lead to stalemate, or too much debate might lead to inaction. But an equal danger is that leaders will become all powerful, resulting in tyranny. Nonetheless, as Americans we believe that for democracy to work, we need strong, executive leadership.

The United States was founded in opposition to the dictatorial aspects of colonial rule. The framers of the U.S. Constitution as well as the drafters of the early state constitutions sought to prevent **tyranny**—oppressive governmental power in the hands of one, a few, or many who use their power to deny the rights of citizens or thwart the public good—by a separation of legislative, executive, and judicial powers. To this separation of powers they added the principle that the city, state, and national governments would serve as a check on each other. For example, cities are under the authority of their state, so a mayor may not become too dictatorial. The power of chief executives is balanced by the power of the legislature and judiciary, and it sometimes can be modulated by the need to stand for reelection.

POWERS OF GOVERNORS AND MAYORS AS CHIEF EXECUTIVES

Skill in the use of political power is of fundamental importance in governing as a chief executive. Our nation's founders never envisioned such dominant roles for government executives as we see today. American Revolutionaries found that the governors sent by King George III of England were often oppressive and unresponsive to local interests and needs. Thus, as the United States established its independence, early state constitutions constrained the powers of governors, using term limits, weak veto powers, few executive appointments, and executive power divided among statewide elected officers such as the secretary of state and the state treasurer.

Over the decades, revised constitutions increased the formal powers of governors and mayors. For instance, the governor's **veto power**, which allows the governor to negate the legislature's enactment of a law, has been strengthened in many states. City charters and state laws often provide mayors and county board presidents with the ability to veto proposed laws and ordinances as well. Veto power sometimes includes the **line-item veto** so the executive can strike individual items from budget bills, and the **pocket veto,** which kills a bill if it goes unsigned by the executive within a specified time. On the other hand, governors and mayors who once appointed employees on the basis of political work rather than merit have seen their authority to do this curbed by courts.[4]

Even where gubernatorial or mayoral powers are considered weak, as in Texas or Chicago, governors and mayors can become strong through political skill and popular support—as in the case of Texas Governor George W. Bush—or through political skill and the power of a political machine—as with Chicago Mayors Richard J. Daley and Richard M. Daley.

Formal and Informal Powers

Elected chief executives possess a combination of formal powers granted to them by constitution or statute, and informal powers that they develop as their political talents allow.

Formal Powers

- Appointing agency heads, board and commission members, personal assistants and other staff.
- Determining policy in all executive agencies and departments.
- Exercising veto power over legislation.
- Calling special sessions of the legislature to deal with specific topics.
- Developing a budget to propose to the legislature.

Informal powers can add considerable clout to the offices of governor and mayor.

Informal Powers

- Domination of political news coverage within a state or city.
- Ready access to the media.
- Perquisites of office, such as entertaining at the governor or mayor's mansion, use of a government plane or helicopter.
- Ability to do favors for lawmakers and local leaders through executive agencies and departments.
- Capacity to raise money for one's own campaign but also for campaigns of other party members.
- Public expectation that the chief executive is the top elected leader.

Governors and mayors generally are prominent in media coverage of state and city news, whereas comments by individual lawmakers are often treated as secondary or are ignored. Skilled governors and mayors can use their access to the media to arouse public support for their programs. They can provide favors for lawmakers and other political allies through the privileges (or "perks") of office, as when offering a ride on the governor's helicopter for a photo op, and they can provide campaign contributions from their own fund-raising efforts. Moreover, they are often seen as head of their political party, which gives them considerable power when dealing with other members of the party.

The Skillful Use of Power

A mayor or governor with abilities and talents can be extremely effective, even on the national scene. This can be especially true if that chief executive also possesses fame and fortune, as with Governor Arnold Schwarzenegger of California and Mayor Michael Bloomberg of New York.

Both chief executives are self-confident, self-made men who also have a reputation for effectiveness. For instance, when the Bush administration rejected the Kyoto protocol on global warming, Mayor Bloomberg's PlaNYC called for a plan to meet its emissions-reduction standards. Likewise, Schwarzenegger signed a bill which set unprecedented high fuel-efficiency standards for California cars.[5]

Such strong, effective mayors and governors are able to make tough decisions in the face of powerful opposition, as when former Governor Eliot Spitzer of New York and Governor Schwarzenegger addressed prison reform.[6]

Nevertheless, the authority of chief executives, whether they are governors or mayors, can be weakened in several ways. First, executive power is often divided, constitutionally or by law, among several other elected executive officials. Second, term limits can weaken an

GOV. ARNOLD SCHWARZENEGGER signing a bill raising the minimum wage.

RAISING OUR **Minimum Wage**

executive: the longer officials stay in office, doing more favors and appointing other officials, the more their power and influence increase. Third, executives are weakened when their state constitutions require public referenda on measures to increase taxes or incur long-term debt. Even if the legislature is willing to enact an executive's tax proposal, the public may vote against it at the polls.

In circumstances such as a local crisis or a weak economy, political skill plays a role in the ultimate authority and power of chief executives. Even with a full panoply of powers, some governors and mayors prove to be weak leaders, while some with weak formal powers become strong and effective executives. For instance, governors and mayors who are skilled speakers can take to the "bully pulpit" to enlist the voters' support for their programs, while simply the threat of a veto can defeat a bill in the state legislature or city council, since overriding a veto can require a majority of two-thirds or more in the legislature. And skillful use of the line-item veto can help a governor or mayor control government spending.

Mayors and governors can exercise a great amount of power, sometimes with little formal authority.[7] When they were governors, Terry Sanford of North Carolina and George W. Bush of Texas operated with few formal powers. Through force of personality and programs, Sanford led North Carolina in the development of a strong higher education system. Bush failed in his efforts to transform public finance for education in his state yet brought about major innovations based on his willingness to charm and work with legislators and key state officials such as the Texas lieutenant governor.

Strong executive power, whether lodged with a governor, county board president, or mayor, can create the bold leadership to resolve problems, implement new creative solutions, and undertake effective actions. There is a unity of purpose and action in a single chief executive that can never be achieved in a legislative body where the members naturally have

A RIBBON-CUTTING CEREMONY Often chief executives cut ribbons for projects. In this case, the actor known for his role as Sulu on *Star Trek* kicks off same-sex marriages in West Hollywood, California.

differing views and can be partisan or contentious. During recessions, governors can guide their states through drawn-out negotiations to enact difficult budget cuts. Such decisions by chief executives are not always right. Many are wrong and properly criticized. But the alternative of doing nothing about an economic meltdown or local crisis is far worse for communities. A positive view of executive power does not deny the importance of checks and balances, which the legislatures, courts, and voters must provide in a democracy to keep executive power from degenerating. But strong executive power is essential in state and local government.

THE ROLES OF A CHIEF EXECUTIVE

Mayors, county board presidents, and governors are expected to be leaders in a variety of ways. They are the ceremonial head of their city, county, or state. They cut ribbons and bestow awards. But they are also leaders who can take action in emergencies like terrorist attacks or natural disasters, and they can provide direction during sustained difficulties like the severe budget deficits that hit many states and cities in 2008–2009.

Leader of the Government

Different historical moments and different political climates call for very different kinds of leadership. There is more than one way to lead. Some distinctions that scholars make are between "democratic leaders" and "authoritarian leaders" as well as "builder mayors," "machine bosses," and "social or civic reformers." There are also distinctions between "charismatic leaders" who can spearhead major reforms and "caretakers" who don't make major policy changes.

Regardless of popularity or leadership styles, not all mayors and governors have the same amount of authority granted them by state constitutions, and some chief executives' power exists because their party also dominates the state legislature. Political scientist Thad Beyle has ranked the governors' power since the 1980s. He examines tenure, budget authority, appointment and veto powers, and whether a governor's party controls the legislature. The results of his 2007 study found that Democratic Massachusetts Governor Deval Patrick had the most clout of the 50 governors.[8]

Legislator in Chief

Governors, mayors, and county board presidents (and the executive branch of government which they head) initiate a lion's share of the critical action in the legislature—the laws, budgets, and appointments to key government posts. True, the legislature can turn down their proposals, but executives have the advantage of initiation in many legislative areas. The saying goes, "Executives propose and legislatures dispose." Executives usually do not propose minor legislation such as putting in stop signs. Chief executives leave many private resolutions—such as congratulating a couple on their 50th wedding anniversary—to the legislators who are closest to the constituents involved. But the city or state budget usually is not initiated in the city council or the state legislature; it is proposed by the executive branch and the elected chief executive.

The deep recession that began in late 2007 proved a supreme challenge for governors, mayors, and other local government executives. Tax revenues plummeted while demands for social services and education and training escalated. Generally, the chief executive rather than the legislature was expected to take the lead in proposing cuts in spending and politically difficult revenue increases. The national government can print money and go deeply into debt to deal with recessions. State and local executives lack that power and must balance budgets. During the recession, many states increased income and sales tax rates as cities and local governments increased fees and property taxes.

Chief executives have a greater mandate from the people to govern, as they are elected by all the voters in the city, county, or state, whereas legislators generally are elected from particular wards or districts. The chief executive can act individually, whereas legislatures must coordinate efforts among the numerous members—such as the 188 state legislators in Maryland, the 201 legislators in Minnesota, or the 51 city council members in New York City. Chief executives also have a greater ability than legislators do to command media and public attention. Using their high office, they can build support for legislation they favor or defeat proposals they oppose. As Alan Rosenthal remarks: "The legislature is no match whatsoever in a communications battle with a governor. Individual legislators may get good press. The institution seldom does, especially when it is under assault [from the chief executive]."[9]

Chief executives set the agenda with their annual state of the city, county, or state address and with the legislation that they and their departments introduce throughout the legislative session. They almost always propose the budget, although the legislative branch may propose amendments. They grant or withhold favors to legislators such as providing government jobs, scheduling an appearance in a legislator's district to announce new government initiatives and generate positive publicity, offering candidate endorsements, and supporting a legislator's legislative proposals and home-district projects such as road building, school improvements, and new parks.

Yet just because governors, mayors, and county board presidents have the power to propose legislation, budgets, and appointments does not mean that the legislative branch automatically approves these proposals. While some "rubber stamp" legislative bodies may approve every proposal of the chief executive with little debate and few votes in opposition, it is just as likely that state and local governments can end in gridlock. This is particularly true, as is frequently the case these days, if the executive branch of government is in the hands of one party and the legislature, or one of its branches, is in the hands of the other party. As Figure 3.1 in Chapter 3 shows, after the November 2008 general elections, Democrats controlled the governor's mansion as well as both legislative chambers in 17 states, and the Republicans had this control in 8 states.[10] Divided government was the case in 25 states, where different parties had control of the governorship and at least one legislative chamber, as seen in Table 4.1. Such divided government provides a check and limit on the power of chief executives, but it often also stalemates attempts to solve major problems because of the differences in party platforms and the struggle for advantage in the next upcoming election.

The veto power gives executives an upper hand in their negotiations with the legislative branch, since it gives them the ability to block legislation they oppose, unless a supermajority in the legislative branch can override the veto. The line-item veto allows some executives to veto an individual item in a budget bill, such as a tax increase or one expenditure item. This lets them shape a law or budget with a scalpel instead of using an axe to eliminate the entire budget.

Table 4.1 **Divided State Government: Party Affiliation of State Governors and Majority Party Affiliation of Legislative Chambers, 2008**

State	Party of Governor	Senate Majority Party	House/Assembly Majority Party	Divided State Government[a]
Alabama	R	D	D	X
Alaska	R	R	R	
Arizona	D	R	R	X
Arkansas	D	D	D	
California	R	D	D	X
Colorado	D	D	D	
Connecticut	R	D	D	X
Delaware	D	D	R	X
Florida	R	R	R	
Georgia	R	R	R	
Hawaii	R	D	D	X
Idaho	R	R	R	
Illinois	D	D	D	
Indiana	R	R	D	X
Iowa	D	D	D	
Kansas	D	R	R	X
Kentucky	D	R	D	X
Louisiana	R	D	D	X
Maine	D	D	D	
Maryland	D	D	D	
Massachusetts	D	D	D	
Michigan	D	R	D	X
Minnesota	R	D	D	X
Mississippi	R	D	R	X
Missouri	R	R	D	X
Montana	D	D	D	
Nebraska	R	Nonpartisan election	Unicameral	
Nevada	R	R	R	
New Hampshire	D	D	D	
New Jersey	D	D	D	
New Mexico	D	D	D	
New York	D	R	D	X
North Carolina	D	D	D	

(continued on next page)

Chapter 4 Executive Power in State and Local Government

Table 4.1	Divided State Government: Party Affiliation of State Governors and Majority Party Affiliation of Legislative Chambers, 2006 *(continued)*			
North Dakota	R	R	R	
Ohio	D	R	R	X
Oklahoma	D	=	R	X
Oregon	D	D	R	X
Pennsylvania	D	R	D	X
Rhode Island	R	D	D	X
South Carolina	R	R	R	
South Dakota	R	R	R	
Tennessee	D	=	D	X
Texas	R	R	R	
Utah	R	R	R	
Vermont	R	D	D	X
Virginia	D	D	R	X
Washington	D	D	D	
Wisconsin	D	D	R	X
Wyoming	D	R	R	X

Total Number: 25

ªGovernor's party affiliation differs from that of majority of at least one legislative chamber.
Source: Council of State Governments, *Book of the States*, www.stateline.org.

Most chief executives use their veto power sparingly. The more powerful they are, the less they have to use the veto. For instance, in the 22 years that Richard J. Daley was the powerful "boss" mayor of Chicago, he never once used his veto power, being able to prevail in the city council without using it. His son, Mayor Richard M. Daley, vetoed only one ordinance between 1989 and 2009.[11] On the other hand, as described in the Struggle for Power feature on the use of the veto in South Carolina, chief executives sometimes have to use the veto power early in their administration to prove their power and dominance. Afterward, the threat of the veto is often sufficient to force compromises from the legislative branch. Of course, being able to block legislation is not the same as being able to pass laws or budgets. In the end, that is the power and prerogative of the legislative branch of government. These differences in the legislative and executive branches of government set up the type of struggles for power about which this book is written.

Party Leader

In addition to their governmental role, local and state chief executives are most often the leader of their political party in their city, county, and state. We no longer necessarily expect them to be a party boss and chief executive at the same time, like Huey Long when he was Louisiana's governor or like many Boston mayors. But elected chief executives provide their

STRUGGLE FOR POWER

Using the Veto in South Carolina

Governor Carroll Campbell had to make extensive use of his veto power during the first year of his administration in South Carolina. He had to establish his power and authority in dealing with the legislature if he was to pass his programs and to prevent laws he didn't approve from being passed. So he used his line-item veto power 276 times that first year and most of his vetoes were sustained by the legislature in later override attempts. While some of his vetoes were successfully overridden by a two-thirds vote of the legislature, the South Carolina legislature soon became wary of opposing the governor and having him veto their legislation. Governor Campbell described the change this way:

"From that point on [after my first year in office] I haven't had to [veto legislation] very often because most

of the time if there is something that [legislators] are in doubt over what I am going to do, they'll come and ask me. Then it is a much better process because they will say, 'Are you going to veto this?' And I'll say, 'Yes, I have to veto it.' So they'll say, 'Well, what will you take, or what do you want?' When you get into that type of process, it is a much better process, but you have to establish the willingness to veto or to stand up when you think something needs to be done. I do that but I don't try to overdo it; but I've done it quite a bit."

Sources: Alan Rosenthal, *Heavy Lifting: The Job of the American Legislature* (Washington, DC: CQ Press, 2004), 177; and Laura A. Van Assendelft, *Governors, Agenda Setting, and Divided Government* (Lanham, MD: University Press of America, 1997), 180.

party with at least a titular leader. They support and provide a voice for their party by participating in party conventions, appearing in mass media advertising, and providing important endorsements for the party's candidates. They work with their party's members and leaders in the legislative branch to pass the programs announced in the party's platform.

Yet most mayors, county board presidents, and governors no longer have the power and political clout to block legislators, judges, and other public officials from getting their party's nomination. As we describe in the later chapter on elections, individuals tend to run candidate-centered, costly, media-heavy primary campaigns and win the party nomination mostly on their own. Afterward, the party may help them get elected in the general election. While the chief executive can provide very helpful endorsements for other candidates, they cannot control all elections. The chief executive is an important party leader but not the czar or boss of the party.

Shaper of Public Opinion

Local and state chief executives are important in shaping public opinion and drawing attention to certain issues and policy proposals. For instance, in Illinois in the 1990s when George Ryan was first elected governor, the vast majority of the public favored the death penalty in capital cases such as first-degree murder. But when 13 people on death row were proven innocent with the help of new evidence including DNA testing, public opinion became more divided on the death penalty.[12] Then Governor Ryan declared a moratorium on executions and appointed a blue-ribbon state commission to determine what reforms were necessary before the state could be sure that innocent men and women were not being put to death.[13] The commission made a series of reform recommendations, which by 2002 were beginning to be adopted. Yet when Ryan studied the individual death penalty cases, he felt compelled to commute the sentences of 160 inmates before he retired from office. Other states are now considering a moratorium or have abolished the death penalty. This is one example of a governor affecting public opinion. Those in favor of the death penalty in Illinois switched from 76 percent in 1994 to 43 percent in 2000 according to polling.[14]

THE HISTORY OF STATE AND LOCAL EXECUTIVE POWER

How much power governors and mayors have and how they use it have evolved over time. What follows is a brief look at executive power from a time of entrenched political machines to more recent reforms.

Political Machines

In the 19th century, a problem arose in local government that the framers of the U.S. Constitution had not foreseen. Powerful, hierarchal political party organizations, called **political machines,** governed big cities, including New York, Chicago, Boston, Kansas City, and St. Louis. Political machines were especially powerful in cities of the Northeast and Midwest. Southern states like Texas and Louisiana were run by statewide parties that paralleled the big-city machines. Political machines run on patronage, precinct work, favoritism, and party loyalty. And they are often marked by corruption. Party "bosses" headed machines in many American cities and states until the late 20th century. The bosses of the famous Tammany Hall machine in New York City mostly ruled through the officials they helped place in government rather than holding office themselves. Huey Long's organization in Louisiana and the Prendergast machine in Kansas City, Missouri, were similar to Tammany Hall.

HUEY LONG OF LOUISIANA at the Democratic National Convention in 1932.

Governor Long of Louisiana took office with relatively little formal power in 1929, yet by his death in 1935, he had acquired dictatorial powers over local government officials, the state university, and even the power of appointment of local schoolteachers.[15] First, Long proposed a good, popular program that the mass of poor and often illiterate voters wanted and needed—and he enacted his program of free school textbooks, night literacy programs, and highways in a state that had few. To pay for the program, he taxed big oil, primarily Standard Oil, which he never failed to cast as "The Devil." Long fought, according to him, against the oil giant on behalf of "the little guy."

In Missouri in 1900, Tom Prendergast and his brother Jim built a political machine that dominated the Kansas City mayor's office, the street and fire departments, and the police department. With this governmental control, the Prendergast machine was able to give out patronage jobs to supporters and contracts to their business friends in return for bribes. Initially, Jim and then Tom served as 1st Ward alderman, but neither ever served in higher office. Tom began to provide services to the city and county from his Prendergast Wholesale Liquor Company and his Jefferson Hotel, where prostitution flourished unchecked by the police. He later became a partner in the Ross Construction Company, which obtained county road contracts, and owner of the Ready Mix Concrete Company, which provided the cement for many civic projects. Paradoxically, the Prendergast machine also slated future president Harry Truman to run successfully for county judge in 1921.

By the late 1920s and 1930s, the Prendergast machine had a patronage army of as many as 37,000 workers and controlled jobs and contracts throughout the state. They provided help to the poor in the form of handouts, to workers in the form of jobs, and to the middle class in the form of political and social clubs in every ward, but they helped themselves more. Although at one time he controlled Kansas City completely and most of the state of Missouri, eventually Tom Prendergast went to jail for tax evasion. The Citizens Reform ticket won local elections in 1940, ending Prendergast's political career and destroying the machine he and his brother had built.[16] The stories of most political machines and most political bosses are similar.

Machines were powerful because they won elections mainly through the use of patronage precinct workers and political corruption; they controlled the executive, legislative, and even the judicial branches of government. Through patronage, they appointed loyal party workers to the key positions in government and even dogcatchers, garbage collectors, and building inspectors. So even though elections were held, machines centralized power in the hands of the party bosses or a single party boss who also was mayor or governor. Long and

"THAT'S WHAT'S THE MATTER."
Boss Tweed. "As long as I count the Votes, what are you going to do about it? say?"

THOMAS NAST cartoon of Boss William Marcy Tweed of New York's Tammany Hall machine.

Prendergast also developed power because they understood the needs, wants, and failings of those around them. Long kept a locked box filled with records and affidavits of failings and wrongs committed by his opponents, and probably his friends as well. "I can buy or blackmail 99 out of a 100 legislators," Long boasted.[17]

Reforms in Executive Power

In reaction to corruption, waste in government expenditures, and failure to take account of the wishes of those who didn't vote for machine candidates, the Progressive movement in the early decades of the 20th century pushed for election and management reforms. Progressives pushed for primary nomination elections to replace party caucuses and for direct election by the voters of U.S. senators. In addition, progressives pushed for the city manager form of government, which divided power between an elected mayor and an appointed professional manager.

Today, governors and mayors have been stripped of much of their **patronage** power, which is the ability of a chief executive to appoint supporters to government jobs based on their campaign work and party loyalty. Because of civil service and other reforms, most hiring for government positions is based on merit. Nevertheless, skilled political executives have replaced their patronage power with the power of money. Campaign contributions can be generated by the award of contracts to do state and city business. With sizable campaign funds, chief executives may be able to secure not only their own reelection but also the election of supporters to the legislature. Still, none has the all-powerful machine of a Tammany Hall or a Prendergast.

THE STATE EXECUTIVE BRANCH

Most state executive branches include a number of statewide elected officials. Another component of the executive branch is the chief executive's staff.

Elected Officials

In Illinois in 1981, Dave O'Neal resigned as lieutenant governor, saying, "There wasn't anything to do." In 1967, former North Carolina Governor Terry Sanford wrote that, except for governor, "the people barely know [that the other state offices] exist, and cannot possibly keep up with their activities." Former Illinois Lieutenant Governor and U.S. Senator Paul Simon echoed that sentiment: "Honestly, I'd have to say that most voters don't know what [state] constitutional officers do."[18] In most states, it is impossible for the lieutenant governor to really capture the voters' interest.

Among the 40 states that have the elected office of lieutenant governor, many of these "lite guvs" are assigned few or no formal responsibilities. Sometimes they are assigned very little office space and few staff members. In Texas, however, the lieutenant governor is arguably as important as the governor, especially when the Texas legislature is in session. The lieutenant governor not only presides over the state senate but also appoints the chairs and members of the chamber's committees. In addition, the lieutenant governor co-chairs the legislative budget board. As one observer noted, "Governors who don't get along [with] the lieutenant governors in Texas can find themselves pretty well boxed out of the lawmaking game."[19] So the powers of lieutenant governors vary immensely from state to state, but most are fairly weak.

When states revise their constitutions, as they do every several decades or so, convention delegates have tended to reduce the number of elected state officials in order to reduce the length of ballots and focus more accountability and responsibility on the governor. But many states still have a number of separately elected officials. Most states elect a secretary of state, attorney general, treasurer, and auditor in addition to the governor and lieutenant governor. This divides the powers of the executive branch among many officials, most of whom appoint their own bureaucracy and set their own policies. Some states elect education, railroad, and agriculture commissioners who are also independent of the chief executive. These state offices often become stepping-stones to the governor's office because of the statewide campaign experience and visibility as well as the fund-raising capacity of officials who employ large numbers and whose work can affect business, as with many secretaries of state.

In addition to the mayor, most cities elect separate city clerks and city treasurers, dividing the city's executive structure in the same manner as in the states. In addition to the county board president, many counties elect a half-dozen or more executive officials, such as an assessor, sheriff, treasurer, county clerk, and clerk of the court.

Independently elected officers can make life difficult for their fellow executives. Since the lieutenant governor is legally in charge when the governor is out of state, controversy could result if the governor and lieutenant governor are from different political parties. The state auditor or comptroller keeps the state's checkbook and pays all the state's bills, and some have used the information at their disposal to evaluate and criticize the governor's handling of the budget. The attorney general, who is the chief legal officer for the state, can issue opinions that conflict with those of the governor's legal staff. These multiple officials dilute the power of governors, mayors, and county board presidents.

The Governor's Office and Staff

With 50,000 or more employees and a score or more of major agencies, a major state governor needs a specialized, nimble, and confident staff in the executive office to deal with the hundreds of problems and questions that flow into the office each day. Walter Bagehot, a 19th-century political thinker, observed that the job of a department head is not to run the department but to see that it runs well.[20] That is good advice for any governor or mayor. If you try to micromanage each agency, you will be deprived of the time to focus on your policy agenda and the political efforts necessary to get it adopted. Thus, you must rely on your staff for support in the task of governing.

The principal assistant to the governor is the **chief of staff**, who generally oversees policy formulation and implementation, serves as chief negotiator for the governor, and manages the day-to-day operations of the office. As a former senior aide to an Illinois governor put it, the chief of staff role is a bit like being secretary general of the United Nations because the office and its agencies are often less a tight-knit organization than an "alliance of nations."[21] The chief of staff usually has charge of several subject-matter specialists who are responsible for areas such as social services, health care, transportation, education, and regulatory agencies. They also serve as liaisons to the various governmental agencies and their directors.

A governor's office often comprises 20–100 or more strong-willed, self-confident, often young men and women who have differing views of how to implement the governor's agenda. In addition, a governor tends to have a "kitchen cabinet" of friends and confidants who, though not a part of the governor's office, try to tell the governor and his office how to operate. Then there are the interest groups that helped elect the governor. The interest group leaders believe they should have access to the governor whenever they wish. And many of the state agency directors have been appointed to please various groups. They and their professional bureaucracies are often at odds with one another and with the governor's office as well.

Other offices of the governor include a state budget bureau, legislative staff, legal staff, press office, personnel office, and scheduler. The **state budget bureau,** or office of management and budget, prepares the annual or biennial budget for the governor and monitors spending by the state agencies. Whereas governors tend to bring in their own "political" people for most posts in the governor's office, budget bureaus tend to be composed of a "professional" staff of analysts who continue from one administration to another, with the exception of the budget bureau director and a deputy or two, who are appointed new with each incoming administration.

The governor's **legislative staff** works with legislative leaders and individual lawmakers, pushing the governor's agenda, counting heads, and handling lawmaker requests for gubernatorial support with their legislation. The **legal staff** provides advice on the extent of the governor's statutory and constitutional authority and the legal implications of individual decisions. The legal office works with agency lawyers on litigation, especially on litigation that would constrain the governor or commit the governor to new and recurring program expenditures. For example, lawsuits charging state government with failing to hire enough veterans or for making too many appointments based on political consideration rather than merit are monitored and often opposed.

The **press office** includes a press secretary, a speechwriter, and other assistants who respond to inquiries from the capitol press corps, issue press releases about accomplishments, write speeches, and organize press conferences for the governor to announce initiatives. Good press secretaries often serve as political counselors to the governor because of their sense of the mood of the media and public. In the new era of the "permanent campaign" to get reelected, some governors meet with their press secretary and other key advisors such as their chief of staff each morning to plan the announcements of the day and to "spin" the media to get their policies and programs adopted and applauded.

Most governors have a **personnel office** which manages and makes recommendations to the chief of staff and governor on staff hiring and appointments to scores of independent boards and commissions. In addition, some governors respond to requests from their political party's county chairs and campaign supporters to place friends in paid posts with state government.

The governor's **scheduler** has the role of evaluating the hundreds of speaking invitations that arrive weekly and allocating the governor's time for meetings with staff, legislative leaders, interest groups, and community delegations. The scheduler also has to protect some private time or family time for the governor. (See the Insider's View feature for a look at one governor's day.) Most governors would probably agree with the observation of Terry Sanford, a former North Carolina governor, who lamented: "There is no one in the governor's office whose only job is to gaze out the window and brood about the problems of the future."[22] An assistant to former governor Richard J. Ogilvie of Illinois evaluated how the governor spent his time during a 30-day period. Only 19 percent of his scheduled time was devoted to "management of state government," while 16 percent was

AN INSIDER'S VIEW

Governor's Day

A day in the life of a governor may begin at 7:00 a.m. and follow something like the following schedule:

- Reading staff memoranda, speech drafts, and newspapers.
- Staff meeting with chief of staff and press secretary.
- Meeting with social service directors.
- Photo with visiting school group from governor's hometown.
- Three consecutive 10-minute meetings with state legislators to hear their requests for support.
- Luncheon talk to state conference of realtors.

- Unscheduled meeting with director of emergency services on tornado that just devastated two counties; governor declares them "disaster areas" and schedule is rearranged for trip there next morning.
- Trip in state plane to New York City with budget director for dinner with credit rating officials reviewing the state's credit rating.
- Return home by midnight.

Source: Based on Ronald D. Michaelson, "An Analysis of the Chief Executive: How the Governor Uses His Time," *Public Affairs Bulletin*, Southern Illinois University, September–October 1971.

devoted to legislative relations during a legislative session. Governor Ogilvie, like most governors, allocated by far the most time (27 percent) to "public relations," which encompassed speaking engagements before political, civic, and interest groups as well as time spent with members of the media.[23]

In addition to these in-state staff members, most major states and cities have government offices in Washington, D.C., which monitor federal legislation that affects state funding and program flexibility. The head of the D.C. office also lobbies on behalf of the governor and state agencies for grants and increased funding from hundreds of federal programs that are administered by state government. This head of the D.C. office also serves as liaison with the state's congressional delegation. Members of both parties in the state delegation often work together with the governor's office for federal funding or state and local projects.

Governors all have different styles or strategies for managing their staffs, ranging from hands on to hands off. Some governors prefer hierarchical, army-like chains of command in which requests and proposals are routed through the chief of staff. Others want to meet directly and immediately with those who have requests, proposals, and suggestions without many buffers in between. Some governors spread authority and direct access among several staff members. Some appoint department and agency heads who are ambitious and who disagree with each other so that they have to bring issues and problems to the governor to resolve. The corporate and military management styles are not necessarily the best ones in government.

Many members of a governor's staff will be quite young, often in their 20s. Many start in internship programs in state or local government, stay on either in government or in a political campaign, and then gain appointment to a governor's staff. Of course, they are under the authority of older staff members who have been with the governor often for many years, such as the chief of staff and key aides. After several years as staff assistants, the younger professionals decide either to make government management a career, become lobbyists, return to the private sector, or go back to their home cities to run for office. Among the governors who started out as staff members are Jim Edgar of Illinois, who started just out of college as a legislative intern, and Gray Davis of California, who had served as chief of staff to Governor Jerry Brown. Young government aides today may become future chief executives.

EXECUTIVE OFFICIALS IN CITY AND COUNTY GOVERNMENT

On a day-to-day basis, a mayor's job can be tougher than that of the governor. A mayor in city hall is closer to the people than is a governor in a state capital office. Basic city services such as water and sewer systems, police and fire departments, and street maintenance affect everyone directly each day. Mayors also indirectly control other services, such as public education, that affect many citizens. There is not much room for a major error in running a city, and citizens are constantly evaluating the mayor's job performance.

Running a city is not always about high drama. Most mayors have an experience more like Shawn Gillen's as mayor of Monmouth, Illinois, a town of 8,000. Gillen actually served as both mayor and city manager. After the election and just before taking office, Gillen traveled to San Francisco for a meeting of the U.S. Council of Mayors, where he hobnobbed with host mayor Willie Brown and New York's Rudy Giuliani. Upon his return to Monmouth, on his first day behind the mayor's desk, he faced these problems: a resident was shooting arrows at a feuding neighbor; a backyard barrel contained burning human excrement that created an odiferous problem in the neighborhood; an irate couple complained that they had planned to park their recreational vehicle in the city park overnight but had been kicked out because they couldn't show proof of marriage to the city patrol officer.[24] Welcome to city government, Mayor Gillen.

Elected Officials in Large Cities and Counties

Large cities and counties have a plethora of elected executive officials in a conscious attempt to dilute the power and authority of mayors and county board presidents. Many large cities elect at least a city clerk and a city treasurer. The clerk keeps the official records of the city and handles city licenses and permits. The treasurer handles and invests the city's money. This means that a mayor does not directly control the city records, critical licenses, and money. The power is divided as a check on too much power in the chief executive's hands.

Frequently, county government is even more divided. Often the county sheriff, clerk, circuit court judges, and treasurer are elected separately from the county board president. There may be other elected officials such as a clerk of the circuit court, a county assessor, and a board of tax appeals. Usually the county board has some minimal control over the budgets of these offices, but not the conduct of the officials in these offices, who run elections, run the courts, assess the value of property for taxing purposes, handle the deposit of tax funds, keep the official records of land and the records of meetings of the county board, police the county, and prosecute offenders. County board president is still a powerful and important position, but it would be much more powerful and important with direct control of all these officials.

In writing state constitutions, city charters, and state laws governing local governments, the members of the conventions and legislatures have believed that divided executive powers, especially at the local level, are important to curb corruption and to keep chief executives from becoming autocrats and dictators. This has led to fragmented government and stalemate in some cases, but it has also kept executive power in check.

Executive Staff in Cities and Counties

Mayors and most county board presidents (or county mayors as they are sometimes called) don't have the huge bureaucracy that a governor has. Therefore, they don't have as large a personal staff. They don't have as many assistants. Their governing style is simpler. They do have to manage budgets, a city council or county board, and city employees, who may number a couple of thousand in a variety of departments. But it is not much different from managing a small or mid-sized company in their hometown.

However, mayors of large cities such as New York, Los Angeles, Chicago, Boston, Atlanta, and Houston have an office operation similar to the governor's. New York City Mayor Bloomberg, Los Angeles Mayor Villaraigosa, and Chicago Mayor Daley all run city governments with at least 40,000 employees and multibillion-dollar budgets. They can't govern informally as mayors did decades ago. They may have different management styles, different political party loyalties, and different political ambitions, but they are what some analysts call "manager mayors." They tend to be pragmatic, rather than ideological, in their public policies. And they must manage large bureaucracies to get the job done.

Mayor Richard J. Daley of Chicago, for example, appointed top-quality bureaucrats to run the city agencies and to make sure that services got delivered to the voters, but he kept a tight control over the city budget. He had a few direct mayoral assistants to help coordinate the city bureaucracy and make Chicago the "city that works," but he was a hands-on manager. All major projects had to get his personal approval, so there was often a bottleneck in government decision-making. Although he could make building decisions and initiate major public works projects, Daley was unable to handle the social crises engendered by racism and segregation or social problems like poverty, education, and police brutality.

Chicago Mayor Harold Washington had a number of deputy mayors to oversee the various government departments. For instance, the mayoral assistant in charge of public safety coordinated and oversaw the work of the police and fire departments, civil defense, and the city's 911 and 311 emergency numbers. The deputy mayors were often able to force coordination and cooperation between sometimes competing departments, and they were able to implement mayoral ideas and suggestions about new services or procedures.

The organizational chart of the city of New York shown in Figure 4.1 reveals that many departments and agencies report to the mayor and deputy mayors, along with dozens of boards, commissions, and quasi-independent agencies that directly operate the government of New York City. But the mayor doesn't directly control them. Since New York's Mayor Michael Bloomberg came up through corporate business structures rather than city government, he had to adjust to being a chief executive in this environment.

All big-city mayors, like governors, have a chief of staff who oversees the entire bureaucracy and often supervises the mayor's schedulers so that important individuals, interest groups, and bureaucrats who need to see the mayor get to do so. The chief of staff also ensures that the bureaucracy functions to accomplish the mayor's goals. The chief of staff, deputy mayors, department heads, and lower-level bureaucratic officials make simple bureaucratic decisions, but the major goals and direction of the administration have to come from the mayor.

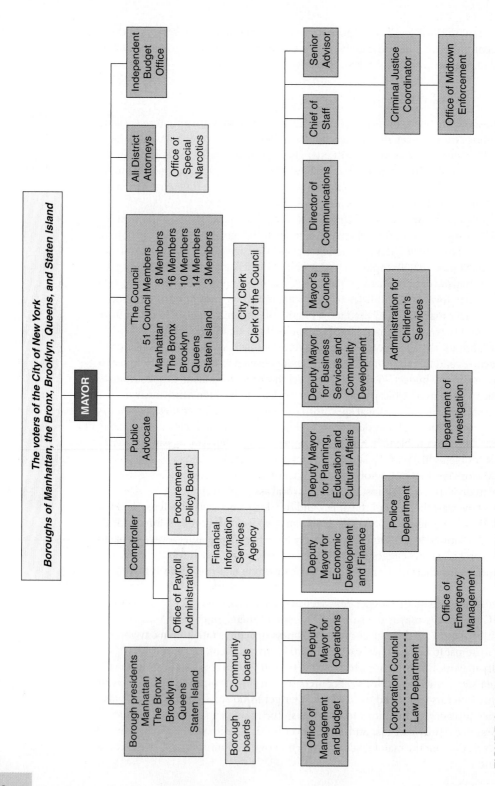

FIGURE 4.1 Organization Chart, City of New York *Source:* City of New York, www.nyc.gov.

Big-city mayors often control as many employees and as large a budget as a governor does, and their daily schedules can be just as busy. Like governors, however, they can't always get their way. But they do have an official lobbying staff in Washington, D.C., and in the state capital to work on getting laws and regulations passed, defeating proposals they oppose, and obtaining federal and state grants to help them provide the services their citizens demand.

MAYORS, COUNCILS, AND CITY MANAGERS

The majority of medium-size cities in the United States employ the council-manager form of government (discussed in Chapter 2). More than 92 million people live in cities and towns with city managers, and over 63 percent of U.S. cities with populations of 25,000 or more have adopted the council-manager form of government.[25] City managers take responsibility for preparing the budget, managing the agencies, hiring and firing personnel. Policymaking, however, still resides with the mayor and council, at least in theory, while management becomes the province of the manager. (See Figure 4.2.) The concept of a professional manager is extended to many county governments and is used extensively as well in smaller forms of government such as school systems that hire a professional school superintendent to run the school under the supervision of an elected school board.

Tension often exists, however, between mayors, city councils, and city managers as they scrap over power and responsibility and frequently cross the blurry line between policy and management. Skilled city and county managers generally keep modest public profiles and work closely and cooperatively with their mayors, city councils, or county boards. They keep these elected officials fully informed of their activities. The average city manager in the United States has spent at least 17 years in their profession, and many serve more than 10 years in the same city manager position.[26]

When Jerry Brown became mayor of Oakland, California, in 1999, the citizens had just abandoned their weak mayor/strong council and manager system and reverted to a strong mayor/weak council system. But Brown kept the city manager who had been appointed by his predecessor. The two developed a strong "Mr. Outside, Mr. Inside" relationship based on shared goals. Mayor Brown operated as a board chair while city manager Robert Bobb acted as the CEO of city government. Brown made it clear that while he would be calling the shots as mayor, he had no intention of usurping Bobb's day-to-day authority. Brown shared City Manager Bobb's chief of staff, and the city manager's policy staff advised Mayor Brown as well as the manager.[27]

Managing big governments where power is often fragmented and overlapping is difficult in good times. Effective management is as much about psychology and sociology as about administration. In the council-manager model, an elected mayor chairs the city council or city commission (which in smaller cities, towns, and suburbs usually consists of 5–15 elected members). The mayor is often elected citywide while the council or commission members are most often elected in small single-member districts called wards. The mayor and the city council (or city commission) enact the laws, adopt the budget, and set the policies of the city.

In smaller suburbs, towns, and cities, mayors are part-time and tend to their own businesses or professions as well as to the city's business. There is a gradual trend toward increasing the mayor's salary and making mayors full-time, but most mayors are still part-time in cities and towns employing city managers.

City managers have the power to hire and fire the other city employees, propose new city programs and services, draft the city budget for council and mayoral approval, and

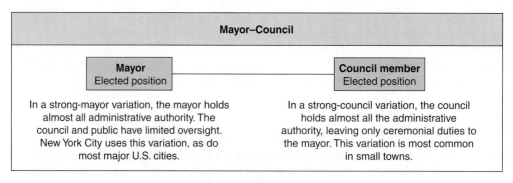

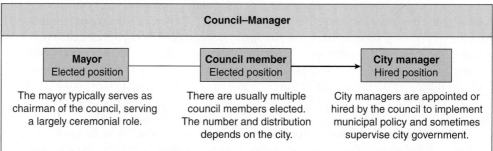

FIGURE 4.2 Diagrams of Council-Manager and Strong Mayor Forms of Government

directly supervise the various departments and agencies that actually deliver city services, like police and fire, garbage pickup, traffic planning, housing, and economic development programs. City managers are taught in graduate school not to meddle in politics and elections. According to the International City/County Management Association, all managers are bound by its Code of Ethics, which states that they "shall refrain from all political activities that undermine public confidence in professional administrators and refrain from participation in the election of members of the employing legislative body."[28] If city managers ignore this code, they will likely be fired when a new mayor or council majority comes to power. Should they back insurgents who lose in their efforts to overthrow incumbents, they will be out of a job as soon as their contract allows them to be terminated.

One example of the mayor/city commission/city manager form of government is found in the city of Hollywood, Florida, a medium-size city of 139,000. Mayor Mara Giuliante, who served until she lost reelection to a fifth term in 2008, chaired the seven-member Hollywood city commission, whose members are elected from individual districts. Giuliante also worked with City Manager Cameron Benson, and Hollywood's 1,700 city employees reported to Benson through department heads. In this system, the mayor and the city commission hold the manager responsible for city government operations. The mayor and commission directly appoint only four city officials: the manager, city attorney, and directors of two special community redevelopment agencies.

Benson described the roles of manager and mayor this way: "The mayor and commission make policy. The city manager implements policy."[29] Mayor Giuliante described their roles almost the same way: "The mayor guides public policy, but she is not involved directly in policy implementation. . . . She is the first point of contact for people. . . . She provides the political leadership that people want. . . . The mayor and the commission are the first or last point of contact to solve problems. They serve as ombudsmen . . . [but they] should let the manager manage."[30]

These distinctions between the powers and functions of managers and mayors are not clear-cut in practice, of course. Mayors and commissioners can meddle in detailed governmental operations, and managers can get involved in politics. This happened with the city manager who preceded Benson. The city commission had split between Mayor Giuliante's faction and a strong opposition bloc, and the city manager had decided that the opposition faction was on the way up and Mayor Giuliante was on the way down in political support. So the manager became antagonistic and tried to undercut her publicly. When Mayor Giuliante and her allies won the next election, the city manager was forced to retire. By contrast, Mayor Giuliante has said of Benson: "The current city manager is a team player. We speak constantly. In some negotiations, the manager and I send a letter with both of our signatures to signal the city administration and the city commission and mayor are united in their positions. . . . It is a matter of building trust. . . . With Cameron we have gotten over the previous disconnect. We avoid turf wars."[31]

As the Insider's View feature illustrates, a city manager has a busy schedule and a wide variety of duties. City managers meet with citizens who have complaints or suggestions about city services, supervise city departments, check up on the delivery of services, work with staff to create the city budget, ensure that spending is within budget resources, attend community events, meet with city planning directors, and prepare reports for the city commission. They attend all city commission meetings to make sure that the commissioners have the information and technical advice they need to make policies. They meet with the mayor in advance of commission meetings to plan the agenda and to respond to problems. They attend numerous community meetings but let the mayor and the commissioners make the speeches. City managers are careful to support the elected officials because, after all, they are the ones who have to stand for election, and they are the ones who hire and fire city managers. It is a position requiring a keen intelligence, tact, and infinite patience. It also means working at least 60-hour work weeks and, often, more than 12 hours a day, from 7:00 or 8:00 a.m. until at least 9:00 p.m. at community meetings in the evening.

This description of the role of city managers and their executive powers is not to suggest that mayors in mayor/manager/commission or mayor/manager/council forms of government are passive. Mayor Giuliante points out that when mayors are first elected, they have to be dependent upon the city manager and staff, but that a mayor soon comes to know the issues and mechanics of governing and can lead with clear directions and with the mandate of the voters behind her. Mayor Giuliante, like the city manager, works 60 hours a week, despite a part-time salary.[32] She meets with developers who want to do major projects but also with her neighbors, even when she is at the grocery store shopping. She meets citizens at neighborhood meetings and official city hearings. She spends the time to ensure that she has the votes she needs to win approval of her programs in the city commission, in which she has by law only one vote and no veto. She ensures that the city manager and department heads are doing their jobs well. She negotiates with neighboring cities, state agencies, and the federal government in Washington, D.C., on projects and to get the funding her city needs.

Mayors have important, but limited, executive powers; they are not dictators who can get any whim carried out. A mayor is a politician who is in charge of a city, but the city

AN INSIDER'S VIEW

A Day in the Life of City Manager Cameron Benson of Hollywood, Florida

8:00 a.m.   Monthly breakfast meeting with Greg Turek, public works director.

9:00 a.m.   New-hire orientation (welcome to all city of Hollywood new employees).

9:15 a.m.   Update meeting with assistant city managers Rick Lemack and Dave Keller regarding operation issues.

9:30 a.m.   Budget workshop review recap with Rick Lemack, assistant city manager; David Keller, assistant city manager; Cynthia McCormack, management and budget director; Macciano Lewis, assistant management and budget director; Gail Rinfeld, human resources director; Gayle Hayes, human resources manager; John Barletta, information technology director; Cynthia Miller, art and cultural affairs director.

11:00 a.m.   Weekly meeting with City Commissioner Sal Oliveri to discuss general city issues.

12:00 p.m.   Monthly lunch meeting with John Barletta, information technology director.

2:00 p.m.   Weekly meeting with Mayor Mara Giuliante to discuss general city issues.

3:30 p.m.   Emergency operations meeting with all department/office directors regarding the budget workshop outcomes.

5:00 p.m.   Urban Land Institute (ULI) panel interview in the city of Lauderhill to discuss the State Road 7 redevelopment initiative.

7:30 p.m.   General Obligation Bond (GOB) presentation to the Beverly Park Civic Association regarding the city's capital improvement needs.

Source: Interview by author with Cameron Benson, March 26, 2004.

manager has day-to-day control of every facet of the work of the city's many employees. For city government to be truly effective, however, the mayor, city manager, and city council must work together.

RUNNING A BUREAUCRACY

Elected officials often have to perform very different functions—politics *and* government, policy-making *and* management—and they may be much better at one function or the other. Often, a good politician is not a good manager able to run a large, cumbersome, and rule-bound bureaucracy. And often a manager who can handle the day-to-day, routine operations of government may not have the vision and charisma to inspire long-term bureaucrats to adopt new policies or different ways of doing things.

In the end, an executive's goal is to run an effective government that can deliver services well. To do that a mayor or other chief executive must run, or at least oversee, a government bureaucracy. While the smallest town or suburb may have only a few dozen employees, many cities, counties, and states have tens of thousands of employees and budgets from hundreds of millions to billions of dollars. While someone who has had a management position in a business, university, or hospital may easily understand how to run smaller towns and suburbs, he or she won't necessarily understand how to run a large state or city government.

Many new mayors and governors face outmoded government structures that were created decades or centuries ago and are no longer up to the challenges of the 21st century.

These massive government bureaucracies were not designed by management consultants or college professors; they have simply grown "organically" and haphazardly. One of your authors, Dick Simpson, has sometimes jokingly threatened his classes on metropolitan Chicago politics and government that they would have only one question on their final exam. They would be asked to design a worse system of local government (short of a totalitarian dictatorship) than the current one. He has never put this question on an exam because it would be too difficult to answer, since there are 540 taxing units of government within Cook County and 1,200 in the six-county metropolitan region. Fragmented government, a history of machine politics, many overlapping, ineffective, dysfunctional administrative units, and low levels of citizen participation characterize Chicago. No one official effectively governs the metropolitan region, and even directing a single unit of government like the city of Chicago with its 40,000 employees and a budget of nearly $6 billion is extremely difficult.

Almost 15 million people work for local governments in the United States, and over 5 million are employed by our state governments, a total of over 20 million, which represents 1 in every 15 adults and children.[33] Hundreds of job classifications exist for the different skills needed to carry out the functions of government and to implement the initiatives of governors and mayors. In addition to administrative and management positions, highly technical backgrounds are needed by those employed in the agencies for environmental protection, nuclear power regulation, agriculture, child protection, highway engineering, public health, and information technology, to name a few.

A key to running a government bureaucracy is to appoint the best people available to head each agency, give them considerable autonomy in running their agency, and hold them accountable for the agency's performance, replacing them if they fail in their task. Like any government executive, however, agency heads, are dependent on their deputies, permanent civil service employees, and other staff. New department heads at the state level, most of whom last only four years or so, can set the tone of the agency, choose good assistants, inspire their staff, and set directions for their agencies. But day-to-day decisions are usually made by the much maligned "bureaucrats" who were there before the current executive and their appointed department heads, and they will be there years after the elected and appointed officials are replaced with new people at the helm of government. As Jane Byrne, mayor of Chicago in 1979–1983, once said: "I gave orders from the 5th floor [mayor's office] and no one responded." The problem of getting bureaucrats whose positions are protected by civil service law to execute your policies and agenda when you didn't appoint them and can't fire them is one familiar to every new chief executive and to many new department and agency heads as well.

Generally speaking, even in the largest cities and states, a chief executive directly appoints fewer than a thousand people. To get their jobs, most government employees pass tests or present educational or professional credentials proving them qualified for their positions—and they can't be refused jobs or fired because of their political preferences. These civil service protections are important in curbing dictatorial and partisan political leaders, and they prevent chief executives from filling every government position with loyal members of one party. Thus, the chief executive's control of rank and file employees is somewhat limited. All employees like to have the favor of the chief executive because they think it will get them praise, raises, and promotions. But they will not carry out just any orders "from upstairs," so managing the bureaucracy can be a challenging task.

Bureaucratic Politics and Policy-making

In a tiny rural county, as few as 20 people may staff the highway department, sheriff's office, and other county offices. For the governor of California, on the other hand, the government can include over 300,000 employees.[34] These staffs are often professionals with a good deal of time on the job and deep, often highly technical experience.

It was not always like this. Half a century ago, the staffs in many states largely turned over if the new governor or mayor came from a different political party than did his or her predecessor. When author James Nowlan was a child in the 1940s, he remembers seeing his neighbor and state policeman Jack Benedict working at the gas station on the corner, out of uniform. When Nowlan asked his father about this, he was told, "Oh, Jack lost his job on the state police. The new Democratic governor fired him and replaced him with a Democrat." In the last century, it was possible for a strong chief executive with the help of a strong political party or machine to appoint almost all officials. This was called "the spoils system," as in "to the victor belong the spoils of office." Government employees at all levels served at the pleasure of their politically elected boss, or patron, and they could not support candidates of their choice without fear of losing their jobs. More important, it was difficult for voters to freely elect the best people to represent them when an army of patronage employees worked the precincts on behalf of government-backed candidates.

Yet slowly throughout the 20th century, nonpartisan civil service, merit-based selection of employees slowly replaced patronage as the predominate method of hiring and retaining government employees. Early in the 20th century, Progressives, political theorists, and civic groups pushed for the elimination of party patronage and for the implementation of civil service personnel codes in government, with education, testing, and interview procedures free from politics as the methods for hiring and promotion. In 1990, the U.S. Supreme Court ruled that public employees could not be hired or fired based on political affiliation or support.[35]

Today, the governor of California can still make about 2,500 appointments to key management positions and boards and commissions, about one for every 100 state employees.[36] The governor of Michigan can make but a handful of appointments. The mayor of Chicago can appoint 1,000 of the 40,000 city employees.[37] And so it goes throughout the nation.

Government employees, sometimes dismissively referred to as "bureaucrats," are often given a bad rap. When author James Nowlan served as director of three Illinois state agencies in the 1970s and 1980s, he found the employees overall to be conscientious, hardworking, and like most of us, interested in being remembered for having done their jobs well, for having made a difference. Think of government employees as members of a 100-piece symphony orchestra. The music the team makes is dependent upon skill, training, practice, teamwork, morale, and the leadership and inspiration of the conductor—the agency director, governor, mayor, or county board president. Generally, the stronger and more respected the orchestra leader, the richer and more responsive the musicians' collective performance.

In 1991, Nowlan surveyed 32 Illinois state agency directors from two gubernatorial administrations, one Democratic and one Republican. The directors were selected for the survey because they also had private-sector management experience. The agency directors found their state employees to be equally as diligent and capable as employees they had managed in the private sector, but 26 of the 32 directors found that it was more difficult in government to hire top-quality personnel and to fire an employee.

The difficulty of terminating a public employee points out that any approach to government personnel management has its strengths and possible flaws. For example, under the patronage system, employees could be easily terminated. But top-quality professionals often eschewed patronage-dominated personnel systems because employment was conditioned on politics rather than professional accomplishment. The civil service system is thus, understandably, more appealing to talented professionals.

On the other hand, civil service employees are protected in their employment by complex termination procedures, and agency heads have to prepare a strong, documented case if they want to fire someone. Employees may be protected as well by strong unions such as the American Federation of State, County, and Municipal Employees (AFSCME), the Service Employees International Union (SEIU), and teachers unions. Rather than fire an unresponsive employee, appointed agency heads often try to isolate the employee, or make life on the job difficult enough that the employee leaves "voluntarily."

Practical Guidance for New Agency Heads

State and big-city governments typically have 10–30 agencies or bureaus that report directly to the chief executive. Some larger cities like New York have 60 or more. Each is typically headed by an appointee of the chief executive, free from civil service testing. These agency directors come in with the newly elected governor or mayor, or not long after the chief executive is sworn in. In some cases, the agency director has one or several assistants who are also appointees of the elected chief executive. For the most part, the agency director is the interface between the political world above and the career civil service world below. A large agency of several hundred or several thousand employees may have half a dozen or more top civil service professionals as associate directors, budget officers, personnel officers, and key program managers.[38]

The governor or mayor has been elected to provide leadership and may have strong ideas about new approaches for running an agency, approaches that may be at philosophical odds with the predominate values of the career executives of the agency. The new governor and his appointed director of human services, for example, may believe in individual responsibility as the guiding philosophy, while the top career executives in the agency may believe in support and nurturing. In environmental protection, the governor may believe in market-oriented approaches for improving air quality, but the bureaucracy may prefer regulation.

If significant value differences exist, the agency director becomes the critical player who must develop a strategy for achieving the governor's policy and performance objectives in an agency whose career leaders are skeptical, even resistant. Yet agency directors should not come into their temporary job, which typically lasts only a few years, with plans to dictate to a civil service that is permanent and potentially unyielding. Instead, a new director needs to develop a working relationship in which mutual respect is created among agency director and career executives, who need each other. The director may need to accomplish, for example, the governor's campaign pledge to stress job training for single mothers receiving temporary assistance, while the career civil servants may want to continue their focus on training the mothers to become better parents. Inspiring the agency to achieve both these goals is the task of the director. The new director may find a way to offer support for the career staff's objectives, and simultaneously may seek their advice about how they would go about achieving the governor's objectives. The career staff might have terrific ideas and become enthusiastic about the task. Surprisingly, many new agency heads fail to ask the career civil servants, who know a great deal, for their advice.

The roles of political agency director and of top agency career executive are both challenging, sometime exasperating, but potentially exhilarating and satisfying. The political appointee has to maintain a constant flow of information to all the elements in the web of politics and government and, at the same time, absorb intelligence from the web. The career executive has to lead the agency orchestra by providing objectives, support, encouragement, professional development, protection from political interference in agency affairs, and whatever else is needed to keep the agency employees involved, interested, and responsive to initiatives that come from above.

Some bureaucrats go into civil service with idealistic objectives about, say, reforming education, lifting the poor out of poverty, and rebuilding dysfunctional families. After a few years of effort, however, some of these employees become frustrated that they will never be able to achieve their objectives, possibly because the bureaucratic system blocks their efforts or because poor, dysfunctional families refuse to cooperate. Over time, some of these employees will shift their efforts from broad societal objectives to more personal ones internal to the bureaucracy, such as protecting their jobs and gaining promotions within the system. The challenge for an agency director is to provide the leadership that supports and justifies idealism and professionalism among agency bureaucrats so they continue to perform well.

CHAPTER SUMMARY

The many state and local government executives in the United States include the 50 governors and myriads of mayors, county board presidents, and chief executives of local boards, villages, and districts. The executive branch also includes all the agencies that carry out the functions of these various governments. The chief executives of these various governments struggle for power during elections and afterward with the other branches and levels of government.

Mayors, governors, and other chief executives possess formal powers, such as appointing top-level officials, exercising the veto power, and calling special legislative session, which are found in their constitutions and charters. They also have informal powers which enhance their ability to govern efficiently; these generally include being head of their political party, having easy access to the media, and being able to raise money for themselves and their political allies. A chief executive who uses both types of power well can be extremely effective.

All chief executives perform various roles. They serve as the leader of their government, legislator in chief, head of their party, and shapers of public opinion.

In the past, much of the power of many governors and mayors came from the political clout gained from running political machines, in which supporters were rewarded with political jobs (patronage) in return for political support and election work. Corruption resulted in the passage of reforms beginning during the Progressive Era at the turn of the 20th century. Today, most governors and mayors no longer have such patronage power, which has often been replaced with the power of money from heavy contributors.

There are several other elected state officials besides the governor. Most states elect a lieutenant governor, a secretary of state, a comptroller, and an attorney general. Chief executives also have a number of support personnel. The governor's or mayor's chief of staff generally runs the executive office, sorts out information, and controls access to the chief executive. Mayors generally deal with the day-to-day functioning of the city and are constantly being evaluated as to the job they are doing. Often, governance is divided among many agencies to keep a mayor or county executive from becoming too powerful.

Although over 63 percent of cities with populations of 25,000 or more have adopted the council-manager form of government (with a nonpolitical manager handling the daily business of government), in states and larger cities policy-maker and managerial roles are combined in one chief executive. Governance depends on how effective they are as politician and executive, or policy-maker and manager. Since a chief executive also has to run the bureaucracy, these skills are important. Many bureaucrats are protected by civil service law, and it is important for the chief executive and agency heads to work well with sometimes skeptical career staffs.

KEY TERMS

bureaucracy The government agencies and offices and their staffs that administer government services and implement policy, p. 74

chief executive The highest elected or appointed official of a state, county, city, or separate local unit of government such as the local school system, p. 73

chief of staff The principal assistant to the governor who generally oversees policy formulation and implementation, negotiates for the governor, and manages the day-to-day operations of the governor's office, p. 87

legal staff Executive office staff who provide advice on the extent of the governor's statutory and constitutional authority and the legal implications of individual decisions, p. 88

legislative staff Executive office staff who work with legislative leaders and individual lawmakers, pushing the chief executive's agenda, counting heads, and handling lawmaker

requests for executive support of their legislation, p. 88

line-item veto The power to strike individual items from a budget appropriation without negating the entire budget, p. 75

patronage The ability of a chief executive to appoint supporters to government jobs based on their campaign work and party loyalty, p. 86

personnel office Executive office that manages and makes recommendations to the chief of staff and governor on staff hiring and appointments to independent state boards and commissions, p. 88

pocket veto The power to veto legislation by not signing it within a specific time, p. 75

political machine A political organization, usually a political party, characterized by patronage, precinct work, favoritism, and party loyalty, p. 84

press office Executive office that responds to inquiries from the press corps, issues press releases about

accomplishments, writes speeches, and organizes press conferences, p. 88

scheduler Executive assistant who allocates the governor's time for meetings with staff, legislators, interest groups, and community delegations, as well as private and family time, p. 88

state budget bureau Executive office that prepares the annual or biennial budget for the governor and monitors spending by the state agencies; also called the office of management and budget, p. 88

tyranny Oppressive governmental power in the hands of one, a few, or many who use their power to deny the civil rights of citizens or thwart the public good, p. 75

veto power The power of a chief executive to negate the legislature's enactment of a law. A veto may be overridden by a special majority vote of the legislature, such as two-thirds or three-fifths of each legislative branch, p. 75

QUESTIONS FOR REVIEW

1. What is a recent example of how a strong state or local executive dealt with a crisis? Was the executive successful or unsuccessful? Why?

2. What are two dilemmas faced in representative democracies that concern how chief executives handle power?

3. What are the formal and informal powers of state and local chief executives? What is a recent example of how a governor or mayor exercised a formal executive power? An informal power?

4. What are the various roles of state and local chief executives? What are some recent examples of each?

5. What is a political machine? How are these old machines different from and similar to 21st-century organizations?

6. What are the roles and functions of (a) a governor's office and staff and (b) other elected state officials in the executive branch?

7. How do the duties of a small-town mayor differ from those of a big-city chief executive? A county board president?

8. What are some of the qualifications that chief executives look for when selecting persons to run the various state and local agencies?

DISCUSSION QUESTIONS

1. *Stalemate* occurs in government when no official or group of elected officials has the power to enact a program or proposal. Yet *compromise* in politics is sometimes seen as weakness or as elected officials compromising away their principles. In the problem below, determine if you would make the compromises necessary to break the stalemate.

   Imagine that you are the top political advisor to Governor Don Smith. It is the post-election November session of the legislature convened to consider gubernatorial vetoes. Smith is completing his second and last term as governor, since he is barred by term limits from seeking reelection. His highest priority—and his policy legacy, as he sees it—is a tax increase required to fund a pathbreaking program of education for children up to 5 years of age and their parents. He has received national acclaim for the policy, but he cannot implement it without the tax increase. With the support of any two of three groups of opposition legislators, he can patch together the necessary majority for his tax increase. He could gain support by in turn supporting one of their proposals. Which of the following would you recommend that he support? Would you recommend the governor get the votes needed from at least two of the groups to pass his tax increase? Which ones, and why those? What are the ethical implications of agreeing with any of these legislative compromises?

   a. Several lawmakers want to authorize casino gambling in the state's largest city, with the revenues to go to that city. Governor Smith has always been an adamant opponent of gambling. Should he forego his principles and now support this casino gambling proposal?

   b. Five veteran "lame duck" legislators (not running for reelection) will vote for the tax increase if the governor will appoint each of them to $100,000 posts in his administration for the remaining two months that he is in office. Because their pensions are based on the highest salary earned in state government, these appointments will allow each of them to more than triple their annual pensions, from about $20,000 a year to $70,000. Governor Smith has decried this practice of his predecessor and has never made such appointments, but should he do so now in order to get his tax increase?

   c. Several members of the minority party have told the governor they will support his tax increase if he will sign a bill enacted by the legislature to ban stem cell research by the state's public universities. While these universities have not been at the forefront of research using embryonic cells, the governor has long stated his support for this type of research because of its possible therapeutic value in curing diseases like Parkinson's. Is it more important to educate children or promote stem cell research, which may not happen in this state to any marked extent anyway?

2. What is your opinion of state and local government employees? For example, are they hidebound bureaucrats who like to give the public a hard time? Or are they smart, dedicated employees trying to better the lives of their fellow citizens?

3. If you were mayor of your hometown or governor of your state, what is the first thing you would do to govern well and effectively?

4. Of the many roles that chief executives like a governors or mayors have to play, which would be easy for you and which would be hard? Which role would you like best?

5. If you wanted to become a city manager, how would you prepare for the job?

6. If you wanted to be elected mayor of your city or town, what would you need to do?

7. If you were ever appointed to head a bureau or agency of government, how would you spend your first days on the job? What would you need to do to accomplish your or your boss's goals?

PRACTICAL EXERCISES

Research the following, using the Internet or local newspaper files in your library. Some of this information will be easy to find and some may require more sustained research efforts. If you cannot find the information on the Internet or in back copies of your local newspapers (which nearly always have either a hard copy index or an Internet index for at least recent years), then ask knowledgeable experts like faculty members, civic leaders, neighbors, or political officials.

1. Determine the name of your governor, county board president, and mayor. Determine to which political party they belong and the vote by which they were elected in the last election. Provide a profile of each: what is their gender, age, education, and previous experience in government? Is he or she of the same party that has a majority in the legislative branch of city, county, and state government?

2. Determine the single most important thing done by the elected chief executives that you have named in Exercise 1. Did they promise in their last campaign to do something that they have not yet been able to accomplish?

3. Research the ways in which these chief executives are leaders. Have they proposed sweeping new programs, or are they caretakers continuing programs of predecessors? Do they dominate too much, or do they share their power with other elected and appointed officials?

4. Name the state or city executive branch officials other than the governor and mayor. How many executive branch officials are elected in your state and city? What are their powers?

5. Find out whether your governor and mayor have strong or weak formal powers, according to the state constitution or local charter. Using newspaper stories on these chief executives, determine if they are considered by the public to be strong or weak leaders, and why.

6. Determine if your city or county has a city or county manager. What are the specific duties and powers of this official?

WEB SITES FOR FURTHER RESEARCH

For information about the governor as chief executive, visit the National Governors Association site at www.nga.org. For the U.S. Conference of Mayors, go to http://usmayors.org.

Mayors of both large and small cities have sites on the Internet. The Web page for Mayor Antonio Villaraigosa of Los Angeles is http://mayor.lacity.org, providing links to the vast resources of the city government. To get to the Web page for Wausau, Wisconsin, Mayor James E. Tipple, you have to follow the "City Department" links on the Wausau homepage at http://www.ci.wausau.wi.us. Does your mayor's office have a homepage, or is it part of a larger town or city Web site?

Governors have their own Web pages. Check out www.scgovernor.com/ for the South Carolina governor's activities, or www.colorado.gov/governor/ for Colorado's governor. Of course, your own governor's Web page is the one that would be most useful to you.

SUGGESTED READING

Beyle, Thad. "Gubernatorial Power: The Institutional Power Ratings for the 50 Governors of the United States." www.unc.edu/~beyle/gubnewpwr.html. Updated to keep current, this site ranks the power of governors in several categories.

Dorsett, Lyle W. *The Prendergast Machine.* New York: Oxford University Press, 1968. A historical account of two powerful political bosses and their machine.

Nowlan, James D. *Inside State Government in Illinois.* Chicago: Neltnor House, 1991. A comprehensive political and historical account of Illinois government.

Rosenthal, Alan. *Heavy Lifting: The Job of the American Legislature.* Washington, DC: CQ Press, 2004. An account of inside politics of state government and the exchanges between state legislative and executive branches.

Sanford, Terry. *Storm over the States.* New York: McGraw-Hill, 1967. A former South Carolina governor's inside view of the complications of running a state government.

Simpson, Dick. *Rogues, Rebels, and Rubber Stamps: The Politics of the Chicago City Council from 1863 to the Present.* Boulder, CO: Westview, 2001. Historical background on Chicago city politics, including the power held by various mayors.

Williams, T. Harry. *Huey Long.* New York: Knopf, 1969. A descriptive account of a famously powerful Louisiana governor.

Lawyer arguing a case before a judge and jury.

JUSTICE AND CORRECTIONS

T he judicial branch of government is the primary guardian of the rule of law in the United States. In that exceptional role, the judiciary is often perceived as being above the struggle for power that is central to the legislative and executive branches. Yet the courts are not removed from the struggle. Groups and individuals compete for the opportunity to define the law, one case at a time, for example. And judicial appointments and judicial elections bring individual judges into the political struggle for power and influence.

Note: Katherine Parkhurst, a Civic Leadership Fellow at the University of Illinois, conducted much of the research and provided the draft text on which this chapter is based.

Judges are elected to their positions in 39 states, and nearly all district attorneys are elected; thus, in most states, voters directly participate in bestowing and distributing power in our justice system.[1] The democratic nature of this process gives the justice system legitimacy and strength, yet it makes the system vulnerable to quirks of public opinion. As we will see, even popular television crime shows such as *Law and Order* and *CSI* affect the public, from which juries are selected.

The right of the accused to seek vindication from wrongful charges is central to the rule of law in the United States. Most individuals who are charged with a crime choose to negotiate a plea bargain, working out a compromise with the prosecuting attorney to avoid a trial. A judge in the case usually accepts the plea bargain and sentences the defendant to either a lesser crime than the one initially charged, or to a reduced or suspended sentence. The actual time served in prison, however, can be affected by public policies regarding prisons. If the state legislature has been reluctant to fund prison space, a person may serve less time than the sentence calls for, to relieve prison crowding.

To understand the justice system, we must first understand the basic structure of the judicial branch of government. Then we need to know what individuals figure in the justice system, and how they move through the various judicial processes. We also need some knowledge of the corrections system and its goals. And we need to understand the distinction between criminal and civil cases. In exploring these topics, this chapter will also illuminate areas of controversy or struggle such as judicial campaigns, plea bargains and discretion of prosecutors, and gaps between sentences and actual time served. The people and processes of the justice system have tangible impacts on the outcomes and what the system is supposed to achieve. But don't expect the experience to be the same as we see in television dramas or films, as the Insider's View feature demonstrates.

Police make an arrest.

STRUCTURE OF THE JUDICIAL BRANCH

State courts are the backbone of the justice system in the United States. They make decisions on virtually every kind of case; more than 90 percent of criminal and civil cases are decided at the state level. State courts handle over 100 million cases each year—an impressive load, given that there are 300 million people in the United States. Over half of the cases that courts handle are very straightforward, such as traffic violations. Criminal cases, civil cases, domestic cases, and juvenile cases can be more complicated. Each year, state courts oversee 20 million criminal trials, 17 million civil actions, 5.6 million domestic cases, and 2.1 million juvenile cases, according to the National Center for State Courts.[2] Understanding the structure of state courts is useful, as most people interact with these courts several times over the course of their lives.

County Courts

The Cook County Criminal Courthouse—the biggest and busiest felony courthouse in the nation—sits in a Mexican American neighborhood on Chicago's southwest side. It's a boxy limestone structure, seven stories high and seven decades old, with Doric columns, Latin phrases carved into the limestone, and eight sculpted figures above the columns, representing law, justice, liberty, truth, might, love, wisdom, and peace. None of them is visible from the back of a police wagon.

A paunchy, balding white officer is behind the wheel of the first wagon to dip down a ramp and dock at the rear door this evening. He opens the wagon's tail, and a chain of 15 men, a dozen black and three white, winds its way out. The prisoners are rumpled and rank—the wagon-tossed, wretched refuse of a major American city, arrested on Martin Luther King Day. The driver follows the prisoners to the door, balancing over one shoulder a clear bag hold-ing 15 smaller clear bags that contain keys, combs, lip salve, cigarettes, lighters, beepers, eyeglasses, belts, shoelaces. In his other hand is a sheaf of arrest reports and rap sheets.

A navy-shirted sheriff's deputy slides open the barred door, and the prisoners file into the courthouse basement. About 1,500 prisoners pass through this doorway weekly on their way to a bond hearing—78,000 men and women a year accused of violating the peace and dignity of the State of Illinois. . . .

Later tonight the prisoners will be escorted through a tunnel to the quivering elevator that will carry them up to Courtroom 100 for their bond hearings. At evening's end the lucky prisoners who can make bond or who get an "I-bond" (a no-money-down individual-recognizance bond) will walk out of the courthouse's front door. The rest of the prisoners—about two-thirds of them, if this is a typical night—will ride the elevator back downstairs and march through a longer tunnel to jail. . . .

[T]he deputies' tone. . .expresses their sentiments about the prisoners. Those marched into the basement are cloaked in the presumption of innocence—in theory. The pre-vailing wisdom down here, however, is: if they're so innocent, why'd they arrive in handcuffs?. . .

The initial group of prisoners is occupying one of three bullpens down another hallway. The benches are full, the overflow on the floor. . . .Thomas, [the sergeant in charge] tells everyone that standing isn't allowed,. . . the idea being that it's harder to throw a punch while sitting. . . . When more prisoners come down the hall, [Sergeant] Thomas splits them between bullpens one and two, jam-ming those chambers. . . . The female prisoners are [temporarily] being held in an anteroom near the base-ment entrance. . . .

When the bars at the entryway are slid shut for the evening at six-thirty, the basement chambers hold 77 prisoners—65 men and 12 women. Those are typical numbers for the night bond court. . . . At seven p.m. the deputies prepare to move prisoners up to the first-floor lockups behind Courtroom 100. [Sergeant] Thomas barks out the directions, and the prisoners pair up and link arms—like first graders, except for the handcuffs the ser-geant clicks on. They follow a deputy through the tunnel to the elevator. . . .

Upstairs the prisoners are parceled out into three lock-ups behind Courtroom 100—one for 12 women, one for the 36 males accused of felonies, the third for the 29 males charged with misdemeanors or who have outstanding war-rants. This last lockup is fetid before the prisoners arrive because of a toilet that backed up earlier today. . . .

Private attorneys are on the way for three of tonight's prisoners, who reached them through a call from the district station. The other 74 prisoners will be represented by two public defenders. . . .

Whether they're bonding out or going to jail, these defendants are beginning a journey. They are setting off into America's criminal justice system. . . . The men and women toiling in this machine work hard—but ultimately they are gears in a machine. . . .The system is run by peo-ple, but, as with many systems, it often seems the other way around. . . . What is typical about a courtroom . . . is about how justice miscarries every day, by doing precisely what we ask it to.

Source: Steve Bogira, *Courtroom 302: A Year Behind the Scenes in an American Criminal Courthouse* (New York: Vintage Books, 2005).

Trial Courts

State **trial courts** are the lowest and most basic level of the judicial system. They are often called the "court of first instance" because they are the starting place of virtually all civil and criminal proceedings. There is one trial court of general jurisdiction for every judicial district or circuit. These districts usually comprise more than one county, though more populated areas have separate systems for each county.

Trial courts have seen a steady increase in their caseloads. Between 1994 and 2003, state courts experienced an 11 percent caseload increase, partially due to population growth. The increase has not, however, been met with an increase in judges. State court systems are instead creating systems that improve court efficiency.

One approach for improving efficiency is creating courts of limited jurisdiction—courts that specialize in only a certain kind of case. For example, one type of court may conduct preliminary hearings for felony cases; another may handle misdemeanors, ordinance violations, and traffic tickets. Many states have "small claims courts," where civil cases can be heard and a plaintiff may seek only a fixed maximum amount for damages. Limited jurisdiction courts usually focus their work in one city or county. These courts often have simplified procedures that move cases through promptly.[3] Quickly handling the many cases that come before the courts has become one of the principal goals of the judicial system, even if absolute justice is sometimes lost in the process. For one judge's improvements in the system, see the Insider's View feature.

Many states have instituted specialized courts of limited jurisdiction in recent years. Forty states now have drug courts; family and juvenile courts are also increasingly widespread.[4] As more courts of limited jurisdiction became operational, their caseloads increased 16 percent between 1994 and 2003. This takes up some of the burden from courts of general jurisdiction, which hear more serious criminal and civil cases. Largely thanks to the expansion of these specialized courts, courts of general jurisdiction have seen only very small increases in their caseloads. In fact, courts of limited jurisdiction have become so common, they now handle two-thirds of the 100 million cases handled by state trial courts each year.[5]

According to one Indiana judge, there is no way to communicate how hectic a trial court judge's day is until you see it for yourself. He says, "Hearing to hearing to hearing, and among those you have [someone] coming in with a really important case with a really deep thoughtful legal questions, that he spent a lot of time on, and you're supposed to be up to speed and be able to ask probing questions when you've been spending a 60 hour week just with your hearings. It's hard to get people to understand. In most areas, judges are working long and tired hours without law clerks." A civil litigation defense attorney adds that these heavy loads in trial courts lead to more appeals. Where lawyers can, they consider the caseloads in various trial courts as they choose a jurisdiction in which to file their claim. None of this furthers thoughtful, careful, and just decision-making.[6]

Trial courts have many staff members who keep things operating smoothly. A court clerk, often an elected official, is responsible for docketing cases, collecting fees, summoning potential jurors, and maintaining court records. It is common for each judge to have a secretary and a legal clerk. A court recorder will write a transcript of every word said during the proceedings, so that it can be used as a reference in the event of an appeal. There are usually several judges serving one geographic jurisdiction, and there may be several dozen in highly populated areas. On average, there is one general jurisdiction judge for every 25,000 residents.[7]

A Productive Judicial and Legislative Relationship

Over his 20 years as part of the Hennepin County District Court of Minneapolis, Judge Kevin Burke worked with the legislature to make his court more efficient and fair. First Burke convinced the legislature to allow more cases to be heard in small claims courts. He also brought in a new system in which one judge hears all aspects of one case, eliminating the need for multiple judges to become familiar with the background of cases. As a result, the average speed at which a case went through the court went from three years down to six months. The improvements in efficiency did not come at the expense of fairness. To the contrary, Judge Burke helped to save the indigent defense system from going bankrupt by lobbying the state legislature to fund public defenders through state general tax revenue instead of local property taxes. Increasingly, judges must not only be fair and knowledgeable of the law; they must be able to work within the political system to make improvements for their courts.

Source: Allen Greenblatt, "Kevin S. Burke, Court Reform: A Keen Focus on the Fundamentals of Justice," *Governing Magazine*, November 2004.

Intermediate Appellate Courts

Appellate courts have a very specific function. They only determine if legal or procedural errors occurred during a trial. They do not investigate the facts of a case, determine a criminal defendant's guilt or innocence, or determine a civil defendant's liability. Thirty-nine states have intermediate appellate courts—a level between trial courts and higher courts of appeal. Several states have an intermediate appellate court to handle criminal appeals and another to handle civil suits.[8] Intermediate appellate courts do not necessarily have the final say in all state court cases, but for many cases this is the end of the line.

In dramatizations of courtroom proceedings, one might see a lawyer stand up during a trial and shout "Objection!" If the case is appealed, an appellate court judge will read the court record, note where the lawyers stated "Objection!" and decide if a mistake in the application of the law was indeed made where the lawyer pointed it out. For example, an appeals court may be asked if the prosecution is allowed to present certain evidence at a trial, especially if there is some question about how the evidence was collected or presented. If an appeal is *remanded*, as most appeals are, it is handed back to the state trial court; the appellate court does not decide the final outcome of the case. Where the decision is either *reversed without remand* (negated) or *affirmed* (upheld), the decision is final, unless there is an appeal to a higher court.

In 2003, approximately 281,000 appellate cases were filed in the United States. This is a large number, but small in comparison to the 100 million cases handled by trial courts. Most appellate courts hear 87 cases per 100,000 residents in their area of jurisdiction. Over time, there has been a steady increase in appellate caseloads without a corresponding increase in judges. Again, courts have relied on simplifying and streamlining processes to achieve greater efficiency. Among the streamlining methods are having pretrial settlement conferences or requiring lawyers to present only an oral or written argument instead of both.[9]

Appellate courts hear many mandatory appeals, plus other appeals at their discretion. In many states, certain types of cases will automatically be heard by an appellate court. Appeals are usually automatic in potential death penalty cases. Depending on the specific state, anywhere between 25 and 100 percent of the cases an appellate court hears are mandatory appeals. Trends in state law have increased the number of cases where mandatory appeals apply. Overall, there has been a 6 percent nationwide growth in appeals over the past 10 years, but this growth has been uneven across states. Seven states have experienced increases of more than 40 percent.[10] In cases where plaintiffs or defendants petition an appellate court for a discretionary review of a case, they are unlikely to have their case heard.

Appellate staffs are similar to trial court staffs, but there are notable differences. Most states that have intermediate appellate courts have at least nine judges serving on the court. Judges tend to rely on more legal staff and less clerical staff. Each judge usually has two or three legal clerks to assist with researching precedents and opinions.[11]

State Supreme Courts

State supreme courts are **courts of last resort**, the final form of state appellate courts. They have the highest level of authority in the state judicial system. This is because a state judicial system is largely independent of the federal judicial system. States are bound by federal court decisions only on matters of U.S. constitutional questions. More than 95 percent of criminal cases arise under state criminal laws, and state courts handle most family law cases, landlord-tenant relations, traffic violations, creditor-debtor disputes, personal injury suits, and commercial interactions.[12] Most ordinary court cases are solely in state court jurisdiction. If you go to a court as a juror, witness, plaintiff, or defendant, you will probably go to a state court.

Professor G. Alan Tarr clarifies the structure of federal versus state courts with three clear principles. First, federal law is supreme, and all conflicts between federal and state laws are resolved in favor of the federal law. Based on this principle, federal courts may occasionally demand to hear cases originally presented in state courts, but this is very rare. Second, each system of courts has the authority to expound its own body of law. This means that federal courts must accept the interpretation of state laws as pronounced by the highest court of that state. Federal courts have no power to interpret or strike down state laws when there is no conflict with federal law. Third, courts recognize autonomy. When a state ruling is based on "adequate and independent state grounds," it is immune from review by the U.S. Supreme Court.[13]

Courts of last resort and intermediate appellate courts are not all alike. Some courts have only three judges, but others have as many as nine. Most state supreme courts have seven judges. Most state supreme courts have discretion to hear cases, meaning they can let lower court decisions stand without granting review. Some state supreme courts must hear every case that is appealed to them. You will have to check the laws of your state to determine the number of judges and rules governing your state supreme court.

PEOPLE IN THE CRIMINAL JUSTICE SYSTEM

In contrast to the public, community-wide focus of the executive and legislative branches, the judicial branch focuses on individual cases. While executive agencies and legislatures must enact policies based on what is appropriate for the community as a whole, the justice system revolves around unique people in unique circumstances. The people that make up the justice system are important because their individual actions collectively determine the fairness and effectiveness of the system overall. Policy-makers, judges, defenders, prosecutors, defendants, and juries are all an integral part of the process.

Policy-makers

Enforcement choices are always trade-offs. Law enforcement draws on finite resources, so someone must decide where police and other government employees should direct their attention. These decisions are often made by policy-makers and, less directly, voters. The decisions have real impact on quality of life in communities, from the time it takes a police officer to respond to a distress call, to gang violence in schools and on the streets, to the availability of treatment for drug offenders. Communities have different problems and different priorities, and thus different enforcement policies.

Policy-makers at many levels of government have an impact on how communities deal with crime. The federal government, for example, affects policy by providing grants to fund different initiatives. Governors influence actions of police and state courts more directly. Mayors, city councils, and police chiefs have the greatest influence of all, even though they operate at lower levels of government. When William A. Johnson Jr. was mayor of Rochester, New York, he focused the community's attention on juvenile crime. Under his guidance, the Rochester police department included a Youth Services Section, comprised of an education component that intervened with troubled students in schools and a crime unit that dealt with youth crime and probation programs. A corresponding Youth Violence Initiative, targeted at people ages 13–21, caused recidivism rates in this age group to plummet.[14]

When Los Angeles had trouble curbing gang violence, the city recruited high-profile, former New York chief of police William Bratton to head the police department. Bratton had won national attention for his "broken windows" program to reduce crime. The program focuses on crime prevention, uses better data systems to track crime trends, and seeks to catch minor criminals before they graduate to more serious crime. New York City had a 39 percent decline in its homicide rate on Bratton's watch. In Los Angeles, he had fewer police officers to work with and more territory to cover, but in Bratton's first year the homicide rate fell by more than 20 percent. The greatest improvements were seen in areas affected by gangs. By 2005, three years after Chief Bratton took over, gang membership had dropped by one-third.[15]

City councils also have an impact on crime and incarceration policy. In Minneapolis, police were frustrated that the city's 455 miles of alleys created a convenient escape for criminals on the run. The city council assisted police by creating a city ordinance that barred people from walking down alleys unless on their own property.[16] In Omaha, the city council passed an ordinance making a third-time prostitution conviction a felony offense, rather than a misdemeanor, to increase offenders' time behind bars.[17] The legitimacy of these laws and policies can be debated, but the central point remains: local and state elected and appointed officials set policies that directly affect the criminal justice system.

Judges

Judges have diverse responsibilities that are not always obvious to a courtroom spectator. Most people know that a judge oversees the trial and explains the process to the jury. Judges also establish a court schedule, rule on motions, ensure that a trial's outcome is filed in the public record, and supervise probation services, among other duties. The job can be difficult and emotional. In most states, 1,000–2,000 cases are filed per trial court judge per year, with trial court judges receiving much heavier caseloads than judges of intermediate appellate courts or supreme courts. Judges who retire early frequently cite stress on the job, court politics, staff issues, and low salaries compared to what they could earn in the private sector as factors contributing to their decision to leave.[18]

But for many, becoming a judge is a rare opportunity. Older judges tend to view their position as the culmination of their legal career, while younger judges may focus on name recognition and prestige. Experience as a judge is often helpful in private legal practice and in shaping a political career.[19] Judges are most often successful white men, particularly in appellate courts. In 2002, only 22 of 361 state supreme court judges were African American. And only about one-quarter of all appellate judges are female.[20] As with the legislative and executive branches, this is changing as more women and minorities go to law school and become lawyers, and as more nominating commissions realize the importance of diversity on the bench.

Almost all courts have a **chief judge** who may hold the seat through appointment, nomination by peers, or seniority. The chief judge typically manages the court building and non-judicial employees. One of the most important responsibilities of a chief judge is to assign cases to individual judges. In making assignments, the chief judge usually considers an individual judge's preferences and abilities. Judges in trial courts may be reassigned to a different specialty every year or two, as some judicial assignments are emotionally difficult to manage. Judges in family courts, for example, may find it difficult to hear child custody battles and bitter divorces over a long period of time. But rotating judges too often causes logistical problems, and it may cause plaintiffs to file their cases only when they would be heard by a judge likely to be favorable to them.[21]

Many states have terms for judicial service, similar to a governor's term in office, after which the judge leaves the position or may compete for reappointment. State judicial terms are usually 6 years, though they vary from state to state. Rhode Island's judges, for example, are selected by the governor and may keep their seat for life. Some states have a term limit, such as 10 years, after which judges cannot seek reappointment. Some states also have age limits, fearing that elderly judges may not always be able to fulfill their strenuous responsibilities. During their time on the bench, judges may also be subject to mandatory continuing legal education.

Judicial Conduct

Judges are expected to respect an ethical code. The American Bar Association adopted a Model Code of Judicial Conduct in 1972, which was subsequently given the force of law by most state legislatures. Judges may not comment on how they would rule on a case or an issue before they hear it. They are not allowed to be members of any organizations that discriminate against minorities. During a trial, judges are prohibited from *ex parte* communication, where they speak with one party in the case without including or informing the other. If judges have any reason to be biased in a case, such as personal knowledge of the case before trial or financial investments that would affect their judgment in civil cases, they are expected to *recuse* themselves; that is, they must withdraw from presiding over the case.[22]

All states have judicial conduct committees that investigate potential violations of the judicial ethical code. Citizens may file complaints with these committees. Ultimately, the committees have the power to sanction judges and make recommendations for a judge's removal. However, the committee cannot change the outcome of cases where impropriety is proven or suspected. This can only be accomplished by appealing the case to a higher court.

Judicial Selection

States use three common methods to decide who will serve on the bench: partisan elections, nonpartisan elections, and merit selection (see Table 5.1). Each of these has philosophical and practical strengths and weaknesses. Most judges get their seat on the bench through some form of election: 39 states use elections for at least some of their judicial positions, and 9 out of 10 state judges are elected to office.[23]

Some judges gain their seats through partisan elections, running as a Democrat, Republican, or other party candidate. Party preference is not supposed to color a judge's legal opinions when hearing cases about legislative redistricting, abortion, medical malpractice, or labor rights. However, voters may expect that when they elect a judge in a partisan election, that judge will decide cases according to the values associated with a particular political party. Voters may expect a judge running as a Republican, for example, to be socially conservative. One problem with partisan elections as a method of judicial selection is that they blur the line between law and politics. Justice is supposed to be beyond partisanship or political ideology.

TABLE 5.1 Judicial Selection in the States: Appellate and General Jurisdiction Courts, Initial Selection Methods

Merit Selection through Nominating Commission[a]	Gubernatorial (G) or Legislative (L) Appointment without Nominating Commission	Partisan Election	Nonpartisan Election	Combined Merit Selection and Other Methods
Alaska	California (G)	Alabama	Arkansas	Arizona
Colorado	Maine (G)	Illinois	Georgia	Florida
Connecticut	New Jersey (G)	Louisiana	Idaho	Indiana
Delaware	New Hampshire (G)	Michigan	Kentucky	Kansas
District of Columbia	Virginia (L)	Ohio	Minnesota	Missouri
Hawaii	South Carolina (L)	Pennsylvania	Mississippi	New York
Iowa		Texas	Montana	Oklahoma
Maryland		West Virginia	Nevada	South Dakota
Massachusetts			North Carolina	Tennessee
Nebraska			North Dakota	
New Hampshire			Oregon	
New Mexico			Washington	
Rhode Island			Wisconsin	
Utah				
Vermont				
Wyoming				

[a]The following states use merit plans only to fill midterm vacancies on some or all levels of court: Alabama, Georgia, Idaho, Kentucky, Minnesota, Montana, Nevada, North Dakota, and Wisconsin.

Source: American Judicature Society, "Judicial Selection in the States: Appellate and General Jurisdiction Courts," 2004, www.ajs.org.

CAMPAIGN POSTERS for Mississippi Supreme Court candidates.

Nonpartisan elections are a common method of judicial selection. The judicial candidate runs without party identification, but the campaign may be funded and supported by a particular political party or interest group. This method has a philosophical advantage over partisan elections in that it is a democratic way to select judges that is not overtly partisan; however, it has few practical differences from partisan elections.

One defense attorney from Indiana insists that he would not be happy working in a judicial system where judges faced elections. "Almost every lawyer in Indiana doesn't agree with every decision, but they never have to worry that they lost a case for illegitimate reasons. . . . I don't feel like I've ever won or lost a case for any reason other than what a case was about. . . . I can't imagine what it would be like to lose a case knowing that three of the five Indiana supreme court justices or two of the court of appeals ruled because they felt some allegiance to some interest group [or party]."[24] The Struggle for Power feature provides another view of the election of judges.

In a merit selection system, usually a governor or selection committee chooses a judge from a list recommended by a nominating committee. Supporters of this method believe that better-qualified people end up on the judicial bench than is the case with judicial elections, where campaign funding and partisan politics can be more important than the candidates' judicial skills. Eleven states, mostly in the East, have systems where judges are selected by governors, special selection committees, legislative bodies, or some combination of those.[25]

Merit selection may not be immune from politics, as a 2006 ruling by the New York court of appeals seems to demonstrate. Lower courts had ruled that the State of New York must pay $4.7 billion to New York City public schools to meet its obligations under the state constitution, but after Governor George Pataki asked the court of appeals to propose a lower

STRUGGLE FOR POWER

Rendering Justice, with One Eye on Re-election

In April 2008, Wisconsin voters elected a judge. The vote came after a bitter $5 million campaign in which a small-town trial judge with thin credentials ran a television advertisement falsely suggesting that the only black justice on the state supreme court had helped free a black rapist. The challenger unseated the justice with 51 percent of the vote. . . . This left many observers with the question of whether elections were the best way to select judges.

The question of how best to select judges has baffled lawyers and political scientists for centuries, but in the United States most states have made their choice in favor of popular election. The tradition goes back to Jacksonian populism, and supporters say it has the advantage of making judges accountable to the will of the people. A judge who makes a series of unpopular decisions can be challenged in an election and removed from the bench. . . . Nationwide, 87 percent of all state court judges face elections, and 39 states elect at least some of their judges, according to the National Center for State Courts . . .

The new justice on the Wisconsin supreme court is Michael J. Gableman, who has been the only judge on the Burnett County circuit court in Siren, Wisconsin, a job he got in 2002 when he was appointed to fill a vacancy by Gov. Scott McCallum. . . . The governor, who received two $1,250 campaign contributions from Mr. Gableman, chose him over the two candidates proposed by his advisory council on judicial selection. Judge Gableman . . . went on to be elected to the circuit court position in 2003. . . .

A working paper from the University of Chicago Law School [in 2007] tried to quantify the relative quality of elected and appointed judges in state high courts. . . . It found that elected judges wrote more opinions, while appointed judges wrote opinions of higher quality. "A simple explanation for our results," wrote the paper's authors Stephen J. Choi, G. Mitu Gulati, and Eric A. Posner, "is that electoral judgeships attract and reward politically savvy people, while appointed judgeships attract more professionally able people. However, the politically savvy people might give the public what it wants—adequate rather than great opinions, in greater quantity.". . .

Still, judges often alter their behavior as elections approach. A study in Pennsylvania by Gregory A. Huber and Sanford C. Gordon found that "all judges, even the most punitive, increase their sentences as reelection nears," resulting in some 2,700 years of additional prison time, or 6 percent of total prison time, in aggravated assault, rape and robbery sentences over a 10-year period. . . .

In the recent election, Judge Gableman's campaign ran a television advertisement juxtaposing the images of his opponent, Justice Louis B. Butler Jr., in judicial robes, with a photograph of Ruben Lee Mitchell, who had raped an 11-year-old girl. Both the judge and the rapist are black.

"My position historically has been that there is something to be said for the public to be selecting people who are going to be making decisions about their futures," [said] Justice Butler [a graduate of the University of Wisconsin law school who served for 12 years as a judge in Milwaukee courts]. "But people ought to be looking at judges' ability to analyze and interpret the law, their legal training, their experience level and, most importantly, their impartiality," he continued. "They should not be making decisions based on ads filled with lies, deception, falsehood, and race-baiting. The system is broken, and that robs the public of their right to be informed."

Source: Adapted from Adam Liptak, "American Exception: Rendering Justice, with One Eye on Re-election, *New York Times*, May 25, 2008.

---

amount, that court ruled that $1.93 billion would be sufficient. All four judges that favored the decision had been appointed by Pataki, and the two dissenting judges had been appointed by his predecessor, Mario Cuomo.[26] Many observers concluded that this was a political decision rather than an unbiased one.

All three methods of judicial selections are often followed by periodic **retention elections**, where citizens vote on whether to keep a particular judge on the bench. When a retention election is used in conjunction with merit selection, it is referred to as the Missouri Plan, since Missouri was the first state to use such a system. One problem with retention elections is that voters do not pay much attention to them and are not very informed about the judges or related issues.[27] In larger metropolitan regions, there may be too many judges on the ballot for voters to be well informed about each one. Usually, judges are retained in retention elections.

One reason for controversy over judicial elections is their expense. Judges and their supporters may run unregulated issue advertisements, and special interest groups may fund the campaigns. In 2000, state supreme court candidates broke fund-raising records in half of the states with judicial elections, raising $45 million, a 61 percent increase from the previous campaign finance record set two years prior.[28] In West Virginia in 2004, a coal company raised money under a lobbying group called "And for the Sake of the Kids," a group created to unseat a specific state supreme court justice. Yet any campaign finance reforms must be crafted to ensure that they do not inadvertently give special interests even greater power, and so that judges still have the financial ability to make voters aware of their name and their ideas.[29]

Defense Attorneys and Defendants

Today, in 8 out of 10 felony criminal cases in state courts, the defendant is represented by publicly financed attorneys, or **public defenders**. The Sixth Amendment of the U.S. Constitution protects our right to have "assistance of counsel" for our defense in a trial. In 1963, the U.S. Supreme Court determined in *Gideon v. Wainwright*, a decision based partially on the Sixth Amendment, that people charged with serious crimes who could not afford legal counsel were entitled to indigent defense services at the government's expense.

States and localities have different ways of providing indigent defense services. Public defender programs are used in most cases, assigned counsel programs are used in some, and contract attorneys are used in very few cases. In a public defender program, a nonprofit organization or a government office provides defense services. A court may appoint attorneys in private practice on a case by case basis, or the government may hire contract attorneys. Contract attorney systems are sometimes criticized as a cost-saving measure that results in inadequate legal representation for defendants. Public defense is provided in the majority of state criminal cases, especially for violent, property, and drug crimes.[30]

Inadequately funded public defense systems struggle in their mission to provide justice for defendants. In California, the state paid a small firm $400,000 one year to provide defense for 5,000 cases. The firm had only two associates to take on 250–300 cases per month. Of the 5,000 cases, only 20 were taken to court—the lawyers encouraged their clients to accept plea bargains for the rest. In one case, an associate noticed that the police might have improperly handled evidence. When she asked the judge for a continuance but was denied, she refused to proceed on the grounds that she could not provide adequate counsel under the circumstances. The other associate at her firm took over the case and the initial attorney was fired.[31]

Research shows that indigent defense systems nationwide face staff and budget constraints. In 1973, a National Advisory Commission on Criminal Justice proposed that public defenders not exceed the following annual limits in any one of these categories: 150 felonies, 400 misdemeanors, 200 juvenile cases, 200 mental commitment, or 25 appeal cases.[32] The American Bar Association is concerned that these standards are not being honored in most states. Rhode Island exceeds the standards by 35 to 40 percent, and it exceeds felony standards by 150 percent. In Pennsylvania, the public defender's office has twice as many cases today as it did in 1980 with the same size staff.[33] In New York, Chief Judge Judith

Kaye commented that she had "not seen the word 'crisis' so often, or so uniformly, echoed" as in the report about the indigent defense system.[34] The American Bar Association is concerned that the number of cases is not a good measure of workload because many cases today are more time-consuming and complicated than most cases at the time the case workload guidelines were established in 1973.

Despite these constraints, outcomes for cases with private attorneys and public defenders are similar. In federal and state courts, people with appointed counsel are found guilty just as often as people with private lawyers. For defendants who are found guilty, those with appointed counsel are about 20 percent more likely to be incarcerated than those with private lawyers, but for shorter terms. The average prison sentence for those with publicly financed attorneys is 2.5 years but 3 years for those with private counsel. Differences in incarceration rates and sentence lengths may be due to the fact that people who already have criminal records are more likely to use public defenders.[35]

Indigent individuals are not guaranteed public counsel for some misdemeanor charges that do not involve potential incarceration, and in some cases counsel is not necessary. Almost one-third of people in jail for misdemeanors represented themselves in court: half of all jail inmates had court-appointed counsel and 15 percent had a private lawyer.[36]

Prosecutors

Gary Walker, a veteran prosecutor in Michigan, once noted: "As prosecutors, we are blessed to occupy a position in which we get to make a difference in society, and we get to do so on a broader canvas than most. We are in a position of trust and honor."[37] According to the National District Attorney's Association, the goals of a prosecutor are: to promote the fair, impartial, and expeditious pursuit of justice; to ensure safer communities; and to promote integrity in the prosecution profession and effective coordination in the criminal justice system.[38]

There are different kinds of prosecuting offices in the justice system. Defendants being tried in a federal court will be prosecuted by a United States attorney. The majority of criminal cases, however, are tried in state courts by local prosecutors. Often called **district attorneys**, these local public officials represent the government in the prosecution of those accused of a crime. Local prosecutors' offices may also go by the name commonwealth attorney, state's attorney, or county attorney. The chief prosecutor may be elected or appointed, and may represent a county or a judicial district. All states have an **attorney general**, the state's primary legal advisor and chief law enforcement officer. The attorney general also has prosecutorial powers.

Like public defenders, prosecutors have heavy workloads. About 2,300 prosecutors' offices collectively handle 2.3 million felony cases and 7.5 million misdemeanors (including minor cases such as traffic violations) on a median annual budget of $355,000. This is the equivalent of prosecuting the entire population of Sweden every year. Prosecutors and defenders thus have a shared incentive to end as many cases as possible with plea bargains, which are used in 95 percent of all cases. Adding to the pressure of the work environment, more than 40 percent of prosecutors receive work-related threats of assault each year.[39]

Jurors

Jury trials are conducted only in trial courts, never in appellate courts or supreme courts, where judges decide the outcome. Only 5 percent of court cases ever get to a stage where trial by jury is necessary, but juries still play a vital role in the criminal justice process. And three-quarters of people polled say they would rather be subject to a trial by jury than a trial with judges alone.[40] Even in cases that are not brought before a jury, plea bargains are often based on how lawyers believe a jury would decide the case.

Juries fulfill an important symbolic role as well, providing ordinary people an opportunity to participate directly in the judicial process. This democratization of the justice system has real consequences. Historically, juries have refused to convict violators of unpopular laws, such as fugitive slave laws in northern states before the Civil War, and laws barring the sale of liquor during the Prohibition era of the early 20th century. In what is known as **jury nullification**, the jury acknowledges the accused violated a law but declares the defendant "not guilty."[41] Not only in such extreme cases but in most cases, a jury system usually results in outcomes perceived as fair. Very few judges express dissatisfaction with the jury system, reporting that jury verdicts are "correct" over 90 percent of the time.[42]

However, one major problem with the jury system is absenteeism. Though 20 million people receive a jury duty summons each year, up to half will be exempted, disqualified, or excused from serving.[43] Many that are asked to serve simply do not show up. One study in Dallas County, Texas, found that less than 15 percent of expected jurors reported to the courthouse.[44] Absenteeism makes running the jury system more expensive and more inconvenient for jurors and administrators alike. Many states now exact significant penalties for jurors who fail to show up when summoned.

Judith Kaye, chief judge of the State of New York in 2000, reformed her state's jury system, from the selection process to the service protocols. Before these reforms, courts took weeks to assemble juries for civil cases, and all juries for criminals trials were sequestered, or isolated, for the duration of the trial. Kaye did away with mandatory sequestering and increased jury pay from $15 to $40 per day.[45] Simple reforms can improve the jury system and are being instituted in many jurisdictions.

Convicted Criminals and Correctional Populations

In 2005, 7 million people were in the criminal justice system, and about 2 million of these were behind bars. State and local governments are responsible for most incarcerated prisoners; less than 10 percent of the incarcerated population is in federal custody.[46] The **incarceration rate**—the number of people being detained in jail or prison in comparison to the population—is 432 per 100,000 persons in the United States, higher than in almost any other country. The U.S. Bureau of Justice and Statistics reports that if incarceration rates remain unchanged, 1 in 15 people will serve time in prison or jail at some point in their lifetime.[47] Those incarcerated are likely to be male, young, and from a minority racial or ethnic group.

Criminal or **correctional populations**—a broad category that includes all individuals under some form of court supervision, including probation, prison, jail, and parole—are overwhelmingly male. Of the general population in the United States, 123 of every 100,000 women are in jail or prison, but 1,348 of every 100,000 men. This correctional population is over 90 percent male; 5 percent of all men in the United States are under correctional supervision. Racial disparities in corrections are glaring. The incarceration rates among males in 2004 were: white males 717 per 100,000; Latino males 1,717 per 100,000; black males 4,915 per 100,000. Almost 10 percent of all African Americans are under some form of correctional supervision compared to 2 percent of Caucasians. Minorities make up more than 60 percent of jail and prison populations.[48]

Criminals also tend to be young people facing difficult life circumstances. Over half of all people in prison and jail are less than 35 years old. The median age of felons convicted in state courts is 30. The Federal Bureau of Investigation reported in 2005 that 15 percent of violent crimes and 26 percent of property crimes were committed by minors. Roughly 40 percent of criminals lack a high school diploma. Many grew up in a foster home or institution, and almost half were raised in single-parent homes. Incarceration affects communities and families. More than half of incarcerated adults have young children at home. In the United States, 2 percent of all children have a parent in jail or prison, and 7 percent of African American children have a parent in jail or prison.[49]

About one-quarter of inmates are behind bars for violent crimes, most typically aggravated assault. One-third of inmates are convicted of property offenses such as burglary, larceny, or fraud, and one-third are there for drug-related charges, predominately drug trafficking.[50] Most of the people who are convicted of crimes and incarcerated eventually return to their communities. More than 95 percent of the people in jail or prison will be released to be reintegrated into society.[51]

PROCESSES OF THE CRIMINAL JUSTICE SYSTEM

From the first filing of criminal charges to the day of a release from incarceration, the criminal justice system consists of many participants exerting power. The prosecutor, representing the government, has a wide degree of latitude in deciding which path a particular case will travel. If the case goes to trial, both sides—prosecution and defense—put on a presentation for the benefit of the jury, as the jury has the power to determine the facts of the case and decide guilt or innocence. If the jury agrees the defendant is guilty, a judge has the power to punish, or sentence, the criminal. At the end of the process, the guilty may be incarcerated in jail or prison to serve out the sentence imposed by the judge.

Charges, Pleading, and Plea Bargains

Filing the charges is the first significant event in the criminal justice process. The prosecutor typically has broad powers to decide the crime with which a defendant will be charged. Different crimes may entail vastly different punishments for the defendant. Of course, the prosecutor is bound by state statutes, defendants' rights enumerated in the U.S. Constitution and in state constitutions, and judicial authority. Yet, in practice, the prosecutor has the power to change the defendant's life.

An important distinction in charges is whether the crime is a misdemeanor or a felony. For misdemeanors, an offender who is found guilty will generally serve a maximum of one year in a local jail. Misdemeanors include numerous minor offenses with typical sanctions of fines, probation, or community service. A felony has more serious implications. A convicted felon who is incarcerated will be sentenced to at least a year in prison, which is a very different environment from the local jail. Once a person is convicted of a felony, many aspects of his or her life are affected. Employers typically ask applicants if they have ever been convicted of a felony, thus making it difficult for felons to find jobs on their release.

Felonies and misdemeanors are treated differently in the justice process. Misdemeanors may be processed in courts of limited jurisdiction, which usually simplify the procedures.

Felonies are handled by general jurisdiction trial courts. Serious offenses may have mandatory appeals, so cases can take a long time to resolve.

Many trial decisions are not automatic or strictly guided by state law. The same event may be described different ways. For example, several college students who had been throwing "Molotov cocktails" in an empty street late at night were initially told they would be charged with using "weapons of terrorism," a felony. The students begged the prosecutor to think of this incident as a mere "disturbance of the peace" misdemeanor. The students eventually pled guilty to a misdemeanor charge involving some jail time, community service, and fines. Generally, as in this example, a prosecutor has the power to file charges, reduce charges, or drop cases altogether.

Is broad prosecutorial discretion good for the justice system, or is it harmful and unfair? One argument in favor of broad prosecutorial discretion is that it makes the justice system more efficient. In approximately 95 percent of criminal cases, the defendants will not have a full trial to determine guilt or innocence but instead will reach a **plea bargain**—the prosecutor will reduce the charge or recommend a minimum or suspended sentence if the defendant will plead guilty.

One study of prosecutors dealing with domestic violence cases in California found that the prosecutors frequently categorized new crimes by repeat offenders as "parole violations," a less serious offense than a new crime. One prosecutor explained that this reduces the procedural burdens of handling a case: "It's easier to make our case. It doesn't get me anything going to trial and risking an acquittal. The standard of proof [in a parole violations case] is preponderance, rather than beyond reasonable doubt. It lowers our burden—puts it in front of a judge rather than a jury, and makes it easier to [get a judgment that] they were in violation."[52]

Prosecutorial discretion and plea bargaining to determine a defendant's charges improves the system's efficiency while attempting to maintain its fairness. Since many states have guidelines demanding relatively uniform sentencing for all defendants found guilty of the same crime, the charging phase may be the only opportunity for the justice system to grant leniency.[53]

The U.S. Department of Justice explains that discretion in the justice system is necessary: "Legislative bodies have recognized that they cannot anticipate the range of circumstances surrounding each crime, anticipate local mores, and enact laws that clearly encompass all conduct that is criminal and all that is not. Therefore, persons charged with the day-to-day response to crime are expected to exercise their own judgment within limits set by law."[54] The American Bar Association says plea bargaining is prevalent because defendants can avoid the time and cost of defending themselves, and both defendant and prosecutor are spared the uncertainties involved with trials.[55]

One Indiana attorney defends the practice of plea bargaining this way: "I believe that the role of state trial court judges is to see that the parties resolve their disputes." This attorney knew one family court judge who intentionally gave a hard time to family members who brought their cases to court, encouraging them to settle their dispute outside of court. "His message was, you don't want me doing this. You want to figure it out for yourself. . . . If people perceive the judicial system as a place to go and fight their petty fights, they'll be there constantly."[56]

Others argue that plea bargaining is inherently unfair. The prosecutorial decisions lack the transparency of trials, and there may not be a way to determine if a defendant's rights have been violated. In cases that go to trial, 90 percent of felony cases end in conviction.[57] Given this result, defendants may feel coerced to accept a plea bargain even if they think they deserve to have the case heard by a jury. Some argue this coercion violates the Fifth Amendment right against self-incrimination.[58] This situation is exacerbated in cases where the prosecutor "punishes" the defendant by filing harsh charges if the defendant does not plead guilty.

A U.S. Supreme Court case that upheld the practice of plea bargaining was *Bordenkircher v. Hayes* (1978). Paul Lewis Hayes, the defendant, was accused of forging a check for $88.30, a crime punishable by 2–10 years in prison. Hayes had two prior low-grade felonies on his record. The prosecutor offered him a 5-year sentence if he pled guilty to the forged check. When Hayes refused, the prosecutor officially charged him under Kentucky's "Habitual Criminal Act," a charge that would automatically result in a life sentence if he were convicted. Hayes was found guilty at trial, and his life imprisonment sentence was upheld by the U.S. Supreme Court.[59]

After being charged with a crime, criminal defendants appear before a judge for **arraignment**, a pre-trial appearance where they enter their plea. A judge has the power to reject a guilty plea in some circumstances, as in cases where the judge suspects that coercion led to the guilty plea. If the judge accepts a guilty plea, the defendant does not have a trial and proceeds quickly to the sentencing phase.

The Trial

Defendants may request a bench trial, where all decisions are made by a judge, but most prefer to have the case heard by a traditional jury. The basic steps of a jury trial are jury selection, opening statements, presentation of evidence, closing arguments, and jury deliberations.

Most states require a panel of 12 jurors, but some allow a smaller jury, especially in misdemeanor cases. The prosecution and the defense may both use a certain number of peremptory challenges, meaning they can excuse some jurors for no given reason. Lawyers also can dismiss an unlimited number of jurors "for cause," which means the lawyers must have a good reason, such as the juror is shown to have a bias in the case.

After a jury is selected, the prosecutor and the defense attorney give opening statements. Essentially, each side explains the case they are setting out to prove. In this phase, a judge may explain the **burden of proof** standard to a jury—the burden of the prosecutor to prove the defendant's guilt. The prosecution always has the burden of proof. If the prosecution does not provide sufficient evidence, the defense automatically wins the case. In criminal cases, the prosecution must present strong enough evidence that the jury can conclude the defendant is guilty "beyond a reasonable doubt." This is the highest standard of proof in the legal system. Other standards are sometimes used in civil cases, such as "preponderance of evidence," which means a jury would only have to believe a charge is true, without having absolute certainty.

The heart of a trial is the presentation of evidence. In a pre-trial process called discovery, the prosecution and defense share information about what evidence they will present, who their witnesses are going to be, what the witnesses will testify about, and any other pertinent information. Witnesses are often asked to give a deposition before trial, essentially giving their testimony outside of court under oath in preparation for trial.

Discovery is one reason it often takes months or years to resolve cases in the court system. Discovery takes place directly between the prosecutor and the defendant's lawyers, with minimal oversight from judges, unless a judge is asked to settle a dispute. Each side is entitled to all facts and documents that the other side holds that may support the case to be argued in court. Even in the best of circumstances, this is a 30–45 day process. And the process may be incredibly disruptive and expensive. Lawyers may spend weeks scanning

AN INSIDER'S VIEW

The CSI Effect

Lieutenant John McMurray of the Chicago police department has an unusual and interesting complaint about juries—television is giving them too many ideas. According to McMurray, much of the American public has developed unrealistic expectations of criminal science from shows such as *CSI* (*Crime Scene Investigation*), *Law and Order*, and others. McMurray writes: "In the television world of CSI, all databases are immediately accessible, they are able to use them to compare evidence collected against all databases, and it is done on demand, sometimes even on the scene. This creates in the minds of the average citizen the unrealistic expectation of how law enforcement collects, preserves, analyzes, and compares evidence. In reality, the entire procedure is a long, sometimes tedious affair and under the best circumstances, a DNA test can be analyzed in about 48 hours."

Crime scenes are often devoid of forensic evidence altogether, and investigators can seldom hope to make a DNA match from a single strand of hair. According to a *U.S. News and World Report* article, prosecutors find that as a result of the CSI effect, it is more difficult to win convictions in the large majority of cases where scientific evidence is irrelevant or absent.

Sources: John McMurray, "The CSI Effect and Modern Day Detective Work: Reality vs. the World on the Small Screen," unpublished paper; and Kit R. Roane, "The CSI Effect," *U.S. News and World Report*, April 17, 2005.

---

e-mail accounts, computer hard drives, other electronic information, and other types of information relevant to the case. In complicated cases, discovery may take more than two years.

During the jury trial itself, the prosecution presents its case before the defense. Often the evidence is not as clear-cut as we see in television crime dramas. The Insider's View feature reveals one aspect of this reality.

Closing arguments and jury deliberations are the last steps of a trial. A closing argument summarizes the information given and relates it to a jury's instructions from the judge, who provides instructions on the burden of proof, legal definitions, and the elements of the crime for which the defendant stands accused. The jury then deliberates in private until they reach a verdict. Some courts do not require a unanimous verdict; less serious charges may require only three-fourths of the jury to agree with the verdict. Mistrials can occur when a jury cannot reach a verdict, juror misconduct is found, or an error occurred during a trial which a jury cannot be reasonably expected to overlook in their deliberations. In the event of a mistrial, a new jury is selected and the case is presented again.

Sentencing

Like prosecutors, judges can exercise a great deal of discretion. Judges have the power to set bail and conditions for release, accept or reject pleas, dismiss charges, and revoke probation. Most significantly, they are usually responsible for handing down a convicted criminal's sentence. In almost all noncapital cases, the judge decides the sentence in the case. In the 38 states that allow the death penalty (see Figure 5.1), capital cases often have jury input. This may take the form of a jury being the sole determiner of the sentence, a jury determining a sentence that a judge can override in some circumstances, or a jury recommending a sentence that a judge officially hands down.

Sentences in criminal trials can range from probation or fines to the death penalty; 7 in 10 felony convictions lead to a criminal's incarceration, 41 percent of defendants are sent to

States that have a Death Penalty
States with no Death Penalty

Alaska	Massachusetts	New Mexico	Vermont
Hawaii	Michigan	New York	W. Virginia
Iowa	Minnesota	North Dakota	Wisconsin
Maine	New Jersey	Rhode Island	

and the District of Columbia

FIGURE 5.1 States with the Death Penalty

prisons, and 28 percent are sent to jails. Average state prison sentences are 18 years for murder, 7 years for violent crimes, 3.4 years for property crimes, and 4 years for drug-related crimes. Men and African Americans tend to have longer sentences than women and Caucasians. The average sentence for a violent crime is 6.8 years for a black man, 5.3 years for a white man, 4.3 years for a black woman, and 3.2 years for a white woman.[60]

When judges determine the sentence, they must respect state sentencing guidelines. State guidelines may be detailed, accounting for each category of offense behavior and offender characteristics, such as prior records. Some state legislatures provide advisory guidelines. Other guidelines have the full force of statutory law, limiting judicial discretion.

A sentence handed down, however, is hardly ever the sentence served. Most murderers are released from prison years before their sentences have been completed. Violent criminals serve, on average, 62 percent of their sentences, property criminals serve 49 percent, and drug criminals serve 43 percent.[61] Many state governments have responded to tight budgets by encouraging early release and good behavior programs. In 2003, 13 states changed their sentencing policies to expand early release mechanisms, and 5 states reduced sentencing guidelines for nonviolent offenders.[62] This is a policy outcome directly affected by the degree of prison crowding, rather than by the severity of the offense.

States that are expanding early release programs and reducing sentences are reversing a long-lasting "tough on crime" trend that began in the 1970s. Advocates of tougher approaches argued that programs emphasizing rehabilitation were failing, judges were often too lenient, and indeterminate sentencing in which parole boards would determine a prisoner's release date were ineffective. When budgets for corrections departments were booming and

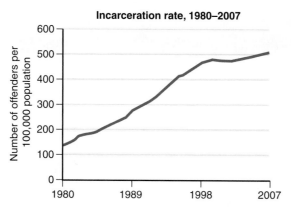

Incarceration rate, 1980–2007

FIGURE 5.2 Incarceration Rate, 1980–2004 Source: Bureau of Justice Statistics, www.ojp.usdoj.gov.

the public was more focused on crime, truth-in-sentencing programs, mandatory minimum sentences, and habitual offender laws reflected the times.

A variety of state programs use **truth-in-sentencing** approaches, which require that convicted criminals serve a certain percentage of their sentence before they are eligible for release. And the federal government's 1994 Violent Crime Control and Law Enforcement Act provided states with federal grants for corrections programs and facilities if state prisoners convicted of violent felonies served at least 85 percent of their sentences, or if the state could demonstrate it was moving toward this goal. Within one year, 57 percent of New York's violent felony prison admissions were subject to "truth-in-sentencing" requirements. Virginia estimated that 100 percent of its violent offender prison population would be meeting "truth-in-sentencing" requirements in just four years.[63]

The federal justice system and 17 state justice systems have sentencing commissions and sentencing guidelines. In many cases, the guidelines established in these systems are mandatory, or beyond a judge's individual discretion. These guidelines often recommend mandatory minimum sentences. Federal sentencing guidelines were established in 1986, and most states with guidelines followed suit throughout the 1990s.[64] Mandatory minimum sentences are most common for cases involving violent offenders and repeat offenders. California became famous for its "Three Strikes, You're Out" policy, which doubled the sentence for a second felony offense and mandated 25 years of incarceration for a third felony offense.[65] Though states vary in how they define a "habitual offender," almost all states have mandatory minimum sentences for criminals with multiple felony convictions.

These policies cumulatively had a measurable effect. Historically, incarceration rates have been around 110 people per 100,000 population, and reached an all-time low of 93 inmates per 100,000 population in 1972.[66] As Figure 5.2 indicates, since the enactment of these "get-tough" measures, incarceration rates have increased, reaching a high of 486 inmates per 100,000 in 2004.[67] This is a higher incarceration rate per capita than any other industrialized country. Incarceration rates also vary from state to state, as shown in Table 5.2.

Incarceration and Release

After being sentenced, convicted criminals are done with the court system unless filing an appeal or petitioning for other post-conviction relief. But they have not made it all

TABLE 5.2	State Prisoner Incarceration Rate in 2006, by State Rank National Rate = 445 Prisoners per 100,000 Population[a]	
Rank	State	Rate
1	Louisiana	846
2	Texas	683
3	Oklahoma	664
4	Mississippi	658
5	Alabama	595
6	Georgia	558
7	South Carolina	525
8	Missouri	514
9	Michigan	511
10	Arizona	509
10	Florida	509
12	Nevada	503
13	Delaware	488
14	Arkansas	485
15	Idaho	480
16	Virginia	477
17	California	475
18	Colorado	469
19	Alaska	462
19	Kentucky	462
21	Ohio	428
22	South Dakota	426
23	Tennessee	423
24	Indiana	411
25	Wyoming	408
26	Maryland	396
27	Wisconsin	393
28	Connecticut	392
29	Montana	374
30	Oregon	367
31	North Carolina	360
32	Pennsylvania	353
33	Illinois	350
34	Hawaii	338
35	New York	326
36	New Mexico	323

(continued on next page)

TABLE 5.2 State Prisoner Incarceration Rate in 2006, by State Rank National Rate = 445 Prisoners per 100,000 Population[a] *(continued)*

Rank	State	Rate
37	Kansas	318
38	West Virginia	314
39	New Jersey	313
40	Iowa	296
41	Washington	271
42	Vermont	262
43	Utah	246
44	Massachusetts	243
45	Nebraska	237
46	North Dakota	214
47	New Hampshire	207
48	Rhode Island	202
49	Minnesota	176
50	Maine	151

[a]From December 31, 2007, to December 31, 2008.

Source: "Prisoners in 2006," *State Rankings 2008*, U.S. Department of Justice, Bureau of Justice Statistics.

the way through the corrections process. They still must serve their time and then return to society.

We have all seen prison movies, but what is incarceration actually like? The experience is inconsistent; no two correctional facilities are the same. Conditions are generally crowded, however. Jails typically operate at 95 percent of capacity, and state prison systems are estimated to be between 1 percent below and 15 percent above capacity.[68]

In the 1990s, author Jim Nowlan taught a college-level American history course at the Hill Correctional Center in Galesburg, Illinois, a "high-medium" security prison built in the 1980s and housing murderers, drug law offenders and white-collar criminals. The prison looked like a small college campus, with dormitories, classroom buildings, vocational training center, cafeteria, gym and extensive weight room, and infirmary, all arranged around a central courtyard. Nowlan had 30 students: one-half were black, one-quarter Hispanic, one-quarter white, and one was Menominee Indian. About half the class seemed interested in the subject matter; the others were happy to receive one day of good time for each class they attended. Although maybe a quarter of the students showed real promise as students, most of the students had very weak, undeveloped writing skills.

At the end of the first class period, Nowlan was approached by a slender white man who was shaking like a leaf. "You must know something about the law," the student blurted out. "I've just been transferred in here from a federal prison. Can you help me get out; you've *got* to help me get out!" Though there are no comprehensive national data on violence in prisons, which the prisoner who approached Nowlan feared, anecdotal evidence is not encouraging. A former Mississippi warden told a blue-ribbon commission studying prison

AN INSIDER'S VIEW

Difficulties for an Ex-Convict

Sharon, a young black woman from the south side of Chicago, was released from prison eight years ago after serving her sentence for drug possession and prostitution. She is now a wife and a mother of two children. She obtained a job at the Food Service Department of Chicago's Bethany Hospital after successfully completing a job-readiness program for ex-felons. The job pays $10 an hour with benefits. Sharon wants to become a certified respiratory care practitioner and she

has already passed the qualifying examinations. However, her chosen profession requires a license that she is barred from receiving due to her ex-convict status. Her case illustrates both the potential effectiveness of training programs and the barriers to employment for ex-felons.

Source: Protestants for the Common Good, "Faith and the Metropolitan Challenge: Race, Economic Justice, and Envisioning the Beloved Community," An Interfaith Discussion Series, Faith and Metropolitan Challenge Curriculum, http://www.thecommongood.org.

conditions, "I've had to negotiate no fewer than eight hostage situations, deal with riots, et cetera." A former New York superintendent admitted that he couldn't protect women prisoners from "being sexually preyed upon."[69]

Prisons no longer seem to serve effectively to rehabilitate prisoners and funding for education and meaningful vocational training are frequently cut back in times of major constraints on state budgets. Prison overcrowding presents problems for safety and health. Medical care is also an issue in prisons. Generally, prisoners have a greater need for medical care and receive less medical care than the general population. According to one study, 1.5 million people are released from prison each year with a life-threatening contagious disease, including 100,000 cases of HIV and 40,000 cases of AIDS. As many as 350,000 prisoners have a serious mental illness. Yet medical attention may be scarce. A doctor in California complained that some prison facilities with 5,000 inmates only have two or three doctors.[70]

Educational opportunities in prison are often thought to be an important component of reducing recidivism and rehabilitating criminals. Educational programs are still generally available in state prisons; high school and general education courses are available in over three-quarters of state prisons and vocational training is available in about half. About one-quarter of state prison inmates say they completed their GED, the equivalent of a high school diploma, while incarcerated.[71] But few prisoners receive the education in prison that they need to succeed in our high-tech, 21st-century society.

Upon release, ex-convicts encounter numerous obstacles. Unemployment rates for ex-convicts are eight times higher than unemployment among the general population.[72] This reflects a difference in skills and education, but also that employers are hesitant to hire ex-convicts. In some states, ex-felons are barred from welfare benefits, food stamps, and public housing, making the burden of unemployment more difficult. The federal government and some state governments bar ex-felons from receiving student loans or financial aid.[73] Ex-felons may not be allowed to vote, they may not be eligible for public employment, and they may be barred from receiving certain business licenses, as the Insider's View feature shows. Some classes of offenders, such as sex offenders, may be barred from living in certain areas. For example, a law in Miami, Florida, bars sex offenders from living within 2,500 feet of schools, parks, or other places where children might gather. The law resulted in sex offenders seeking shelter under a causeway bridge—one of the few areas of the city that meets the 2,500 distance requirement.[74]

OUTCOMES OF THE CRIMINAL JUSTICE SYSTEM

How do we know if we have a "good" criminal justice system? One approach is to look at outcomes such as deterrence, clearance of cases, and rehabilitation. Each of these is discussed below.

Deterrence

Deterrence is any crime-fighting method that seeks to prevent crime, such as increasing the potential costs of committing a crime or the probability of getting caught. Deterrence policies may involve longer sentences, harsh punishments, or community policing. Studies show that simply having more police officers on the street reduces crime rates. Many police forces have developed a "community policing" approach that, according to the U.S. Department of Justice, "seeks to address the causes of crime and reduce the fear of crime and social disorder through problem-solving strategies and police-community partnerships." In just two years, from 1997 to 1999, the percentage of officers designated as "community policing officers" increased from 4 percent to 21 percent.[75] Federal funding for community-oriented policing programs was $1.3 billion in the 2002 budget.[76]

Many theorists argue that punishment will deter crime. The Heritage Foundation, a conservative policy research institute, cites a 1975 study that each execution may deter up to 28 murders.[77] Many people believe that longer sentences also discourage crime. This is one reason the federal government created mandatory minimum sentences.

Debate on whether harsh punishment deters crime becomes especially heated in regard to the death penalty. There is evidence to support both sides of the argument. Murder rates in states without the death penalty are lower than murder rates in death penalty states.[78] Death penalty supporters point out that murder rates have declined while execution rates have gone up, which they say indicates that the death penalty does deter murderers.[79]

The deterrent effect of the death penalty may be mitigated by the fact that a very low percentage of murderers actually receive the death penalty. Also, the death penalty is usually not administered until many years after the murder.[80] In 2005, the U.S. Supreme Court ruled that the death penalty could not be administered to murderers who committed their crime before turning 18. Ten percent of all persons arrested for murder or nonnegligent manslaughter are minors.[81]

Crime rates declined dramatically throughout the 1990s as sentencing policies got tougher and prison populations increased. Between 1993 and 2001, violent crime rates declined by 50 percent, and property crime rates declined by 47 percent.[82] At the same time, state prison populations increased 20 percent between 1995 and 2001. Federal prison populations and jail populations also increased during this time.[83] Some argued that because criminals were serving long sentences, they could not repeat their offenses, and thus "get tough on crime" policies succeeded in deterring crime. Others argued that because the economy improved during this time, providing more jobs, there was less need to resort to crime.

Clearing Cases

One way of evaluating a criminal justice system is to look at the number of crimes that have been solved. Solving a crime usually means catching and punishing the criminal. Typically, police are better able to solve more violent crimes, such as murder and assault, than property crimes, such as theft. Police commit more resources to solving violent crimes, and there are usually more witnesses available. The Federal Bureau of Investigation publishes statistics on the number of cases solved, or "cleared," every year (see Figure 5.3). **Clearance statistics**—the number of crimes

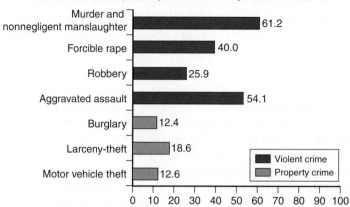

Percent of Crimes Cleared by Arrest or Exceptional Means, 2007

Crime	Percent
Murder and nonnegligent manslaughter	61.2
Forcible rape	40.0
Robbery	25.9
Aggravated assault	54.1
Burglary	12.4
Larceny-theft	18.6
Motor vehicle theft	12.6

Violent crime
Property crime

FIGURE 5.3 **Percentage of Crime Cases Solved** Source: Federal Bureau of Investigation, www.fbi.gov.

committed compared with the number of crimes that resulted in the arrest of the parties involved—are relatively constant, with little change over the past decade.[84]

Many more crimes are committed, however, than appear in the clearance statistics. Not all victims of a crime report the crime to police. Approximately 50 percent of victims of violent crime report the incident, and only 37 percent of victims of property crime.[85] In a U.S. Department of Justice survey, victims who did not report crimes to police commonly offered the following reasons: they felt the incident was a personal matter; they reported the incident to some authority other than the police; the offender was unsuccessful in carrying out the crime.[86] Some authorities opt not to charge an individual with a crime if that individual is already incarcerated for other crimes. These cases are "solved," but they are not considered cleared in the Federal Bureau of Investigation's statistics. Unfortunately, crime is more likely when offenders can reasonably expect to evade punishment, and low rates of crime clearance can be demoralizing to communities and to individual victims.

Rehabilitation

Rehabilitation rates are another way of evaluating the criminal justice system. Zogby International, an independent polling organization, found that 87 percent of Americans favor rehabilitative services for prisoners as opposed to punishment only. Currently, the criminal justice system is not succeeding in rehabilitating criminals in a majority of cases. Over 50 percent of criminals are reincarcerated within three years of their release. This rate of reincarceration is known as the rate of recidivism. **Recidivism** is the tendency to commit another crime and be returned to prison.

Prisons are not a comfortable or nurturing place to live. They often have high rates of disease, with tuberculosis, HIV, and staph infections being some of the most serious common diseases. In California, a federal judge intervened with the state prison's medical system when an average of one inmate a week died due to curable conditions that were neglected or incompetently treated. Also, many prisons are in rural areas, usually far from an inmate's home. This makes it harder for families to visit, even though strong family connections are important to preventing recidivism.[87]

Thus, most prisons are not designed to make inmates better people, and they may even make them worse. State Senator Kermit Brashear of Nebraska, an advocate of drug rehabilitation programs, notes: "Oftentimes you put a nonviolent substance abuser in prison and a violent substance abuser will come out."[88] The Bureau of Justice Statistics reports half of drug-dependent or drug-abusing state prison inmates received at least three prior sentences, and a fifth of inmates in state prisons are sentenced for a drug crime. About two in five state prisoners with drug dependence or abuse issues say they have taken part in treatment programs since their incarceration.[89] Solving their drug problems seems key to rehabilitating many prisoners.

Budget constraints have created an odd partnership between fiscal conservatives and social liberals who promote prison reform focused on rehabilitation. This is particularly true in cases involving drug offenders. Some states, such as Michigan, Texas, and California, are experimenting with programs that involve less time behind bars and more time under correctional supervision programs requiring treatment and drug tests. These programs typically cost much less than incarceration.[90] In a discussion held by the National Conference of State Legislators, Representative Lawlor of Connecticut noted, "From the budget point of view, the do-nothing option—in other words don't change any rules or practices—is probably the most expensive. In Connecticut, we know from the last 15 years that if you don't rethink some policies you will keep spending more to build and run prisons. And it is not really clear what good it does."[91]

CIVIL CASES

Though this chapter concentrates primarily on criminal cases in the justice system, civil cases are another vital component. State courts process 17 million civil cases in a year, though only 3 percent of these cases actually go to trial.[92] A civil case might be defined as any case in the justice system that does not involve a crime. Civil cases include tort actions, contract claims, property disputes, and family matters. The person who brings the case to court, known as the plaintiff, usually claims the defendant is financially liable for some harm the plaintiff has suffered. Most civil cases involve an individual suing another individual or a business. Less often, businesses sue each other or individuals. In about 5 percent of cases that go to trial, an individual or business sues the government.[93] Highly visible civil cases have changed our society. *Brown v. Board of Education* (1954), for example, led to desegregation of public schools.

Civil trials are very similar to criminal trials; most of the people and the processes are identical. There are judges, sometimes juries, plaintiffs who bring the case forward as a prosecutor would, and defendants. One notable difference is the motivation of the lawyers. Prosecutors in criminal courts seek to achieve justice, which may not be consistent with winning the toughest conviction possible. Plaintiff's attorneys in civil trials seek to maximize the amount of money obtained for their clients and themselves and to increase their ability to get future clients. Also, there are no publicly provided defense attorneys in civil actions comparable to indigent defense services in criminal courts. Some states may afford a right to counsel in civil proceedings for certain types of cases only; for example, Illinois grants counsel in adoption and child neglect cases that can result in the loss of parental rights.

As with criminal cases, the majority of civil cases are settled before they go to trial. In civil cases, this saves all litigants the trouble and expense of retaining legal services for many months. Tort cases that go to trial typically take at least two years to resolve.[94] Given that the majority of tort cases involve individuals, civil cases often involve a great deal of

AN INSIDER'S VIEW

Two Tort Cases

Abner Mikva, a former chief judge of the D.C. Circuit U.S. Court of Appeals, has said that the most memorable case he heard in his long judicial career was a tort case. The case involved Gino Andrews, a young man who was caught urinating against a building in Washington, D.C. When police officer Thomas Wilkins approached him with handcuffs, seeking to arrest him for public urination, Andrews fled. Jumping into the Washington Channel, he attempted to swim across but began to struggle. By the time Wilkins could throw him a floatation device, Andrews was unconscious. A woman on a private boat that had come along told Wilkins, "I need to go in, I've been trained." Wilkins shouted back that she shouldn't go in because "he's an escaped prisoner, and could be dangerous." Andrews drowned, and an autopsy revealed that he was under the influence of drugs and alcohol at the time of his death. His parents brought a wrongful death tort against Wilkins and other officers involved. Though Chief Judge Mikva dissented, Andrews's parents lost their case, as the majority of the court held that Andrews's contributory negligence to his own death barred his parents from recovery.

In 2006, an Indiana Supreme Court took up a difficult tort case involving emotional distress over an incident that occurred five months after the September 11, 2001, terrorist attacks. The plaintiffs, Mr. and Mrs. Cook, were on a flight from Indianapolis to New York City, but before even boarding the plane, they were concerned about a fellow passenger. His eyes were bloodshot and glassy, and he had been detained by airport security. On the plane, he refused to take his assigned seat and demanded to sit near the cockpit. During the flight, he grew angry at airline attendants who tried to make him extinguish his cigarette, and he started shouting loudly in a foreign language about the World Trade Center. The plane was forced to land in Cleveland so the man could be escorted away. The Cooks, and many of their fellow passengers, honestly believed this man could be a terrorist. They had never been more afraid in their lives. They sued the airline for negligent infliction of emotional distress, basically asserting that the airline should never have let the man board the plane. They lost their case in the Indiana supreme court.

Sources: Author's interview with Abner Mikva, summer 2006; *Andrews v. Wilkins* (1991); *Atlantic Coast Airlines v. Cook* (2006).

expense. Often lawyers provide representation for clients on a contingent fee basis, where they claim a percentage of the damage award if they are successful in the case.

Torts make up the largest portion of all civil litigation. A **tort** is a case in which an individual, a business or organization, or the government is accused of committing a civil wrong by violating others' bodily, property, or legal rights, or for breaching a duty owed under statutory law. In a tort, the plaintiff accuses the defendant of inflicting harm, either intentionally or as a result of negligence. Defendants who are found financially liable owe the plaintiff for damages such as lost wages, medical treatment, or property damage or loss. Punitive damages may also be awarded if the defendant was "willful, malicious, or fraudulent" in doing harm. Median damage awards for torts are around $28,000. Over half of all torts involve automobile accidents.[95] The Insider's View feature looks at two other kinds of tort cases.

One type of tort that gets a great deal of attention from both the media and legislators is medical malpractice. While medical malpractice torts make up only one-seventh of tort cases, they are about 16 times as costly as other types. The median damage award for medical malpractice torts is $600,000. Damage verdicts are over $1 million in about one-third of medical malpractice trials. The reason for these large damages is that most medical malpractice cases involve death or permanent injury as a result of mistakes or neglect on the part of the medical provider.[96] Due to the size of average medical malpractice cases, all doctors must purchase expensive malpractice insurance, which puts a strain on the health care system.

In some states, legislators have responded to medical malpractice cases by enacting a damage cap, or a maximum amount that plaintiffs can win regardless of the circumstances of their case. State supreme court judges have frequently demonstrated their public policy-making powers, and therefore the importance of the processes by which they are selected and retained, by striking damage cap legislation on state constitutional grounds. For example, the Illinois supreme court has held that damage caps violate multiple provisions of the Illinois constitution.

While medical malpractice awards are becoming larger, other tort awards appear to be declining. The median damage award for torts today is less than half of what it was in 1992. As potential plaintiffs have seen their expected financial benefit drop, fewer are taking their grievances to court. The National Center for State Courts believes that tort filings have decreased by about 10 percent since 1992.[97]

Contract cases are another major source of civil litigation, representing about a third of all civil cases. A **contract case** is a case in which one party sues another for violating an agreement or making an agreement under false pretenses. Contract cases involve failed or broken agreements between buyers and sellers or other parties to contracts. They include fraud, mortgage foreclosures, employment discrimination, or other employment disputes. In most instances, the litigants are individuals suing businesses, or they are businesses, hospitals, or governments suing businesses. Contract cases are more expensive than torts, with median settlements reaching $45,000. Also, in contract cases, the plaintiff is more likely to win. Torts are won by either party about half the time, but 65 percent of contract cases are decided in favor of the plaintiff.[98]

Civil cases are thus an important aspect of the justice system, making up about one-fifth of cases handled in state court systems. Civil cases protect people from fraud and discrimination, and they right the wrongs of undue suffering.

Our justice system has changed over time. More cases are now processed by more specialized and efficient systems. Policies and prisons have also changed because of changes in political attitudes toward rehabilitation and punishment. All aspects of the judicial system are affected by budgets, and all decisions have a cost. As one judge notes in reference to the justice system, "It's an imperfect system, but without it, it's chaos."[99]

CHAPTER SUMMARY

Most states have a three-tiered judicial system: district courts of original jurisdiction, appellate courts, and a state supreme court, also called the court of last resort. The people in the criminal justice system include policy-makers (such as governors, mayors, and legislators), judges, defense attorneys and defendants, prosecutors, jurors, and convicted criminals and correctional populations.

Many participants in the criminal justice system struggle for power. Throughout the process, prosecutors have a good deal of discretion, as they choose which cases to prosecute and what charges to make. Defenders work to have charges dropped or reduced, or to acquit the defendant if the case goes to trial. Today, 8 out of 10 criminal cases have public defenders, but there is no significant difference in outcomes between cases with private attorneys and those with public defenders, despite inadequate funding of public defenders. Judges have the power to set bail and conditions for release, accept or reject pleas, and decide the sentence in almost all noncapital cases.

Three outcomes of the justice system are deterrence, clearing cases, and rehabilitation. While crime rates have declined in recent years, many crimes are unreported and thus not cleared. While many argue that incarceration is primarily about punishing criminals, many efforts are made to rehabilitate prisoners, because most reenter society after serving a sentence.

Civil cases, another component of the justice system, include torts and contract cases. The person filing suit is the plaintiff, whose lawyer tries to maximize a money award for the client. Like criminal cases, most civil cases are settled before they go to trial. Torts make up the largest portion of all civil litigation. In a tort, the plaintiff accuses the defendant of inflicting harm, either intentionally or as a result of negligence, as in a medical malpractice case. Contract cases involve one party suing another for violating an agreement or making an agreement under false pretenses.

KEY TERMS

appellate court A court in which plaintiffs and defendants appeal the outcome from a trial; if legal or technical errors occurred in the trial, the appellate court may order that the case be reheard in trial court with a new jury, p. 108

arraignment A pre-trial appearance where accused persons are brought before a judge to hear the charges filed against them and to file a plea of either guilty, not guilty, or no contest, p. 120

attorney general The primary legal advisor and chief law enforcement officer of a state, p. 116

burden of proof The burden of the prosecutor to prove the defendant's guilt beyond a reasonable doubt, p. 120

chief judge A judge who manages the court building and employees, assigns cases to fellow judges, and handles administrative matters, p. 111

clearance statistics The number of crimes committed compared with the number of crimes that resulted in the arrest of the responsible parties, p. 127

contract cases Civil cases in which one party sues another for violating an

agreement or making an agreement under false pretenses, p. 131

correctional populations All individuals under some form of court supervision, including probation, prison, jail, and parole, p. 117

court of last resort The highest court to which a case can be appealed, p. 109

deterrence Any crime-fighting method that seeks to prevent crime, such as by increasing the potential costs of committing a crime or the probability of getting caught, p. 127

district attorney Local public official who represents the government in the prosecution of accused criminals, p. 116

incarceration rate The number of people being detained in jail or prison in comparison to the population, p. 117

jury nullification A jury decision acknowledging the accused has violated a law but declaring the defendant "not guilty," sometimes to protest a law that seems unfair, p. 117

plea bargain An agreement in which the prosecutor reduces the charge or recommends a minimum or

suspended sentence if the accused pleads guilty, p. 119

public defender Attorney funded by taxpayer dollars to people without financial means so they may properly defend themselves in legal proceedings, p. 115

recidivism The tendency to commit another crime and be returned to prison, p. 128

retention election Election held to decide whether a sitting judge will remain on the bench, p. 114

tort A legal case in which an individual, a business or organization, or the government is accused of committing a civil wrong by violating others' bodily, property, or legal rights, or for breaching a duty owed under statutory law, p. 130

trial court The court where all cases are initially heard or otherwise settled, p. 107

truth-in-sentencing A policy requiring that convicted criminals serve a certain percentage of their sentence before they are eligible for release, p. 123

QUESTIONS FOR REVIEW

1. What are the different levels of state courts, and what does each do? Why do we need several levels of courts?

2. In what ways are local governments involved in the state judicial system?

3. Who are the people in the state criminal justice system? In what ways do they interact?

4. What are the arguments for and against electing judges?

5. What are the steps in the judicial process as a case moves through the system?

6. What role does plea bargaining play in the criminal justice system? What are the arguments for and against plea bargains?

7. How do deterrence, clearing cases, and rehabilitation outcomes measure the effectiveness of state judicial systems?

8. What are three main differences between state criminal and civil court systems?

DISCUSSION QUESTIONS

1. Which method of judicial selection is best in your opinion? What are the pros and cons of retention elections once a judge holds a seat?

2. Are ex-felons in your state barred from welfare benefits, public housing, or voting rights? Should they be?

3. The American court system relies heavily on resolving cases outside court, between prosecutors and defense attorneys. Do you believe that plea bargaining is a good way to achieve justice? Why or why not?

4. What is the best way to address prison overcrowding if state funds are limited? Would you favor reducing sentence lengths for most offenders, expanding early release mechanisms, or some other approach? Why?

5. What should be the most important objective of the criminal justice system: preventing crime, clearing cases, or rehabilitating convicted criminals? What policies would you advocate in order to meet this objective?

6. Imagine you are a juror in a medical malpractice case in which a person sues a hospital in your state for amputating the wrong leg. You and the other jurors have decided the case in favor of the plaintiff and now must decide how much money to award. What factors should affect your decision? The person's age, occupation, lost wages, needs, other factors? What amount do you think is fair? Is that amount possible under the laws of your state, or does your state have a cap on medical malpractice damages? Are states justified in setting such caps?

PRACTICAL EXERCISES

1. What is the structure of the court system in your state? Which different types of courts are there? How many justices serve on your state supreme court? Are judges elected or appointed? Use a search engine like Google with your state's name as a keyword followed by "courts" to find sites explaining the state's court system.

2. If you live in a state in which some judges are elected and judges are retained on the bench by elections, determine how many judicial elections were on the ballot in the last election. This can usually be found in the ballot or election results section of your city or county board of elections Web site.

3. Interview a friend, colleague, or relative who has served on a jury or been called for jury duty. Or if you were called for jury duty, recall your own experience. Was serving on a jury a good or bad experience for you or for them? Did the jury system work well or badly?

WEB SITES FOR FURTHER RESEARCH

The National Center for State Courts site, www.ncsc.org, provides information about judicial administration, acts as a voice for the needs of state courts, and collects and analyzes data relating to state courts.

The Bureau of Judicial Statistics site at http://bjs.ojp.usdoj.gov provides information and data on cases and courts. The bureau is part of the Office of Justice Program, www.ojp.usdoj.gov, in the U.S. Department of Justice, www.usdoj.gov.

The American Bar Association site is found at www.abanet.org.

Findlaw.com, a free legal Web site, can be used for legal research at all court levels in the United States. Go to www.findlaw.com/casecode/#statelaw for state codes and cases.

SUGGESTED READING

American Bar Association. "How Courts Work: Steps in a Trial." www.abanet.org. A useful online article about basic trial procedures.

Bogira, Steve. *Courtroom 302: A Year Behind the Scenes in an American Criminal Courthouse.* New York: Vintage Books, 2005. A vivid portrayal of the workings of the judicial system of Cook County, Illinois.

"A Call to Action: Statement of the National Summit on Improving Judicial Selection." National Center for State Courts, 2002. www.ncsconline.org/D_Research/CallToActionCommentary.pdf. A report on judicial selection.

Clark, John, James Austin, and D. Alan Henry. "Three Strikes and You're Out: A Review of State Legislation." National Institute of Justice, September 1997. www.ncjrs.gov. Information on "three strikes" provisions in several states, and a comparison with earlier repeat-offender sentencing provisions.

"Confronting Confinement: A Report of the Commission on Safety and Abuse in America's Prisons." Vera Institute of Justice, June 2006. www.prisoncommission.org. A comprehensive report with concrete suggestions for prison reform.

Lynch, Timothy. "The Case Against Plea Bargaining." Cato Institute, 2003, www.cato.org.

Olin, Dirk. "Plea Bargain." *New York Times Magazine*, September 29, 2002.

Zimring, Franklin E. "Imprisonment Rates and the New Politics of Capital Punishment." Sage Publications, http://pun.sagepub.com. An article on capital punishment.

President Obama
meeting with
New York Mayor
Bloomberg and
Governors
Rendell and
Schwarzenegger

FEDERALISM AND INTERGOVERNMENTAL RELATIONS
PLAYING TUG OF WAR

Federalism is the constitutional relationship between states and the federal government. Each level of government has specific powers and authority, and like the different branches of government, the different levels serve to check and balance each other. When local governments are added into the mix, the term intergovernmental relations is used.

Intergovernmental relations are the relationships between and among the federal, state, and local governments. Both types of relationships involve politics and the workings of government, and they often involve a struggle for power.

All citizens of the United States are also citizens of their sovereign states or of a special district like Washington, D.C., or a related territory like Guam or Puerto Rico. Thus, all U.S. citizens vote on local, state, regional, and national issues. Often, we may find that our local needs and outlook on issues conflict with those of the policy-makers in Washington. In addition, there are often differences between the government of a state and the governments of its cities, counties, and other local entities. The struggle for power is seen clearly in such conflicts between these levels of government.

The central themes of this book come into play in the tug of war between these different levels of government. Power and influence by city officials are exercised through legal institutions and highly political processes such as elections and lobbying. All three levels of national, state, and local governments are necessary to provide representative democracy. And the outcomes of government rules, decisions, and actions at every level make a real difference to every citizen in terms of the taxes and fees we pay, the government services we receive, and our ability to affect government actions. Thus, there is a struggle for power between not only individuals and organizations but between different levels and branches of government that is played out in both federalism and intergovernmental relations.

This intergovernmental tug of war is political. The debate changes according to vantage points, circumstances, and public opinion. Officeholders at every level of government argue with each other as to how much power federal, state, and local governments should have. The public changes its views about which government can serve it best, depending on local needs but also crises, scandals, and economic downturns.

Conflicts emerging between state and local governments and the federal government take different forms that affect our everyday lives. In 2005, the U.S. Congress passed the Real ID Act, intending to make it harder for terrorists to get driver's licenses and to board planes or enter federal buildings. The act required that states issue new tamper-proof licenses, verify the identity of their licensed drivers, and share this information with other states. State governments argued that the law would be too expensive and difficult to implement by its May 2008 deadline. Six states passed laws refusing to comply with the federal law, and at least a dozen states passed resolutions calling for the federal government to cover all costs. The National Governors Association, the National Conference of State Legislatures, and the American Association of Motor Vehicle Administrators issued a report saying that the states needed at least another eight years to comply with this law. Following such strong resistance, in January 2008 the Department of Homeland Security gave states nearly five extra years to fulfill the law's requirements, lowered the costs from an originally estimated $14.6 billion to $3.9 billion, and allowed states to apply for an extension to comply with the law. As of April 2008, all 50 states had been granted extensions. Nevertheless, strong state opposition to the law continued and in April 2009 a working proposal for new legislation (tentatively called the Pass ID Act) was introduced in Congress to address states' concerns about the costs and implementation of Real ID. The fight over the cost and the implementation of the Real ID Act is a good example of the ongoing tug of war between states and the federal government.[1]

In the struggle for power and influence, state and local governments are often resentful of regulations that the federal government can impose without providing the funding to pay for them. Yet they also find themselves relying on the federal government to fund programs their voters need and want. This conflict becomes apparent when there is a difference between what our representatives and senators in the U.S. Congress support and what local activists demand, while both claim to represent the people. Community activists also try to influence federal policies but find they often must depend on local politicians to help them.

Such was the case in 2005 when the federal Base Realignment and Closure Commission (BRAC) issued a list of the military bases they recommended be closed. In the base closure debate, some localities, such as Yuba County, California, were able to keep their bases open through citizen activism and by getting the governor involved. Yuba County saved 1,286 civilian jobs within its borders. BRAC also voted to close Ellsworth Air Force Base, which would have cost South Dakota nearly 4,500 jobs. In response, the entire congressional delegation of the state joined together and got the commission to reconsider. As a result, Ellsworth base stayed open, a victory especially for recently elected Republican Senator John Thune, who had campaigned on the promise that he would have the ear of then President George W. Bush in keeping the base in operation. In both situations, citizens working jointly with local and state officials were able to influence the politicians in Washington. Yet sometimes local or state politicians do not act soon enough; Missouri officials blamed themselves for acting too late to save over 3,000 jobs in their state that were lost due to base closures.[2]

Whenever governments interact, they must operate within the institutional structures and political processes that have developed over time. The power of congressional committee chairs is one example of how the interaction of institutions and politics can affect state and local governments. State and local politicians once had significant influence over how a representative in Congress would vote. For instance, members of Congress who wanted an easy reelection had to please the political leaders of the party in their state or district. In return, long-term members of Congress were able to use the *seniority system* to become powerful committee chairs and ranking committee members. Because of their seniority on committees, they could make sure that bills beneficial to their local supporters got passed. Often members of Congress try to guarantee reelection by passing **pork barrel legislation** or adding **earmarks** to legislation, which provide funding for jobs or projects in their particular district or state.

Neither committee chairs in Congress nor local and state politicians exert as much influence over the federal lawmaking process as they once did, but they still help determine the final shape of laws and how these are enforced. In addition, state and local governments make their wishes known in Washington through other political means. For instance, many local and state governments work alone or in coalitions to build relationships with federal officials, the way business organizations or other special interest groups do. Although there is not always policy agreement among the different state and local government organizations, the overall effect of the intergovernmental lobby has been to help state and local governments act effectively in influencing federal policy-makers.[3]

Often federal, state, and local policy-makers differ on the political and governmental outcomes of intergovernmental policy-making. Consequently, government officials on the national, state, and local levels often approach an issue from completely different vantage points. Because the struggle for power does not guarantee that the best policy is supported, sometimes the policy adopted is inferior. For example, while the federal No Child Left Behind Act specified certain educational goals for each school district in the country, some state boards of education, such as that of Connecticut, preferred their own established state educational guidelines, which were often stricter than those of the No Child Left Behind Act. Moreover, state officials did not wish to use the state's education budget for what they considered an unnecessary and underfunded program required by the law. As a result,

Connecticut's Republican governor gave the state's Democratic attorney general authorization to file suit in federal court against having to implement No Child Left Behind.[4] In this case, the conflict between state and federal governmental policy resulted in the governor choosing her state's preferences over a federal policy supported by her party.

HISTORICAL BACKGROUND OF FEDERALISM

From the earliest history of the United States, there have been tensions between the national, state, and local levels of government, and discussions about the power of the national government as compared to that of the states are found in the 18th-century debates over ratification of the U.S. Constitution.[5] Article VI of the Constitution declares that the Constitution is the "supreme law of the land." The opponents of the Constitution—called **Antifederalists**—were concerned that its adoption might result in loss of state self-rule and civil liberties, and that a large republic was doomed to failure.[6] The **Federalists**, who supported ratification of the Constitution, wrote a series of articles known as *The Federalist Papers*, in which they tried to calm their opponents' fear of stronger government by explaining both the workability of a large republic and the checks that the proposed Constitution would provide against the national government limiting state powers and civil liberties.[7] James Madison explained in Federalist Paper No. 39 that the Constitution would provide for a government in which there was neither a supreme national government, nor a confederacy of independent states, but a combination of both.

When the Constitution was finally ratified, disagreements continued over the power of the national government in relation to that of the states. Jefferson's Republican Party preferred a small government, based locally in states, while Hamilton's Federalist Party (not to be confused with the faction that supported adoption of the Constitution) preferred a strong central government to ensure national economic growth.[8] At first, the Federalist ideas were more influential, but due to political backlash from unpopular policies and laws, such as the Alien and Sedition Acts curbing civil liberties during a time of possible war with France, Hamilton's followers lost power. Their party disappeared by 1824. The belief in strong federal authority was maintained by the decisions of a Supreme Court led by the Federalist-appointed Chief Justice John Marshall. The Supreme Court today is still the arbiter of federalism and referees the division of powers between the national, state, and local governments.[9]

After the War of 1812, the primary national debate was between nationalists and sectionalists. Nationalists argued that the Constitution was ordained by the people themselves, rather than the states, and that the national government's authority should prevail over states.[10] Sectionalists argued that states' rights were more important than national needs. The decisions of the Supreme Court under Chief Justice Roger Taney were largely pro-state and pronounced a policy of **dual federalism**, which held that the federal government's powers mainly consisted of national defense, foreign policy, interstate and international commerce, while state governments were responsible for almost everything else. Local governments had only those powers which the state chose to grant them.[11]

During the Civil War, President Abraham Lincoln expanded federal powers to preserve the Union, and during the Reconstruction period which followed, governmental authority slowly became more centralized in Washington, although federal and state finances continued to be clearly divided.[12] This Reconstruction period overlapped with the corrupt Gilded Age of the late 19th century, which saw widespread corruption on all levels of government and business. This period ended with the **Progressive movement** of the late 19th and early 20th century, when fed-up citizens demanded an end to corruption in all levels of society.[13] Since state and local political machines had often controlled the actions of national office-holders, the Progressive reforms resulted not only in a decline in corruption but also in a lessening of state and local influence on national officials. Progressive reforms spread to the federal government when Presidents Theodore Roosevelt, William H. Taft, and Woodrow Wilson called for national policies to replace those that had earlier allowed for more state and local government autonomy.[14]

Constitutional amendments passed at this time also enhanced the national government's power. The ratification of the Sixteenth Amendment allowed a federal personal and corporate income tax, thereby increasing money available to the federal government. Although revenue sources for the states and national government were still separate, the amount of federal money transferred to the state and local governments began increasing, and national involvement in state affairs—particularly those affecting state finance—accelerated.[15] The passage of the Seventeenth Amendment, which called for popular election of U.S senators, was a Progressive reform that curbed most of the influence that state legislatures had over the U.S. Senate.

The next great expansion of federal power occurred when the Great Depression began with the stock market crash of 1929 and resulted in the election of Franklin Roosevelt. His **New Deal** domestic policies brought the passage of federal legislation to provide programs to counter the effects of the widespread economic depression. State and local government no longer had the funds to do so. To maintain support of local politicians, Roosevelt's programs called for state and local governments to carry out the programs initiated by the federal government. Very often, federal New Deal funds would go directly to cities, bypassing the states. This new relationship between levels of government became known as **cooperative federalism** and replaced dual federalism.

The power of the federal government remained strong during World War II and the postwar period and increased again during President Lyndon Johnson's administration in the 1960s. Policy areas that had once had been under the authority of the states—such as civil rights, education, and housing—became highly centralized. By the end of the 1970s, federal grants that had become major parts of both state and local budgets were often tied to highly detailed federal regulations.[16] Many argued, however, that the policies imposed too much regulation, disregarded local differences, and lacked local oversight. These and other criticisms led to a series of reform programs by Republican administrations: Richard Nixon's New Federalism in the early 1970s, Ronald Reagan's New Federalism in the 1980s, and Newt Gingrich's Contract with America in the 1990s.[17]

Although all three of these agendas endeavored to limit federal power, they had different approaches. Both Reagan and Nixon showed a reluctance to keep the federal government in charge of domestic policies, but they had different alternatives.[18] **Nixon's New Federalism** was designed to streamline government. Most funding was passed to the states through **revenue sharing**, which is the distribution of federal moneys to state and local governments, and through **block grants**, which are federal grants given to states for a general purpose, such as community development. These approaches give more leeway to states than do **categorical grants**, which are designated for a specific purpose, such as school lunch programs. **Reagan's New Federalism** was designed to reduce the role of the federal government, especially through shifting responsibility for many domestic programs to the states. Since Reagan's goal had always been **devolution** of authority to the states, federal programs were cut in the belief that this alone would encourage state and local governments to fill gaps in services. Unfortunately, simultaneous federal cuts in programs led to a greater need for state and local governments to finance services. With more responsibilities and less federal funding, state and local governments were forced to increase taxes to pay for programs that citizens valued.

In 1994, when Republicans won control of both houses of Congress for the first time in 40 years, conservatives in the House of Representatives began to pass Speaker of the House Newt Gingrich's **Contract with America**, a plan to transfer governmental powers to the states. Specifically, the new Republican majority in Congress proposed changing entitlement programs into block grants, and severely limiting **unfunded mandates**, which are any federal laws that would require state and local governments to spend their own funds to carry out nationally mandated programs. Although the entire Contract with America was not passed, Congress passed the Unfunded Mandates Relief Act, which was supposed to make it harder for Congress to pass unfunded mandates on state and local governments if their costs were above specified thresholds.[19] States requested more autonomy to deal with the increased responsibility, but rather than returning all authority for social programs to the states, Democratic President Clinton accommodated them through a policy of flexibility.[20] During the administration of President George W. Bush, there were major cutbacks in federal funding to state and local governments and an increase in unfunded mandates, most notably in the No Child Left Behind education act.

STATE AND LOCAL GOVERNMENT RELATIONS WITH THE FEDERAL GOVERNMENT

In trying to get their needs addressed, state and local governments find effective ways to work with each branch of the federal government. The relationships between state and local governments and each federal branch are unique and always adapting to new political circumstances. Each branch is discussed individually below.

Relations with the Federal Judicial Branch

The federal court system has had an enormous impact on how much leeway states and local governments have in carrying out federal legislation. For instance, under the leadership of Chief Justice William Rehnquist, the Supreme Court handed down decisions that gave states more authority in its relations with the federal government. When Rehnquist became chief justice in 1986, he and Justice Sandra Day O'Connor began a reexamination of federal versus state power. Since the New Deal, the Court's decisions on congressional power had favored the federal government. However, the Rehnquist Court's decisions tended to favor the states, reinforcing all the political, economic, and fiscal trends of the Reagan and George H. W. Bush administrations giving government authority to the states. In cases involving the Tenth, Eleventh, and Fourteenth Amendments, and the commerce clause (in Article I, section 8) of the Constitution, the Rehnquist Court generally supported the states, but the Court tended to favor federal policy in cases concerning federal takeover of state and local authority, based on the supremacy clause (in Article VI) of the Constitution. Under the leadership of

Chief Justice John Roberts, the Court has generally ruled in favor of state and local governments, with some exceptions concerning federal preemption of state and local law.[21]

Tenth Amendment Cases

The powers not delegated to the United States by the Constitution, nor prohibited by it to the States, are reserved to the States respectively, or to the people. (Tenth Amendment)

In several landmark cases, Justice O'Connor argued that the Tenth Amendment was a safeguard guaranteeing protection of our liberties from both state and federal governments. O'Connor put forth the **clear statement rule**, which held that a federal law cannot be interpreted to interfere with a state's law unless Congress made a *plain statement* in the law to that effect.

Commerce Clause Cases

[The Congress shall have Power] To regulate commerce with foreign Nations, and among the several States, and with the Indian Tribes. (Article I, section 8, clause 3)

During the post–New Deal era, the Court developed a commerce clause doctrine that allowed Congress to regulate any economic activity within state borders that had a *substantial effect* on interstate commerce.[22] Consequently, cases involving the commerce clause had been decided in favor of the federal government ever since the *National Labor Relations Board v. Jones and Laughlin Steel Corporation* decision of 1937, in which the Court ruled that provisions of the National Labor Relations Act did not interfere with interstate commerce.[23] However, the Rehnquist Court declared in *United States v. Lopez* in 1995 that it would no longer automatically assume the constitutionality of any federal legislation invoking the commerce clause.[24]

The *Lopez* case was the first case since the 1930s in which a congressional act was overturned on the grounds that the issue was unrelated to interstate commerce. It concerned the constitutionality of the Gun Free School Zones Act. Alfonso Lopez had been arrested for possessing a gun near a school. Acting as protector of social policy, Congress had reasoned that guns result in crime, which can affect the national economy—and consequently interstate commerce. The Court rejected this claim 5–4, recognizing that there was a long-standing policy allowing Congress to regulate intrastate activity that "substantially affected" interstate commerce, and that the Court itself had greatly expanded the reach of the commerce clause, but even this "power had outer limits" (*Lopez*). Rehnquist's opinion declared that the Gun Free School Zones Act was a criminal statute that had nothing to do with interstate commerce, and moreover, it did not contain wording that could determine when and if a firearm possession near a school affected interstate commerce.[25] According to Rehnquist, the only basis on which the Gun Free School Zones Act could be upheld under the commerce clause power would be if it regulated an activity that substantially affects commerce. In a concurrent opinion, Justice Kennedy argued that the intrusion on state sovereignty contradicted the federal balance the framers of the Constitution had designed and that the Court had to uphold.[26]

Eleventh Amendment Cases

The Judicial Power of the United States shall not be construed to extend to any suit in law or equity, commenced or prosecuted against one of the United States by Citizens of another State, or by Citizens or Subjects of any Foreign State. (Eleventh Amendment)

The Rehnquist Court also strengthened state governments by interpreting the Eleventh Amendment to give more protection to states from being sued by citizens of another state, a concept known as **sovereign immunity**.[27] In *Alden v. Maine*, it extended state immunity not just from private suits filed in federal courts, but also from suits filed in state courts.[28]

Fourteenth Amendment, Section 5

> Congress shall have the power to enforce, by appropriate legislation, the provisions of [the Fourteenth Amendment]. (Fourteenth Amendment, section 5)

The Fourteenth Amendment provides guarantees of equal protection and due process to all citizens. The Rehnquist Court decided cases concerning the Fourteenth Amendment to extend Congress's **enumerated powers** (the powers specifically granted to the federal government in the Constitution) in ways that affect state and local government. Many of these cases addressed civil rights issues concerning discrimination by state laws and regulations, including age and race discrimination.[29] Nevertheless, throughout U.S. history, the justices' decisions have generally been decided by their views on the particular case at hand, rather than by the possible consequences to federalism.[30]

The Supremacy Clause

> The Constitution . . . shall be the supreme Law of the Land; and the Judges in every State shall be bound thereby. (Article VI)

Despite its state-friendly decisions, the Rehnquist Court's federalism did not favor states in cases relating to federal preemption of state authority.[31] The federal government has claimed the power of preemption based on the supremacy clause in Article VI, which holds that the Constitution and federal laws are to be upheld by judges of all states, "laws of any State to the contrary notwithstanding." It is presumably this federal preemption that allows Congress to pass sweeping laws such as No Child Left Behind, which so profoundly affects education in local schools.

The Spending Clause

> The Congress shall have Power to lay and collect taxes, Duties, Imposts and Excises, to pay the Debts and provide for the common Defense and general Welfare of the United States. (Article I, section 8, clause 1)

In addition to the supremacy clause, the Court left a loophole in the spending clause through which Congress can act as well. Rather than place conditions on states in legislation, Congress can make federal funding available to states *only* if they comply with federal statutes and regulations. The Court had long held that Congress has the authority to do indirectly through grants what it cannot do directly under its expressed powers.[32] For instance, in 2003, the Court refused to hear three cases in which federal appeals courts had ruled that Congress had acted lawfully when it required that state agencies waive their right not to be sued as a condition for federal funding.[33]

Several decisions in 2003–2006 suggest that the Supreme Court has begun giving more leeway to Congress, and it did not overturn a single congressional law on federalism grounds during this period. It also upheld the congressional power under the commerce clause in some cases.[34] To what degree the Roberts Court will rule in favor of states remains to be seen, but questions concerning their decisions on federalism played a large role in the Senate confirmation hearings of Justices John G. Roberts and Samuel A. Alito.

In the Rehnquist Court's conservative justices, states found an audience for their petitions and have taken advantage of the Court's favor in several ways. Some states, such as New York, have challenged federal laws on the ground that they have made inroads on state authority. In recent years, states have sued the federal government over costs of finding illegal aliens, road development, EPA requirements, hazardous waste cleanup, education funding, and changes to the Clean Air Act.[35] Other challenges involve regulations that permit federal agencies to wield undue influence over state policies, such as the Motor Voter Act of 1993. In addition, federal laws themselves were challenged in federal courts on the grounds that they exceeded congressional power to regulate interstate commerce.[36]

The state attorneys general offices in most states have become larger, better financed, and more professionally able to deal with challenges to the federal government in court. States have increasingly begun to file *amicus curiae* (or friends of the court) briefs for each other in lawsuits coming before the federal court. The states have also become more careful in choosing cases they wish to bring before the federal court, and are also appealing more cases, with favorable results.[37]

With favorable Rehnquist Court decisions, states became energized and increasingly challenged both congressional legislation and regulations of federal agencies. To these ends, states continue to work together, file more lawsuits, and maintain a high level of professionalism. As a result, although states are obviously not as powerful as the federal government, they cannot be ignored or dismissed, and losing in Congress encourages states to go to the courts. In addition, state courts are once more interpreting state laws according to their own state constitutions. The general trend in recent years has been to strengthen the state and local governments in their attempts to restrict expanding power of the federal government in intergovernmental policy-making.

Supreme Court decisions can also bring legitimacy to state actions—especially when voters demand something be done in the face of federal inertia. In April 2007, the Court handed down a 5–4 decision in *Commonwealth of Massachusetts et al. v. Environmental Protection Agency et al.*, ruling that the Environmental Protection Agency violated the Clean Air Act by improperly declining to regulate new-vehicle emissions standards to control the pollutants that scientists say contribute to global warming. This decision strengthened the authority of California in enforcing its antipollution laws concerning vehicle emissions. Since California's regulations had been adopted by 10 other states, the decision will make it easier for those states to enforce their antipollution laws as well. It may embolden some wavering state legislators to enact tough antipollution laws in their states.[38]

The opinions of the two George W. Bush appointees, Chief Justice John Roberts and Justice Samuel Alito, have not shown marked diversion from the pro-state decisions of the Rehnquist Court's conservative majority, although their legal reasoning concerning federalism may differ from that of Justice O'Connor and the late Chief Justice Rehnquist.[39]

Relations with the Federal Legislative Branch

The interaction between state and local governments and the U.S. Congress involves state and local officials trying to get members of Congress to pass laws that will benefit their citizens. But intergovernmental relations change with the politics of the period, and we must be aware that congressional lawmaking procedures reflect who holds power. This means that whoever holds power must work with many different factions. To succeed, congressional leaders have to overcome obstacles including ideological opponents, organized interests, and the indifference of other members of Congress. To get a bill passed requires constant contact with its supporters, information from experts, effective performance during congressional debate, presidential support, and support or at least no opposition from key committee leadership.[40] In addition, political power determines congressional action in two ways: First, the majority party determines who chairs committees and sets the agenda. Second, the majority party leadership in Congress often decides which laws get passed. During earlier eras, the chairs of the committees performed this function.

When committees dominate lawmaking, policy-making as well as parties tend to be nonideological and often bipartisan, and committee members, especially the committee chairs and ranking minority members, become experts in the committee's policy fields. For example, the longtime members of the Agricultural Committee become highly knowledgeable on all farm issues, and consequently, other members of Congress generally respect their opinions when a farm bill comes to a vote. However, when the majority party holds policy-making power, the congressional leadership will often guide lawmaking procedures to suit its own agenda. Such changes in congressional lawmaking methods during recent years have led to several negative consequences for state and local governments. First, the number of ways that state and local governments can formally contact federal lawmakers, such as testifying during committee hearings, had sharply declined by 2008.[41] Second, because of the emphasis on congressional voting along party lines, it became more difficult to influence a member of the other party to vote in one's favor. It is too soon to tell whether there will be an increase in congressional bipartisanship during the Obama administration.

Relations with the Federal Executive Branch

Just as state and local government officials keep constant touch with congressional lawmakers, they attempt to maintain positive relationships with members of the executive branch. This includes not only the president but also federal agencies.

The Presidency

The president of the United States holds the most powerful office on earth. Yet the power of the presidency is dependent on the institutional power of the office itself as established by the Constitution *and* the public support received by an individual president. In the struggle for power and influence, the president has the advantage, since except for the vice president, the president is the only nationally elected official chosen to represent the citizens of the entire country.

States and major cities need to get the attention of the president if their needs are to become part of the president's policy agenda. Consequently, it is important that governors and mayors keep in personal contact with the president if they can. One informal method is to maintain friendly ties. Republican Mayor Fiorello LaGuardia was friendly with his fellow New Yorker, President Franklin Roosevelt, a Democrat; LaGuardia was able to get billions of dollars of New Deal funding for the city of New York from meetings with the president.[42] Even in relatively low-key meetings, mayors and governors are always mindful that a president's politics develop into presidential policies. These policies often determine federal funding, which state and local issues will get priority, and the extent of cooperation by federal agencies and departments with individual state and local governments.

STRUGGLE FOR POWER

Viewing Katrina from the Rose Garden

After Hurricane Katrina devastated states along the Gulf of Mexico in 2005, Democrats believed that they could capitalize in the 2006 and 2008 elections on the slow government response to the crisis, arguing that it showed President George W. Bush to be out of touch with the concerns of average Americans. But as time passed and the images of those early days faded, it appeared that Bush would more likely be judged on the ongoing, more deliberate process of assisting the storm's victims.

By that yardstick, it doesn't seem that Bush gained much ground. In a prime-time address from New Orleans's Jackson Square in September 2005, he pledged, "We will do what it takes, we will stay as long as it takes, to help citizens rebuild their communities and their lives." But through the rest of 2005 and 2006, Bush was largely absent from the legislative process of enacting new programs and approving spending for the Gulf Coast recovery effort.... Bush apparently concluded that he had more to lose from being associated with the continuing problems of the Katrina recovery effort—some of them very local in nature—than he had to gain from taking a high-profile role.

Months after Katrina came ashore, Congress was still arguing about Bush's proposals, such as his Gulf Opportunity Zone to lure business investment to the region. But while the president spoke frequently about other legislative priorities languishing in Congress, such as renewal of the USA Patriot Act, he stayed away from the Katrina debate. When it appeared that Congress would approve only a fraction of the money needed to make minimal repairs to New Orleans's levees, a personal plea from Mayor Ray Nagin got the White House to double that commitment. But Bush didn't make the announcement—he left it to Donald Powell, who oversees the federal hurricane response.

When the White House released its year-end list of major accomplishments in 2005, hurricane recovery came last, after bankruptcy reform, tort reform, and the Central American Free Trade Agreement.

The $29 billion that Congress approved for hurricane recovery in December 2005 was mostly a reshuffling of money included in a $62 billion commitment by Congress for Katrina earlier in the year. It did not include a $6 billion plan to restore Gulf Coast wetlands, crucial for protecting New Orleans from a future storm surge. Republicans had tied the wetlands money to a controversial trophy that Bush had sought for five years—opening the Arctic National Wildlife Refuge to oil exploration. Democrats killed the oil-drilling provision and the wetlands spending. Sen. Mary Landrieu, D-La., complained about the political maneuvering: "It is truly unfortunate that hope for the people of the Gulf Coast has been clouded by an unrelated political fight centered on a piece of land more than 3,000 miles away."

Source: Adapted from John Maggs "Katrina, from the Rose Garden," *National Journal*, January 14, 2006, 45.

According to political scientist Richard Neustadt, "*presidential power* is the power to persuade."[43] Despite the institutional power of the presidency, presidents must maintain their personal popularity if they want their policies passed. Immediately after the September 11, 2001, terrorist attacks, a highly popular President George W. Bush was able to get his War on Terror agenda passed through Congress. However, after the federal government's failures in responding to the devastation caused by Hurricane Katrina (see the Struggle for Power feature), the president's approval rating sank well below 40 percent, reducing future support for some of the president's policies.

Federal Agencies

The executive branch is more than just the president. It is by far the largest part of the federal government, because it includes not only the Executive Office of the President, but also the cabinet departments, independent executive agencies, and independent regulatory agencies. Not counting the armed forces, almost 2.7 million people work for the federal government.[44] Since governing has become so complex, state and local governments often find that in order to get needed federal funding, they must comply with hundreds of regulations from dozens of different agencies. Independent executive and regulatory agencies are especially problematic for state and local government, since they are generally immune from the vicissitudes of politics. Yet they are charged with making sure that states and local governments comply with executive orders, congressional legislation, and agency regulations and policies.

One critical aspect of intergovernmental relations is at which level—local, state, or national—money can be used most successfully for any given governmental task. This problem is addressed by **fiscal federalism**, the theory that the costs and performance of particular governmental tasks work best at certain levels of government.[45]

In fiscal federalism, economic issues seem more important than political factors, and different levels of government do indeed provide the appropriate services. They are often aided in doing so by politically oriented private organizations.[46] State and local governments want the federal government to fund redistributive policies in which wealthier taxpayers pay for necessary government services for poorer citizens. If federal funds are not forthcoming, cities would prefer states to pay for redistributive policies. On the other hand, state and local governments prefer to control developmental policy (though they would accept federal funding) because they then can make sure that these policies correspond to the needs and preferences of their citizens. Both states and cities try to convince Congress and the executive branch to pay for some developmental policies and as much of the redistributive policies as possible.

THE INTERPLAY AMONG CONGRESS, STATE AND LOCAL GOVERNMENTS, AND FEDERAL AGENCIES

When Congress passes a law that requires action, it generally authorizes an agency or department of the executive branch to carry out its wishes. The agency then puts together regulations so that the law is uniformly followed. If Congress does not appropriate the money needed to carry out the regulations, or state clearly that so much federal money will be given to the states and cities so they can abide by the new rules, the states and cities still have to obey the law. They have to pay the cost of implementing the law locally, bearing the costs of these unfunded mandates. Congress may also pass laws that have priority over existing state laws or regulations; these are called **federal preemptions**. Similarly, a state law can displace a local law or regulation that is in conflict or inconsistent with the state law. But on the federal level, unfunded mandates and federal preemptions are examples of federal takeover of a state's or city's authority.

When Congress passed the Unfunded Mandates Relief Act in 1995, it said that an unfunded mandate only includes *direct orders* from Congress to the state. However, federal departments and agencies still enforce other categories of unfunded mandates and preemptions that shift the costs of carrying a federal program to state and local governments. Congress passes laws like No Child Left Behind that place rigid restrictions on state and local governments yet provide inadequate funding to meet the requirements of the legislation. States and local governments must find money to pay for administrating and implementing federal laws even when they cancel or preempt state or local statutes. (See the Struggle for Power feature on local governments facing federal mandates.)

Local Governments Face Federal Mandates

In 2003, county commissioners in Utah were coming to grips with yet another expensive and unwanted demand from Washington. They were being ordered to buy new voting machines. It was part of the well-intentioned "Help America Vote Act," passed to prevent a repeat of the 2000 election fiasco in Florida. And it sounded generous at first: The feds put up $1.5 billion nationwide for the new machines. But as happens most of the time, federal money won't come close to covering the cost.

Utah County, which contains Provo and is the second-largest jurisdiction in the state, learned in June 2003 that it would be billed $138,210 for new touch-screen voting technology. This news came just as county officials were placing hourly employees on mandatory one-day furloughs to save $200,000. Utah County could refuse to purchase the new machines, but then it would likely lose all federal aid for voting reform and would still have to buy special audio devices for each of the county's 124 voting places to help visually impaired voters. As County Commissioner Steve White pointed out, that would be a huge expense and would help just 200 individuals among the county's 161,000 voters. . . .

Federal lawmakers have not lost their taste for taking stands and then passing the buck downward. Nearly a decade ago, the federal government promised to pay 40 percent of the cost of the Disabilities Education Act, but its contribution has fallen between $10 billion and $25 billion short.

More recently, the Bush administration provided only a small fraction of the $3.5 billion it promised to deal with homeland security costs—and even the promised amount was more than $15 billion short of what governments claimed they needed. Most of all, however, state officials complain that the No Child Left Behind Act of 2001 has forced big expenditures for testing and other changes in the educational system. The act provides $29 billion, but states say their costs will be an additional $35 billion.

Counting all unfunded mandates together, the National Conference of State Legislatures estimates that states are out $95 billion each year, about as much as the total budget shortfall currently plaguing them. In the end, mandates are irresistible to the federal government for the same reason that parents prefer making teenagers mow the lawn instead of doing the job themselves. It's hugely attractive for governments to plant a new policy flag and leave the heavy lifting—and the bill—to someone else. And on the more positive side, unfunded mandates are a means of pursuing national policy while leaving local governments some flexibility in executing it. In the case of Utah's voting machines, after all, the alternative to a mandate would be a rigid federal program turning local governments into branch offices of Washington. Congress doesn't want that; the locals would like it even less.

Source: Donald F. Kettl, "Mandates Forever," Potomac Chronicle, *Governing*, August 2003.

Political Safeguards and State and Local Government Action

During the 1950s through the 1970s, Herbert Wechsler and other political scientists argued that "the political safeguards of federalism"—meaning the power of state and local political leaders to influence policy-making—would effectively protect the interests of state and local governments, despite the growing power of the federal government.[47] In the 21st century, state and local officials have not always been able to get needed fiscal relief from Washington, and disenchantment with Washington's federal policies has been bipartisan.[48] Despite passage of the Unfunded Mandates Relief Act, unfunded mandates and federal preemptions have increased since the 1990s.[49] Nonetheless, some scholars argue that Wechsler may have been partly right about politics as a safeguard of state and local interests. Regardless of party or ideology, national lawmakers are more reluctant to limit state authority and power and are less supportive of federal mandates and preemptions during the election cycle.[50]

In addition, when state and local governments find themselves unable—or unwilling—to pay for the hidden costs of federal legislation, they have strategies for fighting back. They often use *lobbyists, coalitions,* and *informal political networks.* Methods used with more limited success include *bargaining, passing state laws,* and *filing federal lawsuits.*

Because they are sovereign governments, states have more clout with the federal government than all but the largest cities. Thus, states have power when they bargain with federal agencies. In the case of the No Child Left Behind Act, by 2005 at least 40 states had introduced bills in their own legislatures protesting various aspects of the law. Although some states were fined for ignoring No Child Left Behind guidelines, most of the state bills were geared toward making it easier to go along with the act and as a bargaining maneuver to assert some state authority over education issues.

Filing a lawsuit against the federal government is the most extreme way of protesting a federal law. Sometimes bargaining breaks down, and states and cities use this as a last resort in the tug of war with the national government. In response to their concerns about No Child Left Behind, over 100 state and local governments have supported a lawsuit filed against the secretary of education by the National Education Association (NEA) and some local school districts. The outcome of such action can depend on how a court interprets Congress's wording in a law; the lawsuit that Connecticut filed was dismissed, and the NEA lawsuit is under appeal. In addition, change in administrations can influence state policy-makers to reconsider a lawsuit, as with passage of state laws.[51]

Intergovernmental Lobbying

In addition to their constitutionally protected powers and their traditional political roles, another way that state and local governments try to achieve their goals in Washington is by functioning as interest groups. Like other lobbyists, states can use the techniques of *advertising, direct lobbying of rank-and-file members, direct contact by donors,* and *grassroots efforts.*[52] Further, state and local governments, as well as organizations of government professionals (such as attorneys general or school superintendents), can form a type of coalition known as the **intergovernmental lobby**. Through such a loosely based network, representatives of state and local governments can take advantage of their official status as public officials and work to benefit their own constituents and interests.[53]

Intergovernmental lobbies operate as part of the vast network of interest groups in Washington, D.C., and like all lobbying coalitions, there is not always policy agreement among the individual members. Nevertheless, the overall effect of intergovernmental lobbies has been strongly positive in helping state and local governments work effectively with federal policy-makers.[54] Reflecting the different categories of state and local interests, these lobbies include *generalist associations,* which represent major state or local government coalitions such as governors, state legislators, cities, or counties; and *specialist associations,* which represent specific professional fields in government, such as public schools. Other intergovernmental lobbies represent regional groups, offices of individual states and cities, and even partisan groups such as the Republican Governors Association, if goals seem unachievable through bipartisan lobbying efforts.[55]

Individual State and City Lobbies

Today 37 states and territories have their own offices in Washington to address Congress. Operationally, they are a hybrid between a state government agency and an interest group.

AN INSIDER'S VIEW

Cities Come to Washington

Alan Autry, the mayor of Fresno, California, stepped out of his Washington hotel into a bright, cold winter morning. "It's a great day to beg," he declared. That day he would be begging at high levels, pleading with his state's congressional delegation for money. Autry takes this part of his job seriously. . . . In Washington, he had retained the services of Leonard Simon, one of the many lobbyists who specialize in guiding local governments through the federal policy thicket. Autry understands that cities come to Washington . . . because that's where the money is. In his quest for new revenues, Autry is hoping that Washington will help. . . .

It appears that the agenda of cities and counties is more at odds with that of the federal government than it has been for a long time. "You get the feeling sometimes that the feds are just not on the same map as state and local governments," says Lynn Cutler, who worked in the White House intergovernmental affairs office during the Clinton administration and now lobbies for the city of Cleveland. "And—hello—I thought they all represented the same people."

But if states and localities are finding the feds unresponsive to the larger items on their agendas, earmarked funding for local programs remains a growth industry in Congress, with members eager to support projects that can earn political credit for all involved. If only the strongest dogs are going to get scraps in the present budget environment, it may be more important than ever for local governments to hire their own lobbyists, . . . [because]money buys persistence. . . . It is an article of faith on Capitol Hill that members of Congress respond to repetition—they believe that if a city keeps asking about a project, it must really want it. The more visits, phone calls, and e-mails from Simon that are logged into a congressional office computer, the better Fresno's chances of seeing its projects through to completion. . . .

Or so everyone believes. "If you leave it just to the members and the senators," says Cutler, "they've got a whole lot of folks they have to take care of, and they're picking among their children in a way." Only the largest cities—Chicago, New York, Los Angeles—have their own Washington lobbyists on staff, but hundreds of cities, counties, and transit authorities contract out such services. Most of them rely on small-shop lobbyists . . . or big law firms with municipal practices. . . . It's impossible to guarantee results, however, and some mayors question whether the expense is worth it. . . .

These days, hiring a friend in Washington may be about the only way a city can ensure that it will have one. . . . Issues of the greatest importance to cities and counties barely register as a concern in the capital, even among members of Congress who used to be mayors or county commissioners themselves. When the problems do register, the first instinct often is to punt to the states, who can act as middlemen with the locals. The federal government has a constitutionally based relationship with states, and policy-makers find it easier to keep an eye on how programs are doing in 50 states than tracking them through 19,000 local entities. . . . And so the best thing for a mayor to do in Washington right now may be to avoid the broader policy questions and instead focus on specific projects, coming up with a strategy for converting them into earmarked dollars. Then he can present that strategy on the Hill and work the phones until it bears fruit.

Source: Adapted from Alan Greenblatt, "Squeezing the Federal Turnip," *Governing*, March 2003.

Few cities have such representation in Washington, but most large cities like Chicago, Los Angeles, and New York do. Moreover, states have a relationship with the federal government that is grounded in the federalism provisions of the Constitution, but local governments do not have this status. See the Insider's View feature on cities lobbying Washington.

Advantages of Intergovernmental Lobbies

In some ways, intergovernmental lobbies have advantages over other groups trying to reach representatives in Congress. With issues concerning states or local governments and Congress, they can remind members of Congress of the connections between the needs of their mutual constituents back home and the member's reelection. In addition, acting as representatives of governments and government officials, they not only serve as "special interest" representatives for state and local governments but also promote good working partnerships with officials of the federal government. Intergovernmental lobbies can provide federal officials with contacts to unique local groups whose input they might need or intergovernmental lobbies can often provide an insider's information on the preferences of local and state voters and interests. Finally, as elected officials, some members of intergovernmental lobbies—such as governors— can bring their influence with constituents, local party leaders, local government officials, and even elected colleagues from other states to the bargaining table with federal officials. The same advantages hold for local governments relaying their messages to governors' offices, state agencies, and especially state legislatures.[56]

Limits of Intergovernmental Lobbies

Intergovernmental lobbies are limited in ways that private interest groups are not. For instance, they cannot legally give campaign contributions to federal office holders and must avoid taking positions that might anger their own constituents.[57] Another limitation is that for at least the past decade, lawmaking has become highly influenced by reelection needs of public officials. This development has become known as the "permanent campaign," which refers not only to perpetual fund-raising needs but also to the tendency of lawmakers to pass laws shaped with reelection in mind, rather than the common good.[58] In the permanent campaign, getting in touch with policy-makers is often determined by the size of campaign donations. But state and local governments, by law, cannot give political donations. This hinders some lobbying efforts of state and local governments, although they still have clout with their own representatives, who need their support at home.

In addition, the media do not always focus attention on state or local needs. Instead they often concentrate on nationwide attention-getting issues—such as crime—producing a "something must be done" atmosphere. Prompted by such publicity, the public demands laws that will supposedly benefit everyone, but often the real costs are passed to the states through unfunded mandates or federal preemptions.

THE TUG OF WAR AMONG STATES— HORIZONTAL FEDERALISM

Another aspect of intergovernmental relations doesn't involve the federal government and we don't hear about it very much. The relationships between and among states is known as **horizontal federalism**. According to Article IV, section 1, of the U.S. Constitution, "full faith and credit shall be given in each state to the public acts, records, and judicial proceedings of every other state." This **full faith and credit clause** means that states honor each other's laws, public records, and court rulings. For instance, if you

MAYOR WILLIE BROWN of San Francisco and Rep. Nancy Pelosi at a groundbreaking for the new San Francisco Federal Building, 2002.

STRUGGLE FOR POWER

States Helping States

Hurricane Katrina prompted the largest-ever sharing of resources and personnel among states despite bureaucratic bungling that hindered response efforts. Police officers from New York, chainsaws from Virginia, ambulances from Florida, and National Guard troops from 40 states crossed state lines to help hurricane survivors in Mississippi and Louisiana. Dozens of states sent more than 31,000 people to the hurricane-stricken Gulf region to handle search and rescue, law enforcement, and biomedical waste management.

The assistance was triggered through use of a little-known tool in the emergency response arsenal. State governments legally can share fire trucks, helicopters, employees, and other resources through the Emergency Management Assistance Compact (EMAC), endorsed by all states except California and Hawaii. Hurricane Katrina set in motion the largest deployment in the agreement's nine-year history, said Karen Cobulius, spokeswoman for the National Emergency Management Association, a group that provides administrative support for the compact and is an affiliate of the Council of State Governments. . . .

In Katrina's aftermath, some of the borrowed state workers have assisted with fighting fires, providing transportation and re-establishing communications systems. New York sent 100 buses and 28 National Guard aircraft. Oklahoma foresters are helping with debris removal. Florida sent four Black Hawk helicopters, about 100 vehicles, and 1,847 workers to the affected states, totaling about $71 million worth of aid. . . .

When Govs. Haley Barbour (R) of Mississippi and Kathleen Blanco (D) of Louisiana declared emergencies on August 26, a few days before the hurricane hit, they triggered the emergency compact. State EMAC officers sent requests by e-mail for needed items or personnel. Multiple states respond via the Internet, and a state EMAC officer filters responses and decides which state will fulfill the request.

States are not required to assist other states unless they are able, and states that ask for help are responsible for reimbursing all out-of-state costs and are liable for out-of-state personnel. States in need usually rely on federal funding from the Federal Emergency Management Agency to cover the costs.

Often state governments don't possess the kind of medical help requested by states affected by a disaster and must look to local officials. Some states have shared local assets by enacting special agreements to use certain local employees such as doctors and medical examiners for the hurricane response.

Source: Adapted from Kathleen Murphy, "Katrina Sets Record in State-to-State Help," September 9, 2005, Stateline.org.

decide to get married in New York, you are still married should you fly to Florida for your honeymoon. If someone commits a crime in Michigan but is caught in Arizona, the Michigan courts can ask Arizona to extradite the suspect to Michigan for trial. **Extradition** is the procedure by which one government delivers a fugitive to the legal jurisdiction of another government.

In most cases, horizontal federalism benefits states. We see that when they file *amicus* briefs in federal court or when they form coalitions to petition the federal government. States and cities help each other in a crisis. For instance, after September 11, 2001, fire brigades and police were sent to New York City from all over the country. As the Struggle for Power feature on the aftermath of Hurricane Katrina shows, states share their resources, often through state government associations, independent of any federal aid that might be forthcoming.

However, sometimes horizontal federalism becomes a problem. One way that can happen is when states compete with each other over federal dollars. For example, as Table 6.1 shows, there is a discrepancy in how much money state governments get back for each dollar sent to the federal government. Cities are similarly affected. In 2006, New York City and Washington, D.C., had their homeland security funds cut by 40 percent while cities like Omaha had their

TABLE 6.1 Federal Expenditures per Dollar of Federal Taxes in 2005, by State Rank

National Median = $1.09 Received for Each Dollar Sent[a]

Rank	State	Per Dollar
1	New Mexico	$2.03
2	Mississippi	2.02
3	Alaska	1.84
4	Louisiana	1.78
5	West Virginia	1.76
6	North Dakota	1.68
7	Alabama	1.66
8	South Dakota	1.53
9	Kentucky	1.51
10	Virginia	1.51
11	Montana	1.47
12	Hawaii	1.44
13	Maine	1.41
14	Arkansas	1.41
15	Oklahoma	1.36
16	South Carolina	1.35
17	Missouri	1.32
18	Maryland	1.30
19	Tennessee	1.27
20	Idaho	1.21
21	Arizana	1.19
22	Kansas	1.12
23	Wyoming	1.11
24	Iowa	1.10
25	Nebraska	1.10
26	Vermont	1.08
27	North Carolina	1.08
28	Pennsylvania	1.07
29	Utah	1.07
30	Indiana	1.05
31	Ohio	1.05
32	Georgia	1.01
33	Rhode Island	1.00
34	Florida	0.97
35	Texas	0.94

(continued on next page)

TABLE 6.1 Federal Expenditures per Dollar of Federal Taxes in 2005, by State Rank *(continued)*

Rank	State	Per Dollar
36	Oregon	0.93
37	Michigan	0.92
38	Washington	0.88
39	Wisconsin	0.86
40	Massachusetts	0.82
41	Colorado	0.81
42	New York	0.79
43	California	0.78
44	Delaware	0.77
45	Illinois	0.75
46	Minnesota	0.72
47	New Hampshire	0.71
48	Connecticut	0.69
49	Nevada	0.65
50	New Jersey	0.61
	District of Columbia	5.55

[a]Fiscal year 2005. This table shows how much the federal government spent in each state compared to how much the federal government receives from each state.

Source: Tax Foundation, "Federal Spending in Each State per Dollar of Federal Taxes," www.taxfoundation.org.

funds increased.[59] Although New York and Washington still received some of the biggest grants, totaling 18 percent and 7 percent respectively, the mayors of both cities felt that the federal government had let their citizens down. New York's Mayor Michael Bloomberg said the Department of Homeland Security was forgetting that cities like New York and Washington would need financial help for years to pay for antiterrorism teams. Washington's Mayor Williams agreed. "There are just probably many more examples of where you've had great technology but you haven't invested on the ground in people, and it's been a tremendous flop," he said.[60] Based on past terrorist attacks and the likelihood that their cities would be future terrorist targets, both mayors believed their cities deserved greater federal funding.

Conflicts can arise among states when their laws are so different that the full faith and credit clause and the privileges and immunities clause are severely put to the test. According to the **privileges and immunities clause** in Article IV, section 2, of the Constitution, "the citizens of each state shall be entitled to all privileges and immunities of citizens in the several states." In other words, citizens of one state can move freely to another and expect to be treated as fairly as any other citizen.

When the citizens of two states look at issues very differently, their politicians may pass conflicting laws, as when some states legalize same-sex marriage while others pass legislation prohibiting it. Debates will vary in intensity and in which side prevails according to the culture of each state. As Figure 6.1 indicates, while most states prohibit same-sex

STATE POLICIES ON SAME-SEX MARRIAGE

Six states allow same-sex marriage. Eight states, including Connecticut* and Maine, offer either civil unions or domestic partnerships. Thirty-six states have statutes banning gay marriage and 30 have constitutional prohibitions.

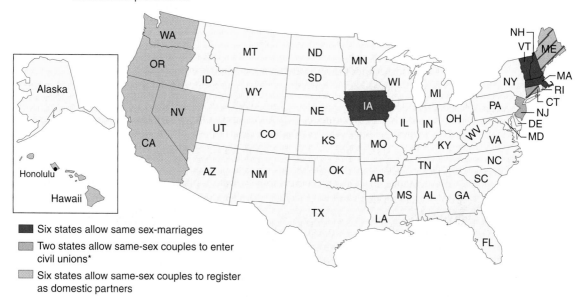

■ Six states allow same sex-marriages

▨ Two states allow same-sex couples to enter civil unions*

▨ Six states allow same-sex couples to register as domestic partners

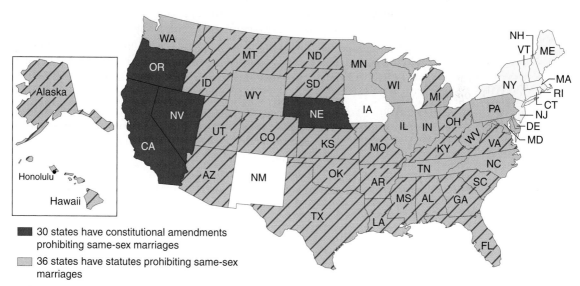

■ 30 states have constitutional amendments prohibiting same-sex marriages

▨ 36 states have statutes prohibiting same-sex marriages

*In Connecticut, civil unions will be offered until Oct. 1, 2010; after that date, couples in civil unions automatically will be converted to marriage.

FIGURE 6.1 State Policies on Same-Sex Marriage

marriage, others have now legalized same-sex marriage or legally recognize civil unions or domestic partnerships. But how will states that prohibit same-sex marriage treat same-sex marriages made in states where they are legal?[61]

THE INTERPLAY BETWEEN STATES AND LOCAL GOVERNMENTS

Power in politics tends to flow downward, from national to state and state to local levels, for reasons of constitutional authority and superior financial resources. Because politics is waged at the precinct level, where local government officials often have intimate knowledge and skills, the political power of mayors, prominent local government board members and school teachers can sometimes flow upward, against gravity.

Thus, whenever a state legislature comes into session, its members must spend much of their time working not only on issues that involve the national government but also on those that deal with local governments. City and county officials see the state government as the source of much of their powers of governance but also funding. Consequently, they send lobbyists to the state capitals for all the same reasons that larger cities, counties, and states send representatives to Washington, D.C.[62]

For instance, local governments lobby in state capitals to get needed funding for their schools, social services, and public works, and to minimize the effects of any state-imposed unfunded mandates and preemptions. The actions of lobbyists from cities and states can minimize the effects of such legislation for their specific city or county, or at the very least, make the state officials aware of the negative political fallout from their actions.[63] State lawmakers and governors worry about how any local interests, including governments, might affect their political bases of support. Any opposition that could erode their bases of power at the margin worries state politicos, which makes the many local governments, their associations, and lobbyists highly influential in the state capitols.

Local governments vary from state to state in how many lobbyists they have in the state capital, the level of organization and sophistication, and the nature of the legislation for which they lobby. For instance, the Texas Municipal League and the Florida League of Cities each has over a half-dozen departments, while the New Jersey Association of Counties has only six employees.[64] When local governments work together, they have more influence in state capitals than they would if they worked alone, just as states have more clout in Washington when they operate through generalist groups like the National Governors Association.

Today many political scientists argue that state and local relations will become more important in the coming years. State and local intergovernmental interest groups have been debating whether they should focus more on promoting effective state and city management and less on what goes on in Washington.[65] State governments will be where people seek out public goods and services. Since most state governments work through local governments, their role will increase as well.[66]

THE POWER OF STATES AND CITIES TODAY: STRONGER IN THE TUG OF WAR?

Many scholars assert that states and cities have been regaining power and have a greater share in policy-making, especially by becoming innovators. Ever since the 1960s, state as well as local governments have became more efficient and responsive, modernizing their institutions through updated constitutions and city charters, fair reapportionments, professionalized and full-time legislators, four-year terms for most governors and mayors, and strengthened judiciaries.[67]

Some observers see the country as polarized along ideological and party lines, divided into red and blue states.[68] But such polarization is not so strong in most state and local governments. Democratic and Republican officials often have to work together regardless of partisan differences, because their states constitutionally require balanced budgets or their state is subject to a court ruling that demands quick action.[69] Bending to the other party's wishes is possible, especially when it comes to money issues. For instance, while governor of Arizona, Janet Napolitano, a Democrat, vetoed bills on abortion and illegal aliens passed by the Republican legislature but compromised on several budget measures.[70] Governors such as Mitt Romney, Arnold Schwarzenegger, and Mike Huckabee, all Republicans, have shown that it is possible to have bipartisan policy-making. Many states are led by a blend of Republican and Democratic officials, creating a sort of purple government. After the November 2008 elections, 25 states had a divided government with the chief executive from one party, and the majority of at least one chamber of the legislature from another.

States have consequently been able to respond more quickly—and often better—than Congress to public support in many issue areas. For instance, 15 states voted to raise the minimum wage in 2006, but Congress was still debating the issue in 2007 and finally passed legislation only in 2008. At present, 30 states have higher minimum wages than the federal government.[71] The bipartisan attitude of state lobbyists in Washington is noticeable when compared to the political attitudes of congressional staff. When interviewed, staff members of state government lobbying offices in Washington said that they worked with officials from both parties regularly, whereas congressional aides did not.[72]

Bipartisanship is also working in cities. While larger cities tend to be run by Democrats, New York and Los Angeles have had Republican mayors, and many suburban governments are controlled by Republicans. Many local independent units of government, like school boards, tend to be bipartisan or nonpartisan. Public opinion approves of such purple state and local governance, and unlike the federal government, many state and local governments have continued the custom of practical, nonideological politics. There is no Democratic or Republican way to pick up the garbage or build roads.

In addition, the power of the federal government is formidable, but when states and other governments form alliances, they can be effective. Governors, especially working together, can exercise a great deal of influence over how a federal policy will be implemented in the states. While they have more weight if they are from the same party as the president, the influence possessed by state and local officials is often determined by public opinion and support for their position. Whoever uses their power and influence best often determines whether states or local governments can make federal policies go their way in Congress, in federal agencies, or through the federal courts.

Although we elect officials to represent us in Washington, sometimes the politics of those we send conflicts with what we want or need. Ideally, when our representatives in Congress ignore the wishes of their constituents, they will be thrown out of office at the next election, but in the real world, incumbents often have such a huge campaign "war chest" and so many other advantages that it is difficult to defeat them. Nevertheless, sometimes citizens become involved and protest what Washington officials have decided. It is beneficial for citizen activists to persuade their congressional representatives to listen to local needs, and when the community can recruit elected officials to sponsor their cause, the chances of success are considerably higher.

Since state and local governments have become more professional, it is no surprise that now statehouses and city halls often come up with new solutions to old problems, despite the reluctance of the national government to follow their lead. Once sleepy city halls and state capitols have been transformed into modern effective governments. In the 21st century, we face significant problems which will eventually require the special resources of national as well as state and local governments to resolve. But the method of cooperation resembles more of a tug of war about power, policy, authority, and money than members on the same team fighting together for victories.

CHAPTER SUMMARY

All citizens of the United States are also citizens of their sovereign states and vote on local, state, regional, and national issues. Yet there are often differences between our local and state needs and outlooks and priorities found in Washington. The struggle for power is seen in such conflicts between these levels of government and is addressed both in the governmental and political arenas. While state and local governments protest federal preemptions and unfunded mandates, they also find themselves relying on the federal government to fund programs their voters need and want.

Because relationships between the levels of government change, theories of federalism develop to reflect the attitudes toward the federal relationship in any era. In the 19th century, the theory that explained this was dual federalism, which held that the federal government's powers mainly consisted of national defense, foreign policy, and interstate and international commerce; state governments were responsible for almost everything else. Local governments had only those powers which the state chose to grant them.

Federal power was greatly expanded during Franklin Roosevelt's New Deal, which called for state and local governments to carry federally funded and initiated programs. This new relationship between levels of government was called cooperative federalism; the national government worked with state and local governments to provide programs for citizens. Policy areas once under the authority of the states became highly centralized, and federal funds became increasingly tied to federal regulations. Disillusionment with cooperative federalism caused the rise of programs devolving more authority to states (called New Federalism) under the administrations of Richard Nixon and Ronald Reagan, and culminating in the Contract with America introduced by Republican Speaker of the House Newt Gingrich in 1995.

Decisions in federal court have also determined the balance of federal power. The Rehnquist Court generally ruled in favor of states over Congress, handing down decisions upholding states' authority in cases involving the Tenth, Eleventh, and Fourteenth Amendments and the commerce clause (in Article I, section 8), but the Court tended to favor federal policy in cases based on the supremacy clause (in Article VI) of the Constitution. The Roberts Court is expected to continue to be supportive of states, but its opinions may not be based on the same constitutional reasoning.

State and local governments often must comply with complex federal regulations to get needed federal funding. Often federal legislation is not accompanied by corresponding funding or the new laws can preempt state laws already in place. In turn, states have imposed state mandates and preemptions on local government. Nonetheless, under the theory of fiscal federalism, state and local governments strive to get a higher level of government to fund their programs, especially for redistributive policies, which can cause undue burden on their citizens. States and local governments have developed strategies to avoid costs of federal legislation, including use of lobbyists, coalitions, informal political networks; and with more limited success, bargaining, passing state laws, and filing federal lawsuits.

State and local governments as well as organizations of government professionals can form a coalition known as an intergovernmental lobby, a loose network whose members can use their status as public officials to remind members of Congress of the needs of their mutual constituents and also promote good working partnerships with federal officials. The intergovernmental lobby has proved effective in getting state and local governments heard in Congress even though its members cannot give political donations and must avoid taking positions that might anger their voters.

The relationships between and among states is known as horizontal federalism. The Constitution's full faith and credit clause demands that states honor each other's laws. Although horizontal federalism usually benefits states, conflicts can arise when states pass conflicting laws, as when some states legalize same-sex marriage while others pass legislation prohibiting it.

In recent years, state and local governments have become more professionalized and consequently more efficient and effective. Because of this, states and cities have been regaining power and have a greater share in policy-making. In the 21st century, we face significant problems which will eventually require the special resources of national as well as state and local governments to resolve.

KEY TERMS

antifederalists The 18th-century faction opposed to ratification of the U.S. Constitution, p. 137

block grant A federal grant given to a state for a general purpose, such as community development, p. 138

categorical grant A federal grant given to a state for a specific purpose, such as school lunch programs, p. 138

clear statement rule Supreme Court ruling that holds that a federal law cannot be interpreted to interfere with state law unless Congress made a plain statement in its legislation to that effect, p. 140

Contract with America A 1994 plan promoted by Speaker of the House Newt Gingrich to eliminate many entitlement programs and transfer governmental powers to the states, p. 139

cooperative federalism A system of government in which powers and duties are shared between state, local, and national governments, p. 138

devolution Returning responsibility for many functions of government to the states, p. 139

dual federalism A system of government in which the federal and state governments are supreme in their own separate spheres of government, p. 137

earmark A provision in legislation that specifies certain spending priorities to a limited number of projects, programs, or grants or funding to be set aside for a special purpose or recipient, p. 136

enumerated powers Those powers specifically granted to the federal government by the Constitution, p. 141

extradition The procedure by which one government delivers a fugitive to the legal jurisdiction of another government, p. 150

federal preemption A federal law that has priority over an existing state law or regulation, p. 145

federalism The constitutional relationship between the states and the federal government, p. 134

federalists The 18th-century faction supporting ratification of the U.S. Constitution, p. 137

fiscal federalism The theory that the costs and performance of particular government tasks are best found at certain levels of government, p. 145

full faith and credit clause Article IV, section 1, of the Constitution, which says that states are to recognize each other's public acts, records, and judicial proceedings, p. 149

horizontal federalism The relationships between and among states, p. 149

intergovernmental lobby A coalition of state or local government officials or professional associations of government officials who lobby in Washington, p. 147

intergovernmental relations Within the United States, the relationships between and among the federal, state, and local governments, p. 135

New Deal The federal programs of President Franklin Roosevelt put in place during the 1930s to overcome the effects of the Great Depression, p. 138

Nixon's New Federalism A plan to make government more efficient by shifting power and authority to state and local officials, p. 138

pork barrel legislation Legislation that specifically benefits a member of Congress's district or state by providing funding for jobs or projects, p. 136

privileges and immunities clause The clause in Article IV, section 2, of the Constitution that states: "the citizens of each state shall be entitled to all privileges and immunities of citizens in the several states.", p. 152

progressive movement A late 19th- and early 20th-century movement that sought to end corruption in corporations and government by regulating businesses and weakening party machines by giving citizens more direct participation in the electoral process and government, p. 138

Reagan's New Federalism A plan to reduce the role of the federal government, especially through shifting responsibility for many domestic programs to the states, p. 138

revenue sharing The distribution of federal moneys to state and local governments, p. 138

sovereign Immunity States cannot be sued by citizens of another state, p. 141

unfunded mandate Any federal legislation that would require state and local governments to spend their own funds to carry out nationally mandated programs, p. 139

QUESTIONS FOR REVIEW

1. What is *federalism* and what is *intergovernmental relations*? How do these two concepts differ?

2. How is dual federalism different from cooperative federalism, and when did each develop?

3. How was the new federalism of President Nixon different from that of President Reagan?

4. What influence does the U.S. Supreme Court have over how laws concerning states are applied? In deciding matters of federalism, how has the Court used the Constitution's Tenth, Eleventh, and Fourteenth amendments and the commerce clause, the supremacy clause, and the spending clause?

5. What are some examples of unfunded mandates and federal preemptions? What effect do they have on state and local self-government and state and local finance?

6. What are the various types of intergovernmental lobbies? What are some of their advantages and limits?

7. What is the full faith and credit clause and what is horizontal federalism?

8. How do local governments in towns, cities, and counties fit into the federalism tug of war?

DISCUSSION QUESTIONS

1. If you were a governor or mayor, would you prefer a block grant or a categorical grant? Why?

2. Some states have legalized same-sex marriages while others have prohibited them. Should there be one standard for marriage nationwide? Why or why not?

3. If a state with no death penalty has been asked to extradite a person accused of murder to a state that has a death penalty, should the state comply? Why or why not?

4. If a neighboring state has received permission from the Environmental Protection Agency to double its toxic emissions, but those emissions will contaminate your state's water supply, what might you do as a governor, state attorney general, or state legislator?

5. What are the arguments for and against the No Child Left Behind Act being an unfunded mandate?

6. If you are a mayor and want federal highway funding for your city, but your governor wants to use these funds for rural highways, what can you do to get this federal funding despite the governor's wishes? What are the political and institutional issues involved?

7. Why is it important that states lobby the U.S. Congress and federal agencies?

8. If you were a lobbyist for a mayor or governor, how would you get help from members of Congress from your district or state who are from the opposing party?

PRACTICAL EXERCISES

1. Using the Internet, investigate whether your state and city have government offices in Washington, D.C., to represent their interests to the federal government.

2. Call the district offices of your representative and senator in the U.S. Congress. Ask if these members of Congress have regular meetings with the state legislators and local officials from the district. Do the congressional offices have a good relationship with their state and local counterparts? Are there any issues on which they work together? Call your city hall and ask if your local government has ever contacted your senator or representative to help solve a problem with a federal agency.

3. Find your state budget on the Internet. (You can try your state's Web page or *The Book of the States*, a publication of

the Council of State Governments at www.csg.org. How much federal funding was received for education, transportation, and agriculture? What percentage of total funding for each issue area comes from the federal government?

4. Find a criminal case in which your state or a nearby state was involved in extradition procedures. (You can probably find such a case through Lexis-Nexis, FindLaw.com, or a search on a local newspaper's Web site.) Describe the circumstances, and name the states involved. Which state initiated the extradition? What was the outcome of the request?

5. Suppose you are a state official of the Great Lakes region of the country, which has access to water in its lakes. Congress has just passed a bill making it mandatory for all states to spend $1 billion to drill for scarce water in the mountains of their state. This is an unfunded mandate. Make a plan for how you would address this problem of what seems an unnecessary expense for your state.

WEB SITES FOR FURTHER RESEARCH

The major intergovernmental lobbying associations all have Web pages:

National Governors Association: www.nga.org.

Council of State Governments: www.csg.org.

National Council of State Legislators www.ncsl.org.

National League of Cities: www.nlc.org.

National Association of Counties: www.naco.org.

U.S. Conference of Mayors: www.usmayors.org.

International City/County Management Association: www.icma.org.

Some generalist associations are the National Association of Towns and Townships, www.natat.org; National Association of Regional Councils, www.narc.org; American Legislative Exchange Council, www.alec.org; and National Civic League, www.ncl.org.

For an examination of federal court cases concerning issues between the federal government and the states, go to www.findlaw.com or www.law.cornell.edu.

Regularly updated discussions, data, and statistics on how unfunded mandates affect states are found in the NCSL's *Preemption Monitor* and *Mandate Monitor*, at www.ncsl.org.

Go to www.stateline.org for current news articles about state governments.

The *Governing* magazine site, www.governing.com, contains timely news articles, features, and blogs about and for state government and its officials. It contains many articles about state-federal, state-local, and state-state interaction.

SUGGESTED READING

Anton, Thomas. *American Federalism and Public Policy.* Philadelphia, PA: Temple University Press, 1989. Argues that the different levels of government make "benefit-seeking coalitions" so that everyone gets what they need.

Beer, Samuel H. *To Make a Nation: The Rediscovery of American Federalism.* Cambridge, MA: Belknap Press, 1993. Discusses the philosophical underpinnings and history of federalism.

Conlan, Timothy. *New Federalism to Devolution: Twenty-five Years of Intergovernmental Reform.* Washington, DC: Brookings Institution, 1998. Describes and analyzes the recent history of federalism.

Derthick, Martha. *Keeping the Compound Republic: Essays on Federalism.* Washington, DC: Brookings Institution, 2001. Gives arguments for states having more power in relations with the national government.

Hamilton, Alexander, John Jay, and James Madison. *The Federalist Papers*, http://thomas.loc.gov. The major primary source in understanding the workings of our federal system; see especially Nos. 10, 39, 46, 51.

Nice, David C., and Patricia Frederickson. *The Politics of Intergovernmental Relations*, 2nd ed. Chicago: Nelson-Hall, 1995. Basic source on intergovernmental relations, including local government.

Peterson, Paul. *City Limits.* Chicago: University of Chicago Press, 1981. Also see Peterson's *The Price of Federalism.* Washington, DC: Brookings Institution, 1995. Describes how fiscal federalism affects all levels of government, especially the local level.

Posner, Paul. *The Politics of Unfunded Mandates: Whither Federalism?* Washington, DC: Georgetown University Press, 1998. Thorough analysis of the origin and effects of unfunded mandates on the states.

Riker, William. *Federalism: Origin, Operation, Significance.* Boston: Little, Brown, 1964. Written during the time of cooperative federalism and the civil rights movement; a strong argument in favor of policies made out of Washington.

Waltenburg, Eric N., and Bruce Swinford. *Litigating Federalism: The States before the U.S. Supreme Court.* Westport, CT: Greenwood, 1999. Details how states have become professionalized in successfully arguing federalism cases in federal court.

Zimmerman, Joseph F. *Congressional Preemption: Regulatory Federalism.* Albany: State University of New York Press, 2005. Discusses the impact of congressional preemption on state and local government.

Obama campaigning for president at a rally.

ELECTION BATTLES

To hold the highest positions in your local or state government, you have to first win elections. As we will discuss in Chapter 8, people who aren't elected officials can have an impact on their community and state. Citizen activists, lobbyists, and journalists, for example, all have important roles to play in society and in government. But it is elected governors, mayors, state legislators, and city council members who pass the laws, determine levels of government spending, and appoint department and agency heads. So elections are critical to implement policies and programs you favor.

Elections involve all four themes of this book. They are one of the prime ways at the national, state, and local levels that we struggle for power. If we succeed in electing a governor, a mayor, a county board president, or alderman, we can change the course of government. We can pass new laws, set regulations, make critical appointments such as the chief of police, build roads, or help to save the environment. One fundamental route to power and influence is through elections.

Elections are central to a representative democracy. In our society, one of the ways people participate in government is by voting to select who will fill the highest positions of governmental power. We choose the people who represent us.

For all their color and excitement, elections are among the most structured of all the methods of achieving power and influence. The laws governing elections are so clear and strict that an election is in many ways like a chess match. Each player gets on the ballot by gathering petition signatures of voters, and each raises campaign funds within strict guidelines. On Election Day, the winner is determined by either a plurality or majority vote, depending on the type of race. A **plurality vote rule** means that the candidate who receives the most votes wins. A **majority vote rule** means that a candidate must receive at least 50 percent of the vote to win the election. Primaries are usually decided by pluralities, but most general elections are decided by majority votes.

Winning elections affects government outcomes. If you favor allowing a woman to decide with her doctor whether she will have an abortion, without outside interference, then you vote for candidates favoring pro-choice policies. If you oppose abortion in all forms, then you support pro-life candidates who want to ban or restrict abortions. The same is true if you favor lower taxes, same-sex marriage, or spending more money on schools—you vote for like-minded candidates who will try to enact the policies you support. Elections are often the fastest way, especially at the state and local levels of government, to get the policies you favor enacted.

GETTING ELECTED

When he was a state legislator in Illinois, Abraham Lincoln observed that there are three basic tasks in a political campaign: canvas the district, identify your voters, and get them to the polls. Nothing much has changed, except for the technology. To win an election, we have to talk to the voters, find out who supports our candidate, and get them to vote on Election Day. The fact that we use e-mail, blogs, paid television and radio ads, public opinion polls, automated phone messages, and direct mail appeals does not change the fundamentals of campaigning.

To be able to accomplish the three tasks Lincoln set forth, four basic resources are required: *time*, *money*, *people*, and *skill*. All are in short supply in most campaigns.

Time

Time is a finite, wasting asset. That is why campaigns seem to start earlier each election cycle. Technically, a campaign lasts only a few months from the time petitions are filed and candidates are placed officially on the ballot until Election Day. During these months but especially in the last two weeks of a campaign, voter attention is most focused. This period is the most intense and crucial to a campaign's outcome. At this stage, there are not enough hours in a day for a candidate to be everywhere, meeting voters in the district, doing media interviews, and rallying the campaign troops. But most campaigns start nearly a year before Election Day, and for a powerful office like president, they really start the day four years before when the last election was completed. Time is a precious resource to be used very wisely.

Money

Money is the fuel for communicating with voters by paid media and for operating the campaign. Simply traveling around a large district or state is surprisingly expensive. A paid staff, office space, mailings to voters, consultants, and specialists—all these components of a modern campaign cost money whether you are running for the school board, a city council seat, or governor. By now, most people know that "money is the mother's milk of politics." Raising money can take a great deal of a candidate's time, especially if the campaign depends on many small donors.

The Internet has made a difference in a campaign's ability to raise funds from grassroots sources. But wealthy contributors are still the major players. For instance, Michigan architect Jon Stryker, a Democrat, gave at least $6.4 million to candidates or political committees in over a dozen states in 2006. Stryker supports social justice and civil rights issues and Democratic candidates. Democrats gained seats in the legislature in all the states in which Stryker supported their candidates. While Republican were once assured of a significant edge with wealthy donors, the Democrats have been catching up.

Rich donors can have a chilling effect on policy-making, since legislators worry that one wrong vote might target them for defeat by these "mega-donors."[1] To overcome the detrimental effects of rich donors on the election process, reformers have developed agendas to counteract them, including limiting campaign contributions, government funding of campaigns, prohibiting direct contributions from businesses and labor unions, and full reporting of contributions.[2]

People

People are a key campaign resource that is less understood than time and money. People include the candidate, the paid staff, and the volunteers who work in the campaign to deliver the campaign message and deliver the vote on Election Day. While people often volunteer enthusiastically, many will not be there when you really need them. They are busy with their lives and careers and can give only so much time and effort. But dedicated volunteers are worth their weight in gold. Volunteers must be recruited—few just show up unasked. They are motivated by a variety of reasons. Some like the bustle of a campaign office, meeting people, and the excitement of a campaign. Some are true believers in a particular candidate.

When your campaign can afford it (and many local campaigns cannot), paid campaign staff will help you use your time more effectively. They can help you get to more key meetings, and they can develop sharp ads and brochures. With their experience and savvy based on earlier campaigns, professionals can help you avoid mistakes. Public relations professionals and other consultants can devise a message and shape a candidate image that are hard to beat.

Skill

The *skill* of a candidate is, of course, critical. A bad candidate who makes innumerable mistakes and errors is not likely to be elected if running against a reasonable opponent. Yet during the course of a campaign, candidates develop and mature. Many candidates start out ill at ease and without a clear message. But as they talk to hundreds of individuals at campaign meetings and events, they develop the capacity to communicate their message in a forceful, succinct, and comfortable style. Before long, the verdict after a campaign appearance is "She really came across well," or "He knows what he's talking about."

These four resources—time, money, people, and skill—help a candidate develop mutually reinforcing elements that determine the success of a campaign: credibility, visibility, and more money. Every candidate has at least a modicum of some of these characteristics when announcing a candidacy, and some have all three. For example, a candidate may have a credible career as a lawyer, high visibility based on service on the school board, and money pledged by friends. For some candidates, one element may help overcome the lack of the other two. For instance, a candidate might be a Congressional Medal of Honor Winner with immense credibility, which can be developed into visibility and money. A wealthy candidate can hire a competent staff, buy paid ads, and get immediate credibility from the press, which brings the visibility necessary to convince the public that she is a serious candidate. The more credibility, visibility, and money you have as a candidate, the more likely you are to win, although sometimes wealthy candidates spend millions of their personal funds on their campaign only to lose the election.

Prospective candidates should hold off on running for office until they have developed some of these characteristics. They will be needed if a candidate is to accomplish the tasks Lincoln set forth: canvas the district, identify your voters, and get them to the polls.

A LOOK AT THREE CAMPAIGNS

Your authors Jim Nowlan, Dick Simpson, and Betty O'Shaughnessy have run successfully for office and have worked on other candidates' campaigns. Nowlan and Simpson also have painfully lost some key elections, while O'Shaughnessy has run only once. All three share their experiences here on how to run and win elections at the local and state levels.

Simpson for Alderman

In 1971, Dick Simpson was a 30-year-old college professor in Chicago who had already established himself politically as Illinois state campaign manager for Eugene McCarthy's presidential campaign and as founder of the Independent Political Organization (IPO). Simpson had helped elect reform candidates to the State Constitutional Convention, Illinois state legislature, and Chicago city council. While those reform officials had passed good legislation and played an important role in state and local government, they hadn't implemented all of the reform agenda. So when a seat opened on the city council, a citizen's search committee drafted Simpson.

Running on his reform credentials, community involvement, and opposition to the Chicago political machine, Simpson promised, if elected alderman, to do the following:

1. Establish a full-time aldermanic service office delivering city services as a matter of right rather than as political favors or payoffs.

2. Vote his conscience, community opinions, and community interests in the city council, even when these differed from the wishes of Mayor Richard J. Daley.

3. Create a 44th Ward Assembly as a "neighborhood government" with delegates elected from every precinct and community organization to direct the alderman's vote in city hall and to undertake projects for the ward's betterment which did not require city hall's approval.

In a bitter battle of political messages in the precincts and the media, Dick Simpson's clean-cut image, reform message, and 500 campaign volunteers defeated the powerful Daley machine and their candidate by a vote of 8,828 to 7,336, or 55–45 percent of the vote. Simpson was reelected four years later against an even stronger political opponent.

Nowlan for State Legislator

James Nowlan had started thinking about running for a seat in the Illinois house in 1963 when he was a 22-year-old college intern with the Illinois Republican Party in the state capital. Now it was 1966 and he was actually doing it. Nowlan originally expected numerous candidates to contest the primary election because the incumbent had retired. If five or more candidates from all counties in the district filed for the nomination, he could win if he carried most of the vote in his home county *and* if he could come in second in the two neighboring counties, which his grandfather had represented 25 years earlier.

The campaign plan was to raise $4,000 (equivalent to about $30,000 in 2009), a lot of money for Nowlan, who was a graduate student at the University of Illinois at Urbana-Champaign earning only $200 a month as a research assistant. He planned to spend a good portion of the money on radio advertising because television was too expensive for this district, which had no central television market. The rest of the money would be spent on billboards, brochures, and travel to meetings (which is surprisingly expensive in rural districts).

Nowlan's campaign theme was simple: "Now is the time for Nowlan." This "time for a change" approach was a good campaign theme because his opponents were all middle-aged while Nowlan represented youth and energy as well as change, especially for his district.

Everything depends on execution of the campaign plan. Nowlan attended all the candidate meetings and debates. He organized friends to canvas key towns in the targeted counties and fellow graduate students to help out. He went door to door personally contacting voters as much as possible. He sought the endorsement of the district newspapers as well as the *Chicago Tribune*, which was highly respected in this Republican district. The *Chicago Tribune* did endorse Nowlan, and with only a few days left in the campaign, he immediately changed the radio ad copy to trumpet this important achievement, which significantly increased his credibility as a candidate. However, Nowlan came in second in the six-candidate race. Sorry but no cigar, no victory celebration, no exciting upset stories in the media.

But this first campaign had other important results. Nowlan had learned how to campaign, something he wasn't taught in the university. He had made many friends in the county Republican Party organizations and had developed considerable name recognition. (See the Insider's View feature for more of what can be learned even in an unsuccessful campaign.) A year later, the incumbent had a chance to move up to the state senate. The house seat was open again. By this time, Nowlan was in the U.S. Army, finishing his second year of a two-year requirement at army headquarters north of Chicago, only four hours away from the district. He ran a second time.

This time, he had much more confidence, more friends and supporters, and stronger support with the Republican Party leaders he had come to know from his earlier campaign. The result was dramatically different. This time it *was* the time for Nowlan.

O'Shaughnessy Runs with a Local Team

Betty O'Shaughnessy had been an active volunteer in her community while her children were at home, but she had concentrated on getting a graduate degree and teaching social studies full time for several years. The 2008 election changed all that. After volunteering, she decided

AN INSIDER'S VIEW

A Lesson from Lou Falletti

In James Nowlan's first try for office in 1966, he paid a call on Louis Falletti, owner of a popular tavern and also a state employee and precinct committeeman for the small community of Italian Americans in his district. A courtly gentleman, Lou Falletti received Nowlan graciously, but he said bluntly that he already had a candidate. Furthermore, Lou said Nowlan would not get a single vote in his big precinct.

Nowlan left Falletti's tavern determined to prove him wrong. He spent more time and money in the precinct than planned, walking much of the precinct door to door to introduce himself, even hanging around the popular Italian bakery next to the tavern to say hello to residents. On Election Day, he did prove Falletti wrong about not getting a single vote in the precinct. Of the 187 votes cast, Nowlan received three.

Two years later when Nowlan ran again, Falletti's candidate had moved up to the state senate. This time Falletti backed Nowlan enthusiastically in a hotly contested four-way race, and Nowlan captured almost all the votes in the Falletti precinct.

From that time on, whenever Falletti wanted help finding a job or needed a favor for one of his constituents, Nowlan worked hard to get him what he wanted. In politics, those who might oppose you during one election, or on one vote in the legislature, might support you the next time. Winning elections and passing laws is a game of addition, not subtraction. You work to get more and more support each time around.

to get involved in government herself. When it came time to select candidates for township trustee, O'Shaughnessy put her name before the township's Democratic Party and was slated to be on the ballot.

The local Democratic candidates besides O'Shaughnessy included the incumbents for township supervisor and clerk, three incumbent trustees, and a new candidate for township assessor. They all decided to coordinate their efforts and run as a team. While each represented different sets of voters in the township, they worked out campaign strategy together with the help of three experienced township and precinct chairs. They held a joint fundraiser, sent out team mailings, and each got supporters to put up yard signs and hold coffees, which they all attended. The candidates also went door to door in their own neighborhoods, asking voters to cast their ballots for the entire team, and they made phone calls to that same end. Although they were outspent by their opponents, the unified team effort got results. O'Shaughnessy and all but one of the others were elected.

DECIDING TO RUN

Elections are composed of individuals and their personal decisions. A candidate decides to run; campaign participants decide to work in the campaign; the candidate and campaign leaders decide which public stands to take; and finally, each voter decides which candidates to vote for. Each individual choice has consequences for the election's outcome. A campaign is won or lost by specific individual decisions. This is what makes election outcomes so uncertain and individual decisions so crucial.

Everyone in a campaign makes a decision to devote time, talent, and money, but for the candidate and the key campaign leaders, these decisions are of a different magnitude. Not only do they risk more of their time and fortune, but they also risk more of themselves. Ordinary citizens provide the support necessary for victory, but the candidate and campaign leaders must launch the campaign and work unceasingly if victory is to be achieved.

Candidates must pay the debts incurred in their campaign, and they risk being ridiculed and sometimes cruelly attacked by the opposition. Most of all, candidates must ask people to elect them to public office. They may find it distressing to stand in front of stores, shaking strangers' hands, or to go to friends and associates to ask for money. Yet candidates are their own best fundraisers and workers. So a candidate must learn to ask people for support in order to run a good campaign. For many candidates, this is the hardest part of campaigning.

No candidate is truly drafted. Friends or a group may ask you if you are interested in running, but you must decide if you will run, and then you must begin to seek the support necessary to win. If you decide to run, the campaign is launched. If you refuse or hesitate, someone else steps forward. The decision of one person to assume the challenge and risk of a bid for public office is the most important of the campaign.

Concrete reasons not to run are many: running will mean time lost from family and may stall a career. And campaigns cost a lot of money. The positive reasons to run can seem terribly abstract: your election will give the community a strong representative and spokesperson; you can make government more efficient; you can pass legislation to improve the community; you can bring integrity, leadership, and dedication to public office.

Personal ambition and ego also enter into the decision. Candidates may run to become famous, to get into the limelight, or to prove that other people love them. Such motives may seem shallow or selfish, but some combination of public and private reasons must overcome all the practical reasons not to run.

Running for office is not limited to experienced politicians. Sometimes, expected political frontrunners decide not to run, and inexperienced newcomers may find themselves running on major party tickets and in important races. See the Insider's View feature for one example. Sometimes a less-favored candidate ends up winning. When three Democratic frontrunners dropped out of the 2006 Hawaii gubernatorial race, Republican incumbent Linda Lingle won handily in a predominantly Democratic state in an election year that favored Democrats.[3]

Once the decision to go ahead is made, the candidate must assemble a staff. Like the candidate, key campaign personnel face difficult decisions. Campaign staff members must expect endless hours of work and separation from their families. Supporting a candidate is one thing, and giving a few hours or donating a few dollars is easy, but serving as a campaign manager or campaign leader requires dedication and personal commitment.

Campaign volunteers are attracted for many reasons, including their concern for other people, dismay at the failures of the political system, and a desire to be a part of crucial decisions affecting our lives. Some people hope to get a job if their candidate is elected, or they may have an interest in getting a particular law passed or repealed. Many people join campaigns because they are a great way to meet people. Campaigns are very exciting, and some people join to be part of the excitement. A sense of civic duty, of being a good citizen, of giving something back to the community also plays a role.

There are so many reasons not to run for office and even not to join a campaign as a volunteer. You are likely to be told: "Politics is such a dirty business! You are too good to get involved with all those liars and cheats!" Or you might hear: "With your job, you just can't afford the time. And you don't know anything about politics anyway."

Aaron Schock Wins Illinois House Seat at Age 23

When he was a 19-year-old student at Bradley University in 2001, Aaron Schock of Peoria, Illinois, ran for a position on his local school board. Going door to door throughout his district, he impressed voters with his energy, sincerity, and ideas for the problem-plagued school system. Running as a write-in candidate, Schock shocked Peoria by defeating the incumbent, who had also been the school board president!

In 2004, when he was 23 and president of the Peoria school board, Republican Schock ran for the Illinois house of representatives against an incumbent legislator. Having established his credibility and visibility, the young candidate convinced the business community to contribute to his campaign. Once again, Schock walked door to door throughout the district. He was indefatigable, walking every day until early evening, which created a buzz of appreciation among residents.

Sensing that Schock might actually have a chance to unseat the incumbent, the state Democratic Party began pouring money into her campaign. This gave the race statewide visibility among political activists and interest groups, which increased interest in Schock and also increased his fund-raising success among Republicans.

As Peoria (population of 120,000) was a television "media center," Schock could use television advertising efficiently. Serious and articulate, he came across to voters as an impressive candidate. According to local political observers, he also benefited from an opponent who had alienated many constituents, even those in her own party, because of her lack of responsiveness to local groups and individual constituents.

Although outspent $800,000 to $500,000, Schock defeated the incumbent by 300 votes. In 2008, continuing his electoral successes, he became the youngest member of the U.S. House of Representatives.

We recommend that you do *not* begin by running for office yourself. For many people, their interest in politics, their passion about important issues, and their belief that they are as well qualified as current office holders convinces them to put their ambitions and their passions ahead of what is realistically possible. If you are truly interested in politics and government, we suggest that you work on another person's campaign and learn the craft of politics before you consider running yourself. Start by working in a campaign to learn about politics, then get yourself elected to local governmental offices in order to learn how to govern well. After these experiences, you will be ready and qualified to run for a major public office.

LAUNCHING A CANDIDACY

One key task in launching a campaign is deciding how many votes it will take to win the election. In most districts, only 60–70 percent of the potential voters are registered to vote. And the winning candidate usually gets only 50–60 percent of the votes cast. Therefore, in a big city council district of 60,000 people, 8,000 votes are enough to win. In a state representative district of 100,000–120,000 people, 15,000 votes may be enough. In a congressional district of 600,000 people, a candidate might win the party primary with 10,000–50,000 votes, depending on which party primary is entered and how tightly contested the race is.

In his book *How to Win a Local Election*, Lawrence Grey discusses deciphering election statistics in order to decide whether or not to run. Grey talks about picking the **target number,** the number of votes needed to win a particular election. It is calculated by dividing the average number of votes cast in previous elections for the seat by two, as long as that number is larger than the largest vote ever received by an incumbent in a contested election for the seat. If you are running for a particular seat, look at the election statistics for the last four or five elections. Find out the results, the number who voted in the race, and the number who voted in similar races. With this information, you can come up with an average number of people who are likely to vote this time. Divide that number by two to determine the target number. But if your opponent is an incumbent, be sure that the target number is more votes than the maximum vote ever received in previous election battles. Once a campaign determines the target number, they must decide how the candidate is going to get that number of votes precinct by precinct.

In addition to voting statistics, the demographics of a district can help determine whether a particular candidacy is viable. For some voters, demographic factors such as race, ethnicity, age, education level, or socioeconomic background can have more weight than a candidate's experience or platform. Polling information on ideology is also important. If a candidate who supports strong gun control laws is running in a district where hunting is popular, that candidate is likely to lose. A conservative running in a heavily liberal district will also be defeated. So it is best to have a candidate who matches, at least to some degree, the demographic and ideological characteristics of the particular district. This is not to say that a candidate shouldn't try to get votes from different demographic groups and different ideological constituencies. But demographics and public opinion in a district have a profound effect on voting, and such voter preference or bias has to be considered at the beginning of a campaign.

DEVELOPING A CAMPAIGN STRATEGY

After picking a target number and studying demographic and public opinion information about the district, a general campaign strategy is developed. Two principal facts need to be considered. First, winning an open seat is easier than running against an incumbent. Second, if there is an incumbent, running a successful campaign is easier in a newly redistricted constituency than in a district that has elected the incumbent many times. An incumbent has greater name recognition than any challenger is likely to have, and citizens who have voted for a person in the past are likely to do so again.

Suppose after collecting all the facts and consulting with advisors and knowledgeable campaign leaders, you still want to proceed with your candidacy. What should you do next? Before rushing to announce your candidacy, determine that a viable campaign can be run and the necessary resources can be raised. These resources include: (1) campaign leaders and staff, (2) endorsements from recognized political and community leaders to lend credibility to the campaign, and (3) money. The Insider's View feature describes these elements in one local campaign.

The number of paid staff members needed to run a campaign varies by the level at which it is run and the campaign's budget. A low-key campaign for a suburban school board seat will probably have no paid staff and a budget of only a few hundred or a couple of thousand dollars—much of which will be supplied by the candidate. A campaign manager who is a spouse, friend, or in-law will serve with only minimal reimbursement.

AN INSIDER'S VIEW

A Local Campaign

Jeff Rosen, dean of professional and continuing education at Loyola University of Chicago, was appointed to fill a term for a trustee who resigned from the Oakton Community College Board of Trustees. However, he had to run for reelection to the board in 2005 when his appointment ended. He spent $241 of his own money for flyers. He turned down a campaign contribution of $3,500 from a college labor union in order to keep his independence as a trustee, even though he was pleased to accept the union endorsement. The college staff union mailed out 4,000 of Rosen's and their other endorsed candidates' flyers and took an ad in the suburban edition of the *Chicago Tribune* for all their endorsed trustee candidates.

Rosen received an endorsement as well from the local community newspaper, *Pioneer Press*, and the endorsements of some former elected officials. One of two winning candidates, he won reelection against strong opponents by a vote of 17,927 to 17,152 for his closest rival. The other winning candidate received 19,311 votes. Hard work, good endorsements, and support from organizations can offset lack of funds and a campaign staff in some local elections. But even then, it is important to put together a solid campaign to win.

In larger local contested campaigns, such as for city council, county board, or state legislature, there are often three paid staff members: a campaign manager, an office manager who also serves as the volunteer coordinator, and a public relations coordinator. These campaigns may cost more than $100,000, but highly competitive ones in more wealthy districts may run more than $500,000. Campaigns for higher office cost over a million dollars and have a number of paid staff, political consultants, and expensive paid media advertising.

Establishing basic resource goals is important early in a campaign. If minimal resource goals can't be met in the beginning, they are unlikely to be met later. Once the campaign is begun, an exact fund-raising plan, public relations program, and voter contact strategy with clear targets for later stages of the campaign must be made. Only campaigns with at least a chance to win are worth running, and it should be clear from the beginning that the candidate has a strong campaign with a real chance to win the office.

A CANDIDATE'S IMAGE

Public relations coordinators create an image for the candidate and the campaign. According to Don Rose, a public relations consultant, "the important thing is to establish early an identity for the candidate, a point of reference for the candidate."[4]

It is certainly true that celebrities sometimes win elections. Wrestler and radio personality Jesse Ventura was elected governor of Minnesota in 1998, and movie star Arnold Schwarzenegger was elected governor of California in 2003 and reelected in 2006. It is also true that some families make a business of politics, and the family name can help in elections, as we have seen with the Kennedys of Massachusetts, the Bush family of Texas, and the various political families of West Virginia such as the Manchins, Warners, and McGraws.[5] Having a famous name or an established public image helps with the name recognition problem, which is a major hurdle in winning elections. But most candidates start without that advantage and must build a public persona to present to the voters.

In any campaign, a candidate's image, however crafted, must be solid and truthful enough to withstand negative attacks. When Michael Dukakis ran for president in 1988, his image as the governor of Massachusetts, the "Economic Miracle State," could withstand neither the Willie Horton advertisements that portrayed him as soft on crime nor the pictures of a polluted Boston Harbor that undermined his image as ecologically concerned. And he is still remembered for looking foolish while riding in a military tank in one photo opportunity.[6] In the face of negative campaign attacks and mistakes made by his own campaign, Dukakis's image as a smart, able chief executive concerned with issues that mattered to the American voters did not hold. By contrast, Bill Clinton, who had scandals and image problems to manage in his 1992 campaign for the presidency, established a campaign "war room" that quickly responded to his opponent's attacks. He did much better in projecting his campaign persona than Dukakis had done. Partly because of that, Bill Clinton was twice successful in his presidential campaigns. Barack Obama had to quickly and decisively respond in his 2008 presidential campaign to attacks linking him to controversial views of his pastor, Jeremiah Wright, about race relations.

Even before the issues of the campaign are joined, a solid image for the candidate must be created. Shaping that image is the task of the candidate, campaign manager, public relations coordinator, and public relations consultants. Failing to do so or allowing an opponent's portrayal to go unanswered will surely lose the election. All candidates must both project and protect a positive image.

A CAMPAIGN'S MESSAGE

Every campaign must have a message that communicates to voters why they should vote for their candidate. A campaign theme unifies the campaign and defines the battleground the way the candidate wants it defined. The theme may be simplified into a slogan on a campaign poster, or it may remain implicit in the campaign literature. In his 2004 campaign for reelection, President George W. Bush's theme was "the world is more peaceful and more free under my leadership, and America is more secure."[7] Critics of Democratic challenger John Kerry's campaign argue that his theme was never very clear.[8] This was a major failing of the Kerry campaign.

The choice of a theme is particularly important *in defining the issues so a majority of citizens identify with a candidate*. A candidate takes stands on many important, separate questions. It is crucial to state the issues and define the alternatives in a way that most citizens will support. A candidate should be distinguished from opponents in strongly positive terms: an honest person vs. a crooked politician; a reformer vs. a corrupt political machine; a peacemaker vs. a warmonger; competent vs. incompetent; independent voice vs. political hack.

While the stress is primarily placed on the candidate's positive characteristics rather than on the opponent's faults, the contrast between candidates is important. In *Campaigns and Elections American Style*, political consultant Joel Bradshaw writes that a campaign theme is "the rationale for your candidate's election and your opponent's defeat." He advises that there can be only one theme in a campaign, not multiple themes, because "You can get [only] one point through to voters who think about this [election] five minutes a week or less." The best theme is a candidate's response to the question, "Why are you running?" Bradshaw concludes that a good campaign theme answering this question has six characteristics: "It must be clear, concise, compelling, connected, contrasting, and credible. It helps also if it can be communicated in an easily understandable and memorable way."[9]

Bradshaw further advises that the secret to a winning public relations plan is the repetition of your campaign theme. The theme is repeated both in words and visually in the campaign literature, paid commercials, debates, speeches, and press releases. Like a lawyer building a case for a client, you keep pounding home the theme in as many ways as possible.

Publicity and Press Conferences

The best campaigns have a single unified theme developed and exploited by publicity. The publicity effort involves a series of careful judgments as to what information is worth communicating, to whom, when, and through what media. Information about marvelous things the candidate has done or stupid things an opponent has said must be released in ways that have the maximum affect in building a campaign, communicating its purpose, and winning on Election Day.

A successful press conference depends on the newsworthiness of what is being said or done there. The most important aspect of news is primarily that action occurs rather than just words are spoken. So the news conference should announce some action, such as the governor is signing a bill into law, the mayor is announcing that the city has received a federal grant of millions of dollars, a candidate for office has filed a lawsuit against the incumbent. Reporters and editors judge newsworthiness by several simple criteria. *Does this action affect the lives of many people? Is this action real or just words?* Everyone feels the impact of new taxes, but slogans or campaign promises are only words. Statements about procedural reforms of government are even less interesting to most news consumers.

A key ingredient in a press conference is a press release read at the news conference and sent to all the media who did not attend. Press releases are often used as well to inform the news media of stories that don't warrant a full-blown press conference but are still newsworthy. The best press release is built around an action: Candidate Smith today called for a federal investigation of his opponent, or Candidate Badillo launched a drive to collect signatures on a referendum petition. *Called* and *launched* are action verbs. After taking an action, such as participating in a protest march, a candidate can explain why she did so.

Figure 7.1 shows a good standard press release from the 2005 "Werkheiser for Delegate" campaign in the 42nd legislative district of Virginia, where Greg Werkheiser was running against incumbent state legislator David Albo. Notice that it announces something Werkheiser did. He raised more money than the incumbent, which is always surprising. The release is in the proper format, gives the date and contact people, and is short. It refers readers to his Web site for further information. Despite his positive publicity, however, Werkheiser lost his election bid. But his good media campaign was almost enough to allow him to unseat a powerful incumbent. And running again for the legislature in 2009, he lost another close election.

The Role of the Media

One timeless way to deliver the campaign message is through the news media, especially the **free media**, coverage that is not paid for by the campaign, such as news reports of a press conferences or other actions or words of the candidate. However, few if any campaigns are

January 19, 2005

Contact: Bergen Kenny
Office (703) 644-0564
Cell (415) 819-0959

WERKHEISER OUT-RAISES OPPONENT
AND LEADS CHALLENGERS STATEWIDE

Greg Werkheiser, candidate for the House of Delegates from the 42nd District in southern Fairfax County, has raised significantly more money to date than his incumbent opponent, David Albo, and leads all challengers statewide in fundraising, according to reports filed with the State Boards of Elections on January 18, 2005.

Werkheiser out-raised his opponent by almost 2 to 1: Werkheiser $81,615; Albo $55,119.

Werkheiser now has twice as much unencumbered cash on hand as his opponent: Werkheiser $50,017; Albo $23,767

Werkheiser's contributions come from 200 individuals; Albo had 8 individual contributors.

Werkheiser leads all challengers statewide in total contributions raised and funds on-hand.

The 42nd House of Delegates District forms the southern tip of Fairfax county. Nearly fifty-two percent of the district voted for John Kerry in 2004, and Governor Mark Warner won the district with similar margins in 2001.

"I am humbled by the support of so many citizens in the 42nd District who are voting with their pocketbooks for a changes in leadership. While we are very pleased with this outpouring of support, I am going to keep doing what I have been doing: listening to my neighbors and working to make our communities in FairFax county a better place to live, work and raise a family," stated Werkheiser.

For more information about Greg Werkheiser's campaign, please visit: www.werkheiserfordelegate.com

FIGURE 7.1 Werkheiser Campaign Press Release

won solely by press conferences and staged campaign events covered by the media. There are a variety of other methods by which a good campaign delivers its message.

The most common media coverage is **paid advertising** such as radio and television ads and direct mail. But most local campaigns can't raise the money for enough paid advertising to win solely by that method. Table 7.1 presents the results of a survey of campaign consultants working for candidates in local elections.[10] It provides a good idea of the mix of techniques used to communicate a candidate's campaign message to the voters, from direct mail to the use of Web sites. Since this survey was taken, the use of Web sites, mass e-mail, and candidate blogs have only increased. But well-funded campaigns still primarily depend on paid media advertising to get their message to most voters.

172 **Part THREE** Elections and Citizen Participation

TABLE 7.1 Tactics Used by Local Campaigns with Paid Consultants

Tactic	Percentage
Direct mail	96
Free media coverage	92
Direct mail speeches to constituents and groups	89
Telephone banks	87
Promotional products such as yard signs, T-shirts, or bumper stickers	83
Radio ads	81
Newspaper ads	79
Literature drops or distribution	79
Canvassing door-to-door in the precincts	77
Participation in candidate debates	73
Public opinion polls	72
Television ads	68
Use of Internet sites	40

Source: J. Cherie Strachan, *High-Tech Grass Roots: The Professionalization of Local Elections* (Lanham, MD:/Rowman and Littlefield, 2003).

Of the mass media, radio is usually a cheap and effective way for local campaigns to advertise. Radio can be targeted to specific audiences such as classical music listeners or talk radio fans, sports fanatics or farmers. To be effective, however, a candidate usually must buy "saturation time" on a few stations. This means at least 10 replays of the campaign ad each week at various broadcast hours and prices. Often a campaign can buy a week's worth of ads on a single radio station for $500–$1,000. Advertising rates vary according to the size of the audience of a particular station.

State and local campaigns also use television ads despite their cost. These ads often provide candidates with the best opportunity to show themselves to voters. According to one study, paid political ads achieve two and a half times more air time than news coverage. In seven Midwest markets, local newscasts aired an average of about 4.5 minutes of campaign ads, but only 1 minute and 43 seconds of election coverage, over a quarter of which was devoted to the big gubernatorial campaigns.[11]

It has become very affordable to buy ads on cable television that are delivered only to homes in the district. The downside is that they reach very few voters on any one cable channel. Like all paid advertising, they have to have an impact on the voters. It is better to pay for a smaller audience that has many potential voters, such as *60 Minutes*, rather than for time on a popular show that does not, such as *American Idol*.[12]

For local races, most mass media advertising may be too expensive, because the media broadcast to a much larger audience than the residents of one district and the bulk of the viewers or readers cannot vote for a particular district's city council or state legislative candidate. Mass media that is cheaper and more affordable for state and local candidates include cable television, community newspapers, and local radio stations that have cheaper advertising rates. See the Insider's View feature on how cable television advertising enhanced a city council campaign in Southern California.

AN INSIDER'S VIEW

Cable Ads in a California Election Battle

Advertising on cable television offers a tremendous opportunity for candidates to get their messages across to voters, particularly those who are running for local office. That point was underscored earlier this year in a city council campaign in Calabasas, a Southern California community at the northern end of Los Angeles County, with a population of about 20,000. A city council campaign in a relatively small community would not seem to lend itself to sometimes-pricey cable advertising. Yet the costs for 30-second cable TV spots run on behalf of Councilman James Bozajian were affordable enough to incorporate into our campaign. While other candidates focused only on direct mail and traditional field operations, Councilman Bozajian's inclusion of cable TV spots presented a full complement to the voters. The result was his landslide re-election, at the same time that two other incumbents were voted out of office . . .

There are three major rules to cable television advertising: time element, product quality, and strength of buy. The time element refers to when a cable TV advertising campaign should begin, and this is an area of some disagreement within the political consulting world. Many consultants and amateur candidates think only in terms of a campaign's final few weeks. This approach is fairly effective in broadcast advertising, which has more immediate reach, but costs a lot more. The world of cable television is a marathon, not a sprint. It takes time for viewers to accumulate substantive viewings of campaign TV spots. Therefore, it is best to begin earlier rather than later. Start advertising 12 weeks before the election so that voters, especially those who will vote by absentee ballot, have enough time to see your TV spots and digest your message . . .

Cable is less expensive than [other media advertising]. As an example, a typical prime-time spot on FOX News cost Bozajian only $17.[13]

Source: From Todd Blair and Garrett Biggs, "Cable Advertising: An Underrated Medium for Local Elections," *Campaigns & Elections*, September 12, 2005.

While direct mail is still widely used to communicate a candidate's message, some political consultants believe it is obsolete, impossible to test for effectiveness, and nowhere near as cost-efficient for fund-raising as Internet and e-mail campaigning.[14] Nonetheless, direct mail is used in 96 percent of local campaigns with paid consultants, more than any other tactic. It is important for campaigns, of course, to design their direct mail so that it actually gets looked at by voters.[15]

Personal Campaigning

In films like *Blaze*, in which Paul Newman portrays Earl Long running for governor of Louisiana, or *The Last Hurrah*, in which Spencer Tracy plays an old Boston party boss attending political rallies and torchlight parades, we see the common images of political campaigns. Campaigning is about the candidate meeting the voters, shaking hands, being seen in parades and meetings, and speaking at large rallies of political supporters. Even in this modern era of mass media, targeted direct mail, and videos on YouTube, a candidate still must make personal appearances to communicate effectively with voters. An important way for candidates to campaign remains going out and meeting the voters through door-to-door canvassing. This method was used by Dan Kotowski, a successful Democratic candidate for the Illinois state senate in a predominantly Republican district,[16] and by Ed Salvatore, who won the 2005 mayor's race in Albion, New York.[17] Both candidates had gone door to door, speaking with potential voters personally. See the Insider's View feature on one such effort. Since a candidate has to be seen positively by the voters and the press, personal campaigning can be vital.

Advice on Personal Campaigning

Never underestimate the power of a handshake. That and a good look in the eye from the candidates themselves seem the key to winning local elections, perhaps even more important than a candidate's party affiliation or record while in office.

. . . David Albanese was a shoo-in for mayoral re-election in the spring of 1998. The village had just received a hearty Canal Corridor Initiative grant, Albanese was among the vocal leaders opposing a landfill expansion in the community, and taxes seemed in check. Albanese also had the backing of the Republican Party. He and his supporters assumed he would easily defeat his challenger.

The incumbent barely campaigned. He believed residents would like his record of accomplishment.

But Ed Salvatore, a Republican who crossed over and ran as a Democrat, believed the village could be doing more to lower taxes, and to hear resident concerns. He vowed "to give the community back to the people." He walked every broken sidewalk in the village and knocked on almost every door.

On Election Day, Salvatore emerged the winner. He has since followed through on a pledge to cut the tax rate. He also got many of the sidewalks fixed. His campaign in 1998 is often cited as the way to win a local election. You can't rest on your laurels in a small town election and assume everyone loves you. You've got to wear out your knuckles rapping on local doors.

Source: Adapted from Tom Rivers, "Memo to Candidates: Press Flesh," *Batavia, N.Y., Daily News*, November 17, 2005, 4A.

A candidate has to be a lot of places, doing a lot of things in the campaign. In author Dick Simpson's aldermanic campaign, on a single Sunday he had to attend a church to meet the parishioners at the post-service coffee hour and attend a strategy meeting at the campaign headquarters; he also held a press conference where other elected officials endorsed him, attended three gatherings to recruit volunteers, and raised money; he then gave an important pep rally speech at a training session for his volunteer precinct workers. The next morning, he was at a bus stop shaking voter hands by 7 a.m., then gave an in-depth interview with a community newspaper reporter, and by the afternoon was walking door to door along with two volunteers to meet voters. And so it went for many weeks in the campaign.

WINNING ELECTIONS

Winning campaigns are often based on solid precinct work, an understanding of absentee and early voting processes, and new technology—from the Internet and blogs to e-mail and e-videos.

Precinct Work

Staffing a field operation with trained workers able to carry out a petition drive, registration drive, door-to-door canvass, and poll watching on Election Day is still the secret to winning most state and local elections. Precinct work provides a much more personal and less expensive way to reach voters, to register them, to deliver a campaign's message, and to get them to the polls.

One way of obtaining the necessary precinct workers is to receive the endorsement of a political party. Loyal party supporters are willing to do the often unglamorous precinct work. Other organizations also provide volunteers—groups like labor unions, concerned interest groups, and religious groups. But in most campaigns, volunteers have to be recruited one at a time.

Many volunteers are recruited at social events called generically "coffees" even if coffee is not necessarily served. These events are often held in the home or apartment of a supporter who has invited neighbors to come hear the candidate.

In high-tech, candidate-centered, modern campaigns, sometimes paid advertising along with specialized direct mail and automated phone campaigns and use of the Internet can defeat precinct work. In an economy where everyone seems to be working long hours, it is hard to find enough people to volunteer to cover all the precincts and reach all the voters. Often, gated communities and high-rise buildings prevent door-to-door canvassing. Yet precinct work remains one of the best ways to run state and local campaigns. In campaigns at higher government levels, it will not be sufficient and will have to be supplemented by other more expensive methods of campaigning, but grassroots campaigning is still critical.

Absentee and Early Voting

Among the methods states use to increase voter turnout are absentee voting and early voting. Both methods allow a voter to cast a ballot before Election Day, and the requirements vary by state. **Absentee voting** allows you to cast a ballot early by mail if you know you cannot get to the polling place on Election Day. Some states require that you go in person to a government office to request an absentee ballot. **Early voting** allows you to cast a vote at a special voting location, such as a county or township clerk's office, for a period of time before Election Day, and you do not need a specific reason or excuse for voting early. Table 7.2 shows that in 2008, 31 states allow no-excuse early or absentee voting, 14 states and the District of Columbia require an excuse for in-person absentee voting, and voters in one state, Oregon, cast all their votes by mail. All states offer some form of at least absentee ballot by mail.[18]

Candidates must adjust to the increasing numbers of early voters who are not susceptible to traditional late-campaign election drives. Early voting is especially critical because 10 percent or more of the votes will be cast before the polls open on Election Day. Campaigns now frequently encourage all their supporters who have signed up on the campaign Web site to take advantage of early voting, which frees them for Election Day duties and helps reduce long lines at the polls.

Good campaigns work hard to mobilize their voters in advance for both early and absentee voting. Most campaigns plan for an advertising push in the last two weeks of the campaign—paid advertising, direct mail, and even free media coverage peak then—but early voters may not have been exposed to these by the time they vote. Since it is almost impossible to identify who is going to vote early, campaign volunteers can't give them a last-minute reminder. Newspaper endorsements may not yet be available to guide voters. Nonetheless, campaigns desperately want to win early votes to increase their margin for victory in a close race.

Campaigns also are aware that absentee voting is susceptible to fraud. Unscrupulous campaign volunteers, for example, may visit elderly constituents in nursing homes who don't know anything about the candidates or issues but will let the "nice" volunteer help them. They sign the request for the absentee ballot and later will vote as they are told. Another problem is that absentee ballots sent to military service members and civilian voters overseas can be lost or delayed too long to be returned in time to be counted, which can affect the outcome of close elections.[19] The military and some states have been looking into online voting as an alternative method.

TABLE 7.2 Absentee and Early Voting, by State

State	Offer No Excuse, In-Person Early Voting	Offer No Excuse Absentee Voting	Offer Absentee Voting—Witness or Notary Required	Offer Absentee Voting—Witness or Notary Not Required
Alabama			✓	
Alaska	✓	✓	✓	
Arizona	✓	✓		✓
Arkansas	✓	✓		✓
California	✓	✓		✓
Colorado	✓	✓		✓
Connecticut				✓
Delaware			✓	
Florida	✓	✓	✓	
Georgia	✓	✓	✓	
Hawaitt	✓	✓	✓	
Idaho	✓	✓		✓
Illinois	✓			✓
Indiana	✓			✓
Iowa	✓	✓		
Kansas	✓	✓		
Kentucky				✓
Louisiana	✓		✓	
Maine	✓	✓	✓	
Maryland				✓
Massachusetts				✓
Mchigan			✓	
Minnesota			✓	
Mississippi			✓	
Missouri			✓	
Montana	✓	✓		✓
Nebraska	✓	✓	✓	
Nevada	✓	✓		✓
New Hampshire				✓
New Jersey		✓	✓	
New Mexico	✓	✓		✓
New York			✓	
North Carolina	✓	✓	✓	
North Dakota	✓	✓	✓	

(continued on next page)

TABLE 7.2 Absentee and Early Voting, by State *(continued)*

State	Offer No Excuse, In-Person Early Voting	Offer No Excuse Absentee Voting	Offer Absentee Voting—Witness or Notary Required	Offer Absentee Voting—Witness or Notary Not Required
Ohio	✓	✓		✓
Oklahoma	✓	✓	✓	
Oregon		✓*		✓
Pennsylvania			✓	
Rhode Island			✓	
South Carolina			✓	
South Dakota	✓	✓		✓
Tennessee	✓		✓	
Texas	✓		✓	
Utah	✓	✓	✓	
Vermont	✓			✓
Virginia		✓	✓	
Washington		✓		✓
West Virginia	✓			✓
Wisconsin	✓	✓	✓	
Wyoming	✓	✓		✓
DC				✓

*Oregon conducts all elections solely by mail ballot.

Source: National Council of State Legislators [[http://www.ncsl.org/programs/legismgt/elect/absentearly.htm.]]

New Technology

Today's campaigns make use of high-tech electronic tools to reach voters. These include the Internet, blogs, e-mail, e-videos, YouTube videos, and automated phone calls.

The Internet

Ever since Howard Dean's campaign to become the Democratic Party's 2004 presidential nominee, the Internet has been a popular tool for informing voters, rallying scattered volunteers to meetings, inspiring supporters, and raising money through online donations. Dean raised $7.5 million from April to June 2003, and two-thirds of his contributions were from the Internet.[20] Even more impressive and later widely copied was Dean's mobilization of volunteers through "meet ups" in communities throughout the primary states.

Many local and state candidates now use the resources of the Internet. All members of the U.S. Congress and most congressional candidates have Web sites. A campaign Web site gives information about a candidate, a message from the candidate, a chance to join the campaign as a volunteer or a contributor, endorsements by well-known figures and local citizens (whose pictures and quotes often change each time the site opens), information on the district, and copies of press releases, press stories, and photographs. A Web page is critical because it allows campaigns to recruit volunteers and campaign contributions almost for free and keeps supporters and the press informed of the candidate's activities.

Today even the most local candidates may use sophisticated numbers-crunching Web software programs that can give them all sorts of demographic information on the voters they need to reach. For instance, during the 2004 election for District 5 city commissioner in Palm Beach, Florida, incumbent Bill Moss as well as his challenger Hank Porcher used such tools.[21] Such programs and the lists they generate can help candidates target voters, contributors, and people who will put up yard signs. Developers of this new technology claim it can make a difference in close elections, while experienced campaigners and old-time politicians see it as a support for the old-time tactics of personal contact with voters.[22]

When the election gets to the gubernatorial level, Web sites become even more professional. Figure 7.2 shows the campaign Web page of Indiana Governor Mitch Daniels when he was running for reelection in 2008. It is easy to find newsrooms, links to campaign videos, biographies, the candidate's campaign message, and links for volunteering and making contributions. Daniels was reelected with 58 percent of the vote. Modern candidates believe that an informative, attractive Web site is a must for winning any election even if they are incumbents with many advantages.

The Blogosphere

One component of any good campaign Web page is a **blog**, or Web log, an online journal or diary. A blog on a political Web site can be updated regularly to give the reader all the latest news of the candidate and the campaign, as well as recent comments about the can-

FIGURE 7.2 Indiana Governor Mitch Daniels' Campaign Web Site

didate in print, broadcast, and electronic media. The Pew Internet Research Group found in 2006 that 14 million Americans used blogs as their primary source of political information, and another 12 million actively maintained their own blogs.[23] A 2007 poll by Media/Zogby found that 55 percent of adults interviewed thought of blogging as an important aspect of American journalism in the future.[24] Since 2009, many candidates have been using Twitter to send Tweets to their supporters and to provide even quicker campaign updates.

Blogging has become standard for all campaigns on all levels. Campaigns hire bloggers, typically given a job title something like "new media outreach coordinator," to build up relationships with other influential bloggers, get the campaign's messages and themes across, and troubleshoot if necessary.[25] Good campaigns even schedule breakfast or other meetings for all bloggers in their district in order to give them access to the candidate and gain positive campaign coverage.

Campaigns also place ads in blogs, which can be of great advantage for candidates for state and local elections, since even popular blogs charge little for ads. A weekly ad can cost $250–$2,900, while a 30-second TV broadcasting spot might cost $100,000 per week in a medium-sized market. While the most popular political blogs only reach a few million readers, according to a recent study by George Washington University's Institute for Politics, Democracy and the Internet, those who read blogs are more likely to participate in the political process.[26] In the summer of 2007, 1,500 bloggers showed up at the Yearly Kos bloggers convention, along with seven of the eight Democratic presidential candidates, who addressed the convention. By the 2008 campaigns, political blogging had become a major feature of campaigns and campaign strategy.

E-mail

Many state and local candidates maintain e-mail lists that allow the campaign staff to send an e-mail message with instructions, information, and encouragement to all volunteers and supporters at once. Campaigns also send e-mail newsletters that relate recent campaign news and have links for the recipient to contribute or to volunteer to work for the candidate's election. E-mail newsletters for candidates often also tell about campaign successes in raising money, up-and-coming fund-raising deadlines, "meet-ups" of volunteers, and opportunities to influence an endorsement by powerful political groups, such as MoveOn.org or the National Rifle Association. Sending campaign news and exhortations via e-mail and Twitter is an inexpensive and effective way to keep in touch with supporters. Internet technology lets candidates inspire supporters, sign them up for volunteer efforts, and harvest financial contributions almost immediately. It costs 40 cents for each dollar raised by direct mail, but less than a penny when it is raised by e-mail.

CANDIDATE LUCY BAXLEY watching her campaign commercial while running for governor of Alabama.

E-videos

E-videos, short videos posted on the Internet, are an inexpensive and innovative way for a campaign to advertise or gain publicity. Candidates and campaign volunteers can easily make their own videos and put them on a campaign Web site or send them to YouTube. Most candidates now film all their commercials and many campaign events digitally so they can be replayed on regular television, cable television, YouTube, Facebook, the campaign Web site, and as e-videos. As the Insider's View feature explains, e-videos were helpful for two local candidates in their race for mayor of Plano, Texas.

By the 2008 elections, e-videos were being produced by most campaigns and many supporters and placed on sites such as YouTube and MySpace. Such e-videos can be downloaded and e-mailed to family, friends, supporters, and other e-mail lists that a campaign can obtain. In addition, any campaign ad, once broadcast, can be found on YouTube. But e-videos are not always helpful, as U.S. Senator George Allen discovered while running for reelection in Virginia. At a campaign rally, when Allen noticed one of his opponent Jim Webb's supporters, he referred to the young man as "macaca," which was seen as a racial slur. A video of the incident was soon being repeatedly viewed on YouTube.[27] National media soon picked it up. Not only did Allen lose the election, but his defeat tipped the party balance in the U.S. Senate. During a July 2007 CNN-YouTube jointly sponsored Democratic presidential debate in South Carolina, the questions for the candidates were selected from those the public submitted by videotape via YouTube.[28]

Automated Phone Calls

According to Mac Hansbrough of the political consulting firm NTS, in the past 10 years, the number of phone calls made in state and local campaigns by professional telephone bank (phonebank) firms has increased dramatically.[29] Both the number of calls made professionally for individual races and the number of campaigns doing the calls have increased. The reasons for this mushrooming volume are many. First, the cost of calling has steadily decreased due to competition among the phone companies, all striving for a piece of the vast national phone call volume. Second, computer-assisted dialing has become available to local and regional political phoning firms. Many of these have become national enterprises working in many states, subcontracting call centers from commercial telemarketing firms. Third, the easy availability of personal computers allows electronic voter files, scripts, and other items to move about with ease among campaigns, political phone vendors, and call centers. Fourth, more campaigns use professional political consultants, who often steer their clients to professional calling and away from less dependable volunteer telephone callers. Finally, updated voter files are now available in electronic format from political data firms, which purchase the raw data from state and local voter registration offices. These factors have given rise to a substantial increase in the number of phoning vendors from a very few in 1976 to a substantial number in 2008.

Twenty-five years ago, voter files were available in only a few states and many of those files were not very current. Today most states and cities make such files available to campaigns, state parties, and often directly to voter file vendors, in fairly up-to-date electronic formats. These files are enhanced, phone matched (through several local, regional, and national vendors), and sold to campaigns and campaign phone and mail vendors at reasonable cost. It is now commonplace for a phone vendor to get an order from a campaign or organization in the morning, have the voter file in the afternoon, and be calling voters in that file in the evening. Computers and the Internet have made all this possible.

AN INSIDER'S VIEW

Local Candidates Score with E-videos

Remember that zany cowboy from *Blazing Saddles* and the Pace Picante Sauce commercials? During the 2006 Plano, Texas, mayoral contest, he was showing up on computer screens, singing praise for Plano mayoral candidate Ken Lambert. Meanwhile, on Google Video, cheering teenagers proclaimed their support for incumbent Mayor Pat Evans, calling her "da bomb diggity." (That's a compliment.)

The run-up to Plano's 2006 city elections brought the familiar influx of yard signs, mailers and Rotary forums. But as voters upgrade to faster Internet speeds, more local candidates are deploying Web videos as a campaign tool.

"There's an intimacy there—instead of reading material or even seeing a print ad, you actually see them and hear them speaking to you," said Michael Kitkoski, a freelance video editor in Rockwall who has produced Web videos for the past four local elections.

Candidates say e-mailing the videos is more immediate and cost-effective than traditional campaign ads, especially if recipients forward the video clips to friends and family.

"The efficiency of doing that is so much better than just more and more printed material," Mr. Lambert said. The Lambert campaign produced two videos for about $6,300—one e-video of the candidate, and a cable TV ad featuring *Blazing Saddles* actor Burton Gilliam that also is making the rounds via e-mail.

Airtime fees for the 30-second Gilliam ad on local cable have ranged from $7 to more than $100 per spot, Mr. Lambert said. But e-mailing a link to the video is free, as was the production of Pat Evans's video. A group of Plano teenagers sparked another Evans video when a 14-year-old high school freshman named Jennifer first met Ms. Evans through a National Junior Honor Society project, and Jennifer invited students from across Plano to record the video at her home. She and her dad edited the spot, posted it on Google Video, and got the word out via e-mail and MySpace.com. It ran 97 seconds and concluded with a message from the mayor, and received over 4600 Web hits by the week before the election.

Source: Jake Batsell, "In Plano Mayor's Race, Web Ads Are a Sure Hit," *Dallas Morning News,* May 11, 2006.

Automated calling, also known as robo calling, began in earnest about 10 years ago. Candidates and their supporters can make and distribute brief recordings cheaply, quickly, and easily, due to the advent of computer dialing. Messages are recorded on recording lines provided by the political phoning firms. The phone vendor takes the target database of phone numbers, puts it with the recording, and sends out the call. Using this technology, it is easy for a campaign to reach 10,000 or even 1 million voters in a single day.

New Methods, Old Dynamics

Although the technology and specific methods of running campaigns continue to develop, perhaps even more rapidly in the Internet, cell phone, and cable television age of the 21st century, the fundamentals of election battles remain unchanged since even before the days of Abraham Lincoln. First, one way or another, campaigns have to canvass a district and get out the campaign's message about the qualities of the candidate and the candidate's position on key questions that matter to the voters. Second, campaigns must discover which voters will vote for a candidate. Third, campaigns must get the voters who do favor their candidate to cast their votes, and must get those votes counted. A campaign may send its message by a volunteer going door to door with a campaign brochure, by an e-video delivered through the Internet, or though a social networking site such as Facebook. A campaign may have discovered who will vote for their candidate through responses to telephone calls or Internet sign-ups. But whether a candidate campaigns in the old-fashioned or high-tech way, the campaign's purpose and principles remain unchanged. It is our democratic faith that free, fair, and vigorously contested election battles will elect worthy representatives at every level of government who will pass public policies that represent the will and the interests of the people who elect them. We can't understand democracy and the struggle for power in cities and states without understanding elections.

Electoral politics, which can decide who holds power in our cities, counties, and states, is only one arena for the struggle for political power in state and local government. Although elected officials of most states and many cities represent millions of people, they are nevertheless more accessible to citizens than are most federal government officials. This means that during election campaigns, personal contact and making connections with local groups and individual citizens can make the difference between victory and defeat. While incumbents have a name recognition advantage, they still must convince voters that the policies they adopted while in office have had a positive effect on their district, ward, or town.

It is important to stress that campaigns are difficult, and most often the incumbent wins. Congressional incumbents who run for reelection win more than 90 percent of the time, and the reelection rate for state legislative incumbents is almost the same. You might think that in local elections there would be greater turnover. However in 2001, 81 percent of the incumbent suburban mayors and village presidents won reelection in Cook County, Illinois. In 2005, the number decreased only slightly to 74 percent. Incumbent and "acting" suburban village leaders defeated challengers in 36 of 49 races, and 36 won with no opponent on the ballot to challenge them.[30] Running for election against an incumbent is a daunting task.

Campaigns for elected office are exciting, difficult, demanding, and rewarding if they are serious and well run. They affect not only who is elected but the directions that state and local governments will take in the future. They are a real opportunity for you while you are still students to volunteer to elect the candidates you favor, to make a difference in the outcomes of the elections and the direction of government policy-making, and to learn the craft of politics. You can participate directly in this struggle for power at the local level.

CHAPTER SUMMARY

Elections, which are central to a representative democracy, affect all government outcomes. In order to get elected, a candidate needs time, money, people, and skill. These four resources help a candidate develop the credibility, visibility, and more money that will be needed in a successful campaign. Candidates also must calculate their target number: how many votes will be needed to get elected.

There are many reasons not to run, and positive reasons such as civic duty often seem abstract. Once the decision to run is made, factors that will help you win are key campaign staff (volunteer or paid), support of a major party, and a demographic match with your constituents, although it is also important to seek votes from other demographic and ideological groups. It is also important to have endorsements from key public figures, media coverage, and money.

A candidate's image must be solid to withstand scrutiny, and a campaign must have a unifying theme which defines the issues in a way that persuades a majority of voters to identify with the candidate. A candidate's message should stand out using strong, positive terms. Media coverage is important; it is best to concentrate on what information is newsworthy and what actions the candidate has taken.

Traditional means of getting your message out include publicity and press conferences, media coverage, paid television and cable TV ads, direct mailings, and door-to-door personal campaigning. Winning the campaign requires precinct work, understanding absentee voting and early voting requirements, and using new technology such as the Internet and Web pages, blogs, e-mail and e-videos, and automated phone calls.

Regardless of using old or new techniques, a winning campaign still needs to (1) canvass a district to get the candidate's message out, (2) find out which voters support the candidate, and (3) get the vote out. The methods used in national campaigns are also successful in even the smallest local elections.

KEY TERMS

absentee voting Voting before Election Day, generally by mailing an absentee ballot. In some states, a voter must provide an acceptable reason for voting absentee, p. 176

blog A Web log; an online journal or diary, p. 179

early voting Voting before Election Day in person at specified locations, p. 176

free media Media coverage not paid for by the campaign, such as news reports of a press conferences or other actions or words of the candidate, p. 171

majority vote rule A candidate must receive at least 50 percent of the vote to win the election, p. 161

paid advertising Advertisements paid for by the campaign in one or

more media, including press, radio, television ads, and direct mail, p. 172

plurality vote rule The candidate receiving the most votes of all candidates running wins the election, p. 161

target number The number of votes needed to win a particular election, p. 168

QUESTIONS FOR REVIEW

1. What are four essential resources a candidate needs to successfully run for office, and why is each necessary?

2. What are the arguments in favor of and against running for office?

3. Why is a *target number* significant in terms of launching a candidacy? How is it calculated?

4. What is the best way to find a good campaign staff? To gain important endorsements? Raise funds for a campaign war chest?

5. Why is a candidate's image vital to a successful campaign? Why is a campaign message essential for a successful campaign?

6. What are the mechanics of holding a press conference? What do media journalists ask themselves when they are trying to decide whether a candidate's press conference is newsworthy or not?

7. What are some of the traditional methods of campaign publicity? Which are the most often used, and which are most successful?

8. What are some of the new technologies that campaigns use to get a candidate's message across? How have these methods changed political campaign strategies?

DISCUSSION QUESTIONS

1. What are the reasons that different people get involved in politics? What would cause you to work in a political campaign?

2. What sort of image would you want your candidate to project? A man or woman of the people? A fighter for justice? To what extent must the image be based on the real person and to what extent can it be manufactured by public relations people?

3. If you were running for city council or to be a state legislator from your home district, what theme or message would you run on?

4. If you needed a 30-second radio ad for a candidate for local office, what would it say?

5. How can you successfully campaign on the Internet? What are the problems or drawbacks to campaigning on the Internet?

6. Why is precinct work so essential to a winning local campaign?

7. Is it ethical to use demographic data and consumer purchase information to microtarget voters and to send them specifically tailored messages by direct mail or the Internet? Should it be regulated? Why or why not?

PRACTICAL EXERCISES

For the following exercises, choose a particular elected official such as a member of your city council or your state legislator. Then research the answers to the following questions.

1. Who is the current elected official from your city council or state legislative district? You can find out your city council and state legislative districts from your voting registration card or by going to the state or local government Web site and entering your address. You can also get information by calling the local League of Women Voters or Election Commission office.

2. Is the elected official a Democrat or Republican? Do voters in the district tend to vote Democratic or Republican? You can determine this by checking the voting records on government Web sites for candidates and results in the last elections if these are broken down by district.

3. What was the vote for the incumbent and the opponents in the last two elections? You can then calculate the target number of votes needed for a candidate to win the election. (Usually, this will be at least one vote higher than the highest vote the incumbent received in a strongly contested election. If the incumbent won by more than 66 percent in one election but won more narrowly in the other election, it will be the number of votes in the closer election.)

4. Based on newspaper stories, is the incumbent likely to run for reelection in the next election? Or is the incumbent likely to retire or leave office due to term limits? When is the next election for this office?

5. When will redistricting occur by law? (This is nearly always after the next general U.S. Census.) How substantially is the district likely to be changed by redistricting?

6. Has the incumbent cast any unpopular or controversial votes that would make reelection difficult? Sources for this information would be newspaper stories, Web sites of various interest groups, and government Web sites.

7. Based on the information you gathered for the preceding questions, would you advise a potential candidate to mount a campaign against the incumbent? If so, on what issues do you think a good campaign against the incumbent could be mounted?

WEB SITES FOR FURTHER RESEARCH

For nonpartisan information on incumbent office holders and candidates running for national or state offices, use the information at Project Vote Smart: www.vote-smart.org.

For information on local candidates, visit the League of Women Voters site: www.lwv.org.

For information about the Democratic Party and its candidates, use the Democratic National Committee site at www.democrats.org. For the Republic Party and its candidates, use the Republic National Committee site at www.rnc.org.

Campaign contribution information for candidates running for federal office can be found at the Center for Responsible Politics site: www.opensecrets.org. Campaign contribution information for state and local candidates can often be found on a state or local board of elections Web site.

For information on political issues as well as campaign coverage, go to media sites like CNN's All Politics, www.cnn.com/politics.

Additional information can be found on local newspaper Web sites. And each candidate running a serious campaign will have a Web site, which can usually be located by doing a web search under the candidate's name.

SUGGESTED READING

Armstrong, Jerome, and Markos Moulitsas. *Crashing the Gate: Netroots, Grassroots, and the Rise of People-Powered Politics*. White River Junction, VT: Chelsea Green, 2006. On the rise of political blogs and the use of the Internet in politics.

Faucheux, Ron, ed. *The Road to Victory: The Complete Guide to Winning Political Campaigns—Local, State, and Federal*. Dubuque, IA: Kendall/Hunt, 1998. A basic how-to book.

Johnson, Dennis. *No Place for Amateurs: How Political Consultants are Reshaping American Democracy*. New York: Routledge, 2007. The best book on the role of political consultants in local and national campaigns.

Semiatin, Richard, ed. *Campaigns on the Cutting Edge*. Washington, DC: CQ Press, 2008. The latest campaign technologies, how they are used, and their drawbacks.

Shaw, Catherine. *The Campaign Manager: Running and Winning Local Elections*. Boulder, CO: Westview, 2004. The nuts and bolts of a campaign manager's job in local campaigns.

Simpson, Dick. *Winning Elections: A Handbook of Modern Participatory Politics*. New York: Longman, 1996. A step-by-step look at local election campaigns.

Thurber, James, and Candice Nelson. *Campaigns and Elections American Style*. Boulder, CO: Westview, 2004. The campaign techniques used in political advertising, campaign consulting, and running modern campaigns.

Demonstration in support of same-sex marriage in California

CITIZEN PARTICIPATION IN STATE AND LOCAL GOVERNMENT

This chapter is about advocacy, about efforts to gain specific outcomes from government. All four themes of the book are illustrated in this chapter. For example, political parties struggle to elect their own members to formal positions inside governments. If they elect a majority of the members in government bodies, they will capture the formal powers of leadership. Yet lobbyists, journalists, and citizen activists in community organizations exercise various degrees of influence, even power, over elected officials.

POLITICAL PARTIES

Political parties are organized around the formal task of getting their own advocates *inside* government, that is, into elected office. Powerful interest groups, such as teachers unions, will sometimes run favored candidates for the state legislature, yet these aspirants do not run as, for example, members of the Missouri Education Association Party. They usually run as Democrats or Republicans with, in this case, the backing of the Missouri Education Association. Since the Civil War, the Republican and Democratic parties have been the two major parties in American state and local politics.

Parties are often thought of as playing three roles in the political process:

1. *The party organization*, such as the Republican State Central Committee of North Carolina or the Democratic Central Committee of Broward County in Florida.

2. *The party inside government*, such as the Democratic majority in the New York State Assembly or the Republican minority in the Massachusetts Senate.

3. *The party in the electorate*: party members as well as those who identify with a party.

The party organization recruits candidates, supports them in their election battles, raises money to conduct party affairs, adopts a party platform of values and positions, and communicates party positions to the electorate.

Party Structure

Party organizations can be thought of as a pyramid (see Figure 8.1). At the base of the pyramid are the party's members who work the precincts. **Precincts** are the smallest election units in each state, comprising about 500 registered voters. Precinct committeemen are generally elected in primary elections. They are expected to communicate party information to voters and recruit and identify like-minded voters. They are also expected to get these voters to the polls on Election Day to support the party's candidates.

Typically, the precinct committeemen elect a county chair and sometimes a ward committeeman from among their members following their own election, generally "voting their strength," that is, casting one vote for each party vote cast in their precinct in the preceding primary election. The county chairs have often in American history been the linchpin of party politics. In large metropolitan counties, the county chair has often been a power behind the scenes of the elected party-in-government officials. In many elections throughout history, the county chair has been responsible for recruiting and electing the officials.

Above the county chair on the pyramid is a state central committeeman, who generally represents a congressional district of more than 600,000 citizens. In some states, the state central committeemen are elected by the county chairs, who vote their strength. In other settings, the state central committeemen are elected at the

National Chair

National Committee

Party State Convention (generally elects Nat'l Com.)

State Chair

County Central Committee (generally one committeeman or woman per congressional district)

State Chair (generally elected by precinct committee by a vote weighted to number of party votes cast in each precinct)

County Precinct Committeemen or women (sometimes called "captains")

FIGURE 8.1 Political Party Organization Pyramid

primary election. State central committeemen elect a state party chair and one or more members to the party's national organization.

Party Power and Influence

The power and influence of a party organization lies in its capacity to control nominations in the primary election and then elect those nominees to office. The more successful the party organization is at those two tasks, the more power it has over the elected candidates inside government.

Legendary urban political party bosses such as Ed (Boss) Crump in Memphis, Tom Prendergast in Kansas City, and Richard J. Daley in Chicago had the power to control primary and general election outcomes, primarily because of their armies of government-job-holding precinct committee people and the money that those committee people and local business people contributed to the party coffers. If a party-backed elected official failed to vote the way "the Boss" wanted, that elected official could be denied renomination to office. That power made political aspirants highly responsive to their patrons.

Today, most political parties are pale shadows of what they were under the famous urban bosses, primarily because of the Progressive movement in the early 20th century. Progressives fought corruption in all sectors of society and sought to weaken the power of party machines through a series of reforms giving citizens more direct participation in the electoral process and government. During the Progressive Era, many states replaced the closed party nominating caucus with primary elections, which opened the nomination stage to a broader electorate. Civil service and merit selection of governmental employees began to replace political patronage job holdings. As patronage slowly ended, so did the interest of thousands of party faithful in serving as precinct workers.

Progressives also sought to wrest political control away from county and urban party bosses by trying to take partisanship out of politics.[1] As a result, many states adopted legislation to take local elections away from the major parties and instead use local parties such as the People's Party and the Citizens' Party for slating candidates for local offices, especially in suburban areas. Another device to take elections away from party bosses was to eliminate party identification in many local elections. In most states today, candidates for offices such as school board run without any party labels in nonpartisan elections. A **nonpartisan election** is one in which candidates run without any party label.

The advent of television in the 1950s provided a powerful communications tool by which candidates could circumvent the party organization and reach voters directly. This further weakened the influence of dwindling numbers of active precinct committeemen. Modern mass media replaced the need for party patronage armies in higher but not all local elections.

As a result of these reforms, county-level party organizations now have difficulty finding people to run for the precinct committeemen. Similarly, precinct committeemen often have little incentive to go beyond the minimum effort of talking up the party's candidates. Since the strength of a party's influence is assessed in its capacity to motivate committeemen as well as to elect party candidates, most parties today are weaker than previously.

Author Nowlan is vice chair of the Stark County Republican Central Committee in central Illinois. In 2007, when a veteran member of Congress decided to end his service to the 12-county district that included Stark, several prospective candidates expressed interest in running for the Republican nomination. The chair and vice chair of the Republican committee in Stark County had to decide whether to endorse and campaign for one of the candidates or stay out of the contest. If the county party endorsed a candidate, worked hard in the campaign, and the candidate won the nomination and election, he or she might be very responsive to Stark County's needs. But if the county party endorsed a candidate who lost, it might be ignored by the winning candidate.

Stark County Republicans eventually endorsed a candidate for the congressional seat and held a fundraiser for him. The county party thus committed itself, for better or worse. Such is the game of possible credits and debits that is central to the struggle for power and influence in party politics.

LOBBYISTS AND ORGANIZED GROUPS

Most often, citizens get involved in politics and government in reaction to a perceived threat, such as a highway planned through their neighborhood or a sudden rise in their taxes. Sometimes the motivation is to elect an outstanding individual to public office or pass meritorious legislation for the general public good. In any case, citizens often find that they need to be part of a group to achieve their goals. Interest groups play a critical role in U.S. politics, especially in state and local government. They are important in articulating people's concerns, for example, and informing and amplifying the views of individual citizens. Interest groups will be discussed more fully later in the chapter when we look at citizen activists, but here we will focus on how interest groups, business and industry organizations, labor unions, environmental protection clubs, and other similar groups lobby to achieve their goals in government.

One example of the appeal of lobbying and the decision to use paid lobbyists occurred in Florida a few years ago. Florida's cities wanted the state legislature to enact a program that would assist in urban development. The legislature was willing to support the program but insisted that a revenue source be found first. The cities and their legislative supporters discovered that the dry-cleaning industry wasn't represented by lobbyists at the state capital. When the urban development program was approved, it was funded by a sales tax on dry cleaning. Within a few days of the new tax being enacted, dry-cleaning businesses jointly hired a lobbyist to protect their interests in the future.[2]

In Illinois, faced with an economic recession, the state borrowed money in 2004 to fund public employee pension funds. They then invested the funds, hoping that the interest on their investments would be higher than the low borrowing costs. This was a risky venture but better than cutting the pension funds. The next year, the governor and the state legislature borrowed over $2.2 billion from the state's legally required contributions to the funds. As a result of shortchanging the pension funds, the public pensions dropped to a ratio of assets to liabilities of 65 percent, the lowest level for public pension funds in the United States. Public employees, especially state university faculty, staff, and retirees, were stunned. How could the state do this to them?[3] The state retirees' newsletter explained that state contributions to pension funds were an easy target to fill gaps in the budget.

The retirees lacked the political clout to stop the state legislature from diverting contributions from their retirements funds. Their response was similar to that of the Florida dry cleaners. They hired the former chief lobbyist of the Illinois Education Association to lobby

BENDIB CARTOON illustrates the way many people see lobbyists.

on their behalf. They also organized "Political Information Committees" to coordinate grass-roots lobbying from each chapter of the State Universities Annuitants Association and began an immediate drive to increase their membership. In addition, they joined a coalition with other state employee unions with similar interests. In the cases of the Florida dry cleaners and Illinois university and state employees, hiring lobbyists was seen as an effective way to get the influence they needed to change state policies.

Lobbying and lobbyists often have a negative image. Many people picture rich guys in fancy suits smoking big cigars while giving out bribes to governors, mayors, state legislators, and aldermen to get favors for their clients. Lobbyists supposedly buy votes through wining, dining, and making big campaign contributions to sleazy politicians on the take. Yet, while such activities do occur, this is not an accurate picture of a lobbyist. Some observers define lobbying as "the systematic effort to affect public policy by influencing the views of policy-makers." Others describe lobbying as "any communication by someone acting on behalf of a group, directed at a government decision maker with the hope of influencing that person."[4] Many honest women and men represent interest groups—including women's rights organizations, environmentalists, labor unions, community organizations, and government reform groups. Lobbying is nothing more than a sophisticated exercise of citizens' constitutional rights to "petition government." Simply defined, **lobbying** is an organized effort to influence public policy on behalf of a particular interest area.

Lobbying can be done by **paid lobbyists**, full-time professionals who are hired by organizations to influence government legislation, regulations, or decisions. But many lobbyists

AN INSIDER'S VIEW

Lobbyists: The Hired Guns of Tallahassee

In 2004, the *Miami Herald* identified a number of top lobbyists of the 2,000 registered lobbyists of the Florida legislature and executive branch:

1. Al Cardenas is a partner in the firm Tew Cardenas Rebak Kellogg Lehman DeMaria Tague Raymond & Levine, which was founded in 1992. His major clients include Walt Disney Co., various hospitals and cruise lines, Bell South, and Florida Power and Light in addition to entire countries like Panama and the Dominican Republic. Cardenas was previously the Republican state party chair in Florida. He says, "We help [our clients] understand the nuances of Florida law." He also helps his clients to pass laws that favor them.

2. Van B. Poole is a partner of Poole, McKinley & Blosser, which was founded in 1993. His major clients include Auto Nation, Florida Marlins, Dell Computers, the Seminole Tribe of Florida, and the Motion Picture Association of America. He previously served as a state legislator in the Florida house, the Florida senate, and on Florida Governor Martinez's cabinet. Like Cardenas, he also headed the Florida Republican Party. He says of his lobbyist role now, "I'm just as effective now as when I was in office." He has lots of access at the state capitol from his contacts and fame from previous elected and appointed positions in government and the Republican Party.

3. Ronald L. Book founded his own lobbying firm in 1987. His firm represents an amazing 150 clients, including Associate Industries of Florida, Miami-Dade County, numerous cities, and Calder Race Course. He is widely recognized as one of the most influential lobbyists in Tallahassee.

4. Rodney Barreto, Courtney Cunningham, and Brian May of Barreto are three principals in the lobbying firm of Cunningham, May, Dudley, Maloy, which was founded in 2000. Their clients include the City of Weston, Zoological Society of Florida, Aarmark Correction Services, and Florida Insurance Council. Barreto is a former police officer and key fundraiser for former Miami-Dade County Mayor Alex Penelas and Florida

Governor Jeb Bush. Cunningham is a former aide to a county commissioner and Brian May is the former chief of staff for Mayor Penelas. They effectively lobby at both the state and local government level.

5. John Thrasher belongs to the Southern Strategy Group of Florida, which was founded in 1999. He has as his clients AOL Time Warner, AT&T, Blue Cross and Blue Shield, and the University of Miami. He was Republican speaker of the Florida house from 1999 to 2000. He says that beyond opening the door to key public officials, he and his firm "become advocates for our clients." Before he was elected to the house, he was general counsel for the Florida Medical Association for 20 years. Campaign fund-raising is an important part of his lobbying, as it is for all the most powerful lobbying firms. His company advises their clients on where to make campaign contributions so as to "provide support for [candidates] who philosophically agree with us," that is, those candidates with a conservative Republican philosophy.

6. George I. Platt of Shutts & Bowen joined his lobbying firm in 1998. He has as his clients Florida Wetlands Bank, WCI Communications, BellSouth, and Coral Ridge Country Club. He has close ties with the Broward County Democratic Party. Platt says, "the best lobbyists have the ability to persuade people through a variety of means: because of friendship and connections, their ability to raise votes because they represent business groups or unions, and their ability to raise money. Being an expert on an issue is also a factor."

7. On the other side from most of these lobbyists is Mark Ferrulo, who lobbies for the Florida Public Interest Research Group (PIRG). He argues: "White-hat lobbyists like me, who don't provide any financial carrots to legislators but just provide the public interest perspective, are almost always at a disadvantage. Without money, there's no influence or access."

Source: Adapted from Beatrice E. Garcia, "The Hired Guns of Tallahassee," *Miami Herald*, March 8, 2004.

---

are unpaid and do not register as official lobbyists under state and local laws. They simply organize the efforts of groups to promote certain government policies and to oppose others. They appear at hearings and meet privately with lawmakers, staff members, and government officials who can grant the group's requests.

A study of the most influential interest groups in state government found that business organizations, teachers' organizations, utility companies, lawyers, and health care organizations are consistently among the most powerful and effective. Such interest groups and their lobbyists principally did five things: they testified at legislative hearings, contacted government officials directly, helped to draft legislation, alerted state legislators as to the effects of bills on their districts, and organized influential constituents to contact legislators.[5] To better understand lobbyists and their efforts, see the Insider's View feature about lobbyists in Tallahassee, Florida.

In large states like Florida, New York, and Ohio, there may be several thousand registered lobbyists, while in smaller states like Alaska, Delaware, Hawaii, and Maine there may be only a few hundred. There are 112 registered lobbyists in the city of Miami; 212 in Cook County, Illinois (in which Chicago is located), and nearly 1,000 in New York City.[6] The requirements for registering as a lobbyist differ widely in each city, county, and state. These regulations as well as the size of a city or state and its budget account for the variation in the number of lobbyists who register.

For most of the 20th century, there were only a few dozen lobbyists at a state capitol and never more than a handful at a city hall or county building, since state and local government budgets were not large and the businesses that were regulated could simply approach the politicians directly without the intermediary of a lobbyist. Today, however, as political scientist Alan Rosenthal argues, "just as individuals turn to attorneys for help on legal and other matters and to financial consultants for counsel on investments, so they turn to other professionals, that is, to lobbyists when they require assistance in dealing with government."[7]

Rosenthal and other scholars of lobbying at the state and local level describe at least five different types of lobbyists:

1. *Contract lobbyists* who have a number of clients ranging from a half-dozen to more than 100.

2. *Association lobbyists* who represent a single group such as a labor union, chamber of commerce, or a manufacturer's organization.

3. *Company lobbyists* who work for a single business.

4. *Intergovernmental lobbyists* who work for single units of government or for associations composed of local governments such as cities or counties.

5. *Cause lobbyists* who are concerned with issues that have no direct commercial, material, or government interest for themselves or their organizations, such as the Sierra Club's concern for the environment or Common Cause's concern for political campaign reform.[8]

The *contract lobbyists* are most often former elected officials; former staff members of legislators, governors, or mayors; or active political party leaders. They develop access over time and use their access with public officials—in both the legislative and executive branch at both the state and local government level—to promote the interests of their clients.

Association and company lobbyists are less likely to have such strong political backgrounds or be former government officials. They usually come from other positions from within a union, company, government, or association although they now specialize in public or governmental affairs. For instance, in 2005 the Silicon Valley Leadership Group, representing 190 firms and employing 250,000 people, presented their proposals to state legislators in California for maintaining economic growth in the region. Specific legislation for which they lobbied included an exemption for local manufacturing and research firms from paying sales taxes on certain equipment, and a $5 per vehicle assessment to pay for projects to relieve traffic congestion.[9]

Intergovernmental lobbyists also specialize in public affairs but represent state and local governments and government officials. Intergovernmental lobbyists can represent generalist groups, such as the National Governors Association or the National Council of State Legislators. In addition, specialist groups or public official associations exist for nearly every field that operates in government, such as school superintendents or finance officers. Most individual states and large cities lobby Washington, and cities and other local government entities lobby the state government. Unlike other lobbyists, state and local government officials in the intergovernmental lobby coalition cannot give campaign contributions, but they have an advantage in that they often represent many of the same voters as the people they are lobbying.[10]

Cause lobbyists come from within movements, such as the environmental, government reform, civil rights, or women's rights movements. Cause lobbyists may be either volunteers or paid. Sometimes a number of cause movements join together to hire a single lobbyist, like Mark Ferrulo, the Public Interest Research Group lobbyist described in the Insider's View feature on Tallahassee lobbyists.

As with governments themselves, the demographic complexion of lobbyists is changing. One California study found that before 1972 there were no women lobbyists, but by 1983 one-third of the new lobby registrants were women. The number and percentage of women lobbyists has continued to increase at a rate that corresponds to the growing number of women in high levels of government. Women-owned lobbying firms are also proliferating.[11] Essentially, as more women become elected officials, government staff members, and party leaders, they move over to lobbying as they retire from their government or party positions, just as their male counterparts do. There are also more African Americans and Latinos lobbying effectively at city hall, the county building, and the state capitol. Three of the top eight lobbyists in Florida listed in the Insider's View feature on Tallahassee lobbyists are African American or Latino, for example.

There is no course of study to prepare someone to become a lobbyist, no degree offered in lobbying at universities, and most people in the field don't start out to have lobbying as their career. They begin as public officials, political activists, leaders in their association or company, or as people in a movement. As they learn the craft of politics and the workings of government from the legislative to the executive branch, they develop the skills and contacts that allow them to become lobbyists and to get paid for their work. Successful lobbyists can earn millions of dollars, depending on the clients they represent. And millions of dollars are spent on lobbying activities each year in each state. Table 8.1 shows how much money is spent on lobbying within each state, as well as the number of registered lobbyists.

Lobbyists exist, of course, because they represent businesses or interest groups. According to one study, the 12 most powerful interests at the state level are:

1. General business organizations.
2. Schoolteachers' organizations.
3. Utility companies and associations.
4. Lawyers.
5. Health care organizations.
6. Insurance companies.
7. General local government organizations.
8. Manufacturers.

9. General farm organizations.

10. Doctors and hospitals.

11. State and local government employees.

12. Traditional labor associations.[12]

The most significant interest groups, like businesses and teachers, are among the most important groups in 40 or more states. On the other hand, some groups like state and local government employees and traditional labor unions are seen as the most important groups in as few as 15 states. But all of these groups have more clout than groups without a legislative agenda and paid contract or in-house lobbyists.

TABLE 8.1 Lobbying in the States

State	Lobby Spending 2006	Number of Registered Lobbyists 2004	State	Lobby Spending 2006	Number of Registered Lobbyists 2004
Alabama	No Total	603	Montana	$6,924,175	536
Alaska	$26, 272, 631	342	Nebraska	$9,993,827	345
Arizona	$2,602,866	800	Nevada	N/A	N/A
Arkansas	N/A	452	New Hampshire	N/A	227
California	$271,680,365	1267	New Jersey	$55,321,166	935
Colorado	$24,396,668	670	New Mexico	N/A	855
Connecticut	$38,419,882	635	New York	$151,000,000	5117
Delaware	$136,200	270	North Carolina	$14,146,337	726
Florida	$121,760,708	2029	North Dakota	$1,875	154
Georgia	$1,202,269	1506	Ohio	$349,417	1401
Hawaii	$4,413,155	312	Oklahoma	$161,652	362
Idaho	$869,664	392	Oregon	$16,148,614	629
Illinois	$1,279,213	2195	Pennsylvania	$54,090,812	355
Indiana	$21,999,739	684	Rhode Island	N/A	201
Iowa	$7,802,064	849	South Carolina	$19,815,024	385
Kansas	$938,745	560	South Dakota	N/A	625
Kentucky	$14,424,699	653	Tennessee	$10,000	512
Louisiana	$1,113,298	531	Texas	$120,215,500	1489
Maine	$3,227,761	279	Utah	$228,668	351
Maryland	$37,085,356	637	Vermont	$5,943,594	425
Massachusetts	$78,960,743	569	Virginia	$15,367,800	987
Michigan	$,22,692,687	1283	Washington	$38,717,055	964
Minnesota	$53,287,186	1385	West Virginia	$329,267	458
Mississippi	$17,697,439	422	Wisconsin	$26,826,964	815
Missouri	$1,074,258	1116	Wyoming	$159,325	365

Source: Reprinted by permission of The Center for Public Integrity.

Lobbyists make money lobbying the executive branch and separate government agencies as well. For instance, Illinois lobbyist Robert Kjellander served for years on the Republican National Committee and was selected as their treasurer in 2005. He received $4.5 million in fees from the Carlyle Group, a global investment firm, as a commission for getting them $500 million in investments from the Illinois Teacher Retirement System pension program.[13]

In what became known as the "Tennessee Waltz corruption scandal," several Tennessee state legislators were arrested in 2005 and charged with taking bribes to pass favorable legislation. They had been caught in an undercover FBI sting operation intended to uncover public officials corrupted by lobbyists wishing to do business with the state legislature. The FBI had formed a fake company and passed around payments to ostensibly make it easier for the company to do business with state government agencies. Seven legislators were caught accepting over $90,000 in bribes.[14]

How to Lobby the Legislature

Corruption is not the norm for lobbyists. Most lobbyists simply help businesses and organizations navigate the maze of state and local government. Some lobbyists defend their work as providing marketing tools for smaller businesses that would otherwise have to spend the money on a sales department.

So when they are honest and there aren't big bribes or illegal actions, how do lobbyists influence state and local government for their clients? See the Insider's View feature for some advice from a professional state lobbyist.

What do lobbyists do in their efforts to pass and defeat legislation, including city, county, and state budgets? These are lobbyists' eight most important tasks, according to a survey of lobbyists and organizations:

1. Testifying at legislative hearings.
2. Contacting government. officials directly.
3. Helping to draft legislation.
4. Alerting state legislators to the effects of a bill on their districts.
5. Having influential constituents contact a legislator's office.
6. Consulting with government official to plan legislative strategy.
7. Attempting to shape implementation of policies adopted by the legislature.
8. Mounting grassroots lobbying efforts.[15]

Lobbyists can't do these tasks unless they monitor the legislative process closely and effectively. They have to know what bills, amendments, and budgets have been introduced, who is sponsoring and pushing them, the effect of such legislation on the interest group they represent, and the likelihood of each bill passing or being defeated. Some of this preliminary work can be done through a computer and the Internet. For instance, the government Web sites of most states and many cities will have a link to a "Daily Calendar" that shows legislation introduced, legislation up for a vote, and scheduled committee hearings. Many government Web sites also link to a summary or text of all legislation that has been introduced.

A lobbyist must also meet with legislators face-to-face, either informally in the lobby of the state capitol when the legislature is in session or formally in a scheduled appointment in the legislator's office. Generally the meetings are 10–20 minutes. If lobbyists need to impress

Advice from a Professional Lobbyist

In Illinois in the 1990s, Mothers Against Drunk Driving (MADD) wanted the state legislature to toughen laws against drunk driving by reducing the blood alcohol content limit from 1.0 to .08. According to Don Udstuen, retired lobbyist for the Illinois State Medical Society, here is what to do if you are lobbying in support of such legislation:

Find your champions—In lobbyist terms, a champion is someone who completely agrees with and is willing to fight for your cause. In this case, you need to find two types of champions.

Go high-profile first—The first champion you want to support your legislation is a high-profile lawmaker in an elected position of power, such as the secretary of state or the governor. Having the backing of a high-profile politician is a must because the media are much more willing to follow an issue that has the backing of a strong politician. Also, lawmakers who may be wavering on the issue could possibly support the decision of someone in power.

Go to the trenches—The next type of champion you will need is a lawmaker, or two, who will fight for you in the trenches, the floor of the legislature. Since you are already an established lobbyist, you should have a pretty good idea who will strongly support you and who will strongly oppose you; it is those who are wavering on the issue that must be convinced. Since this bill will first go through committee, more than likely the Transportation Committee since it has to do with driving regulations, you will need to identify which members of that committee are willing to support you as well. Committee members are the most important in this stage of the process. If the bill doesn't get out of committee, it will never see the floor of the Senate or the House.

Start at the local level—Due to the strength of the opposition, efforts must start at the local level. Without local support, the political system will knock you and your bill out. Furthermore, if you begin by putting this bill in committee, it will surely die a "mysterious death." Therefore, you must activate your base at the local level.

Use local meetings:—By using MADD leadership and membership, organize local meetings all over the state. Invite all local and pertinent state officials, including your champions, to these meetings so that they know there is local support for this bill. If a politician knows a majority of his or her constituency is in favor of a certain bill, he or she will typically vote in favor as well. At these meetings, you will need to use local "horror stories" in order to put a face to the effects of drunken driving. This will increase sympathy for your cause. These horror stories include any person or persons who have been affected by drunken driving. MADD will have the resources needed to find the proper horror stories in each local area of the state.

Use public opinion—For controversial issues like this one in which there is major opposition, you must also line up public opinion on your side. Build a strong profile for this issue by meeting with state and local media editorial boards. If the editorial boards can be persuaded to back your issue and therefore write about it, public opinion can be swayed your direction. With the combination of public opinion and this issue becoming a "hot issue," legislators will be forced to back this issue due to pressure at home.

Use the power of numbers—As a next step in the lobbying process, put together a day trip to the state capitol with around 1,000 local supporters of this bill. This will show there is strong support for this bill, but it will also clog up the corridors of the state capitol with people, forcing legislators to take notice of the bill. Give all your supporters large buttons saying, ".08 Saves Lives" so that everyone supporting you can be easily recognized by the legislators.

In the capitol—Using your legislative champions in the Transportation Committee, especially the committee chairperson if you can win him or her over, you will succeed in getting your bill to the floor of the House of Representatives. At this point, you need to educate legislators who support you so that they will be able to argue successfully for you on the floor.

Ensure no amendments—Do not allow other legislators to start putting amendments on the bill. For example, an opponent to the bill may try to lower the legal limit even further to .02. At face value this looks like a good idea, but even most proponents will not vote in favor of an extreme such as .02.

Go back for support—You now need to go back for more public and media support, especially in legislative districts where the legislator is still undecided. At this point, you have become a choreographer of a musical, telling everyone where their cues are, and how to dance once they get there.

Organize more big grassroots lobbying days at the state capitol, bringing thousands of supporters to clog up the corridors once again. Continue to remind people that this is not an anti-drinking bill; this is an anti-drinking *and driving* bill.

To the Senate—Once the bill passes successfully in the House, it will enter committee in the Senate. The entire process starts from scratch now.

Source: Adapted from an interview by University of Illinois student Andrew Flach with Don Udstuen.

a legislator, they bring along two or three influential members of the group they represent from the legislator's district. Most often, lobbyists simply discuss the legislation that concerns them and ask for the legislator's support in passing or defeating it. And of course, lobbyists will ask what else is happening in the legislature that might affect their concerns.[16]

It is usually impossible for a lobbyist to contact the entire city council, county board, or state legislature unless these have few members. In larger city councils and in state legislatures, lobbyists concentrate especially on making contact with legislative leaders (the speaker, president, or party floor leaders) and the heads of the committees to which the pertinent legislation has been assigned. The leaders of the legislative body and the committee chairs control the procedures by which legislation will get a hearing and be killed or passed.

In the Chicago city council, legislation is usually not called up for a hearing unless the mayor wants it to pass. The process of forcing a vote on the issue without it first having a committee hearing requires a two-thirds majority vote by the entire council, so unless a committee hearing is called, the legislation is effectively dead. In less hierarchal political environments, all legislation gets a hearing, has a chance to be perfected in committee debate, and gets to the floor for a vote even if the committee recommends voting against it. However, if a governor, county board president, or mayor wants particular legislation to pass, the odds of it getting a timely hearing, a good committee recommendation, and passing are infinitely better. Conversely, legislative leadership wanting to kill legislation can do so in hundreds of ways. So most lobbyists begin with a top-down strategy of obtaining the help of the legislative leaders.

Nearly all legislation that is going to pass must go through a committee process, which is why lobbyists list testifying at committee hearings as one of their primary duties. The most important aspect of testifying is to have your lobbying message, just as in an election campaign one of the important things is to develop a powerful campaign message or theme.

As Mooney and Dyke-Brown advise in their lobbying handbook: "The most important part of preparing your testimony is defining your lobbying message. Identify *a single sentence with no comma* that sums up your arguments. For example, HB2365 will improve health care for the rural poor in your state. Or SB 583 will cause some small businesses to close. Use this lobbying message as the theme for organizing the rest of your testimony." They further recommend demonstrating the impact of the bill on the people, businesses, or groups from the committee members' districts, and that your testimony should be written as well as verbal. For the oral testimony, they recommend:

1. Don't read prepared testimony; just speak directly about the virtues or faults of the legislation.

2. Keep your testimony brief and to the point, ideally less than five minutes. Supplement with written materials and personal lobbying of committee members.

3. Be sincere, open, and honest. Be yourself. Don't posture, become emotional, or engage in melodramatics.

4. Be novel in your approach. Use posters or other visual aids; conduct a brief demonstration of an activity.

Chapter 8 Citizen Participation in State and Local Government **197**

5. Remember that if you haven't done the preparatory legwork before the committee hearing, even the finest oral testimony is unlikely to change anyone's mind.[17]

Normally, neither testimony in committee nor debate on the floor changes many votes. Votes are obtained by getting a good sponsor for the legislation, lining up group and public support, showing that the legislation will either help or hurt constituents in legislator's districts, and by getting a commitment from each legislator on the committee on the legislation before the hearing is held.

It is important for lobbyists to know the legislators they hope to lobby successfully. Previously the best opportunities were social rather than formal meetings in legislator's offices. Both legislators and lobbyists hang out in state capitol cities like Austin, Texas; Tallahassee, Florida; or Albany, New York. They are far from home, staying either in hotels or rented apartments. In many of the smaller cities that comprise most state capitals, it is natural for legislators and lobbyists to go to dinner together or out for drinks after a legislative session. After all, lobbyists are very interested in politics and government just as the public officials are. In cities, lobbyists lobby city council members or city officials in the same place where everyone lives year-round. However, there are still lots of restaurants or bars for lunch or dinner. Traditionally, the lobbyists picked up the check for the meal and drinks—but with new lobbying regulations forbidding gifts of anything of value, or at least requiring all expenses for meals and gifts to be reported publicly, this is less often the case. Wining and dining public officials is no longer the chief method used by lobbyists. All free lobbyist-paid hunting trips or vacations for legislators are also out.

Official parties, especially campaign fundraisers held by legislators, mayors, and party officials—where the price of admission ranges from $75 to several thousand dollars—are one sanctioned way for lobbyists and organization leaders to meet public officials and curry their favor through campaign contributions. Of course, the socializing at receptions is less useful than the private dinners or drinks after legislative sessions, since so many people mill around and contacts are brief. But at the campaign fundraisers and official receptions, lobbyists get social contact with the legislators they will soon visit to ask for support. Officials who get large campaign contributions from lobbyists will not forget them. Their phone calls will be answered promptly, and they will have an easy time scheduling meetings with legislators.

Lobbyist professional associations set forth ethical standards for lobbying. They include straightforward rules like not to lie and not to bribe public officials. The American League of Lobbyists has among their rules lobbying with honesty and integrity, complying with all laws, and avoiding conflicts of interests among their clients (such as representing clients on opposite sides of the same issue).[18] The American Society of Association Executives' "Standard of Conduct" includes the admonition that while keeping the clients interest primary, lobbyists should have a proper consideration for the general public interest and protect confidences of clients and elected and appointed officials.[19] For a more detailed set of practical rules, see the Insider's View feature on the ten commandments of lobbying.

Grassroots Lobbying

Beyond making contact with legislators, developing relationships with them, and asking for their support, and beyond any formal testimony at hearings, there is often a need for a grassroots lobbying campaign to gain passage of significant legislation. **Grassroots lobbying**

AN INSIDER'S VIEW

The Ten Commandments of Lobbying

1. Never lie or mislead a legislator about the facts of an issue, its relative importance, the opposition's position or strength, or any other matter.

2. Look for friends in unusual places. In politics, a friend is someone who works with you on a particular issue, whether Democrat or Republican, liberal or conservative. A friend on one issue may oppose you on every other issue.

3. Never cut off anyone from permanent contact. Don't let a legislator (or another lobbyist) consider you a bitter enemy just because you disagree. Today's opponent may be tomorrow's ally.

4. Don't grab credit. What you and your group want from the process is public policy in line with your interests. Legislators and others may want the public credit. Let them have it. Nothing is impossible if it doesn't matter who gets the credit.

5. Make your word your bond. Don't make promises you aren't positive you can keep.

6. Don't lobby opponents who are publicly committed to their positions. It wastes your time, and it alienates them further. It is more productive to support your allies and lobby legislators who claim to be keeping an open mind.

7. Always notice and thank everyone who has helped you. People like to be appreciated and it costs nothing to say "Thanks!" A person who feels unappreciated will probably not help you again and may even go out of his or her way to hurt you.

8. Don't gossip. Knowing legislators' peculiarities and peccadilloes is one thing; talking about them is another. If you get the reputation of telling everything you know, you'll soon find that no one will tell you anything.

9. Do your homework. There is no excuse for not having the facts and figures to support your case when you need them. It makes you look unprofessional and reduces your credibility.

10. Be there. You can know your opponent; you can develop imaginative and reasonable compromises, you can burn the midnight oil to digest all the arguments. But you have to be in the right place at the right time to win the day.

Source: Christopher Mooney and Barbara Van Dyke-Brown, *Lobbying Illinois* (Springfield: University of Illinois at Springfield, 2003), 41, adapted from the Illinois State Support Center, *Manual of Public Interest Lobbying in Illinois*, Springfield, Illinois, 1984, 7.

is lobbying that arises (or appears to arise) from the concerns of many individuals rather than being undertaken by representatives of an organization. A traditional method of indicating grassroots support for an issue is a letter-writing campaign in which legislators are deluged with mail from their constituents urging them to vote for or against a particular piece of legislation. Or voters are urged to telephone the legislator's office about a particular piece of legislation.

More and more often, grassroots lobbying campaigns are coordinated through an e-mail alert sent to all members of a group such as a labor union, industry executives and their suppliers, or cause groups. Some groups like MoveOn.org specialize in notifying their supporters by e-mail blasts, asking them to send money to buy television ads, to call particular legislators, or simply click a particular button to e-mail a letter to their congressional representatives or the president. This can create a flood of e-mail messages overnight. Better yet, an e-mail blast notifying supporters can create hundreds of phone calls to key legislators, council members, or commissioners. But at the state and local level, this is not usually the most effective technique, because it is not targeted to legislators whose minds are not made up. In addition, if these e-mails are impersonal form letters, legislators give them less weight as a measure of public opinion.

AN INSIDER'S VIEW

Lobbying in Miami

In the 1990s, Humberto Hernandez was Miami city attorney and later a city commissioner. He eventually was convicted and sentenced to jail on charges of vote fraud. However, as city attorney he helped negotiate a $271 million garbage recycling contract and as commissioner he passed an ordinance to give the winning firm a two-year extension to build a compost plant to handle the garbage. The 30-year contract went to a New Jersey firm, Bedminster Seacor Services Miami Corp. It was represented by a team of lobbyists (earning $2 million for their services) including Maurice Ferre, a former mayor with extensive city hall contracts and lobbyist Julio Gonzalez Rebull Sr., who was a close friend of many Miami city commissioners. When Ferre ran for a seat on the county commission, he turned over his lobbying efforts to his sons.

Xavier Suarez, who was mayor of Miami at the time the contract was voted on, later suggested there was a better way to select firms for city contracts: "In an ideal world, we would have a lot less lobbying, certainly of this [Miami City] commission, on behalf of all these groups."

The contract negotiated that year was a very good deal for Bedminster. Under its original terms, the city would pay Bedminster to dispose of 204,000 tons of waste—even if the city did not produce that much garbage. The small print of the contract virtually guaranteed that the city would come up short. Bedminster would not have to accept any "tree stumps or limbs over two inches in diameter." But the city said it had more than 56,000 tons of yard waste every year. A good deal of it is larger than two inches.

A retired county executive, Dennis Carter, says he did his damndest to kill this contract. Joe Pinon, an assistant city manager who later reviewed the contract, concluded: "This is not the best contract . . . how the hell did we get into this?" The journalists who broke this story concluded: "Lobbyists often play key roles in shaping public policy to fit their clients' needs, capitalizing on their relationships with the politicians they helped elect. Taxpayer needs come second."

Source: Manny Garcia and Tom Dubocq, "Political Ties: How Miami's Garbage Contract Could Cost Millions Extra," *Miami Herald*, December 9, 1997; reprinted in Dick Simpson, ed., *Inside Urban Politics: Voices from America's Cities and Suburbs*, New York: Pearson Longman, 2004), 205–9.

Letters, phone calls, e-mails, and visits from constituents do have some impact. Legislators understand that they were elected to represent their constituents and that their reelection depends on keeping their constituents satisfied. As a Chicago alderman, author Dick Simpson never received more than five letters or phone calls on any piece of legislation pending in the city council. He was able to develop a sense of the opinions of his constituents and their problems by attending community meetings and holding his own representative ward assembly meetings. Nevertheless, every constituent letter and phone call counted and was considered in proposing or voting on city legislation.

The best grassroots lobbying technique by far is a delegation of concerned citizens who travel to a legislator's office to talk about a particular issue or proposed legislation. Most effective of all is to meet legislators in their district office or have them come to a meeting in the community. The face-to-face contact with their own voters and contributors helps to gain the support of legislators even if wealthy and powerful lobbyists are working against you. Beyond that, most people find meeting with their own legislator to be an exciting and personally rewarding experience, especially when at the end of the meeting the legislator vows to support them.

Legislators as Lobbyists

Another general rule of lobbying for any interest or group is that the best lobbyist is a legislator. This is most easily seen in the case of the legislator's own legislative body. A group that can convince a city council member to support its cause has a huge advantage, since the friendly city council member can introduce and sponsor the legislation, convince fellow council members and maybe the council leaders or mayor to cosponsor it, notify the group of when hearings will be, help plan testimony, provide a list of the aldermen who have to be convinced to vote for the legislation if it is to pass, and handle the legislation on the floor of the city council. Council members and state legislators can do this better than most outside lobbyists because of their knowledge and their floor privileges.[20]

Lest you think all major lobbying is done at the national and state level, the Insider's View feature describes how a team of well-connected lobbyists and former public officials were able to push through a 30-year garbage recycling contract which appears to have had greater advantages for their client than for Miami and its taxpayers.

JOURNALISTS

Journalists play a different role from lobbyists, but like lobbyists, they can influence elected officials. Journalists who most often play a politically influential role are investigative journalists and political reporters. There is, for example, the well-known story of Bob Woodward and Carl Bernstein at the *Washington Post* who in 1972 broke the story that a burglary at the Democratic Party headquarters at the Watergate Hotel in Washington, D.C., had been supervised by White House aides to President Richard Nixon, who was facing a reelection campaign. The reporters' stories on the intrigue and "cover-up" led to the resignation of the president.[21]

Political reporters and investigative journalists believe that an important part of their job is to uncover corruption and improper actions on the part of public officials. Another task of the news media is to tell us what is going on in election campaigns—who the candidates are and what they stand for, and to report on what laws and regulations are being considered that might have profound impacts on our lives. Often the reports of journalists can cause candidates to win or lose elections, laws to pass or fail, or public officials to go to jail. Journalists provide us the knowledge we need to be effective citizens. They are among our watchdogs of government.

Pressures on the Press

Harris Meyer, former editor of the *Daily Business Review*, published in Miami, Florida, points out that the press can't really affect politics and government decisions unless there is a constituency.[22] For example, President George W. Bush's Social Security privatization proposals in 2004 and 2005 were effectively covered by the mass media because retirees and the elderly are a powerful constituency. They buy newspapers and watch television news programs, and they have a high voter turnout rate at elections. But for many problems and policies, there isn't a constituency. One example is the difference in reactions to the stories which Meyer's paper did on voting machines and the problems of touch-screen voting machines that leave no paper record to allow pollwatchers and others to check the accuracy of the vote count. His paper presented evidence that votes were miscounted in several previous elections in Dade and Broward Counties. In Miami, with a new reform mayor, the city switched to scanned ballots, which allow for recounts and even alert voters if they have marked too many names in a ballot category or not enough. Adjacent Broward County, which has less of a reform movement, kept the same flawed voting system despite the demonstrated problems. Thus, the attitudes of the politicians in office and the degree to which there is an active and powerful constituency can determine which issues the media will cover well and whether anything will be done by the government to respond to the information that the media provides.

Another problem is that many local newspapers or other smaller market media operations cannot afford the specialists needed to cover complicated issues. Instead, they count on their generalist reporters to find and understand often highly specialized information. In Meyer's *Daily Business Review*, reporters are each expected to know five different industries. This is a heavy demand. For example, the airplane and auto industries are vastly different despite both being transportation industries. In covering complicated issues like health policy or Social Security, it takes reporters a long time to learn the issues. Few nonmedical publications have a specialist to write knowledgeably on cancer, for example, or advanced gene research.

It takes time for reporters to learn their "beat" well enough to add significantly to the public's knowledge of important and complicated issues. A lot of expertise is needed to cover state and local politics well. A reporter has to spend the time to learn the major players in government and to learn what has happened in the city, county, or state in the past before he or she can accurately describe the politics and government today.

In addition to specialized or in-depth knowledge, if we are to get better journalism we have to have better journalists, which requires that they be paid better. It requires more education and knowledge to cover stories now. Beyond education and pay scale, journalism has to be seen as an honorable profession and as one that protects our democracy by making sure that the public knows what is happening in politics and government. Currently, business leaders are much more highly regarded than journalists and much better paid.

Another difficulty is the lack of sufficient space and time in the media for covering real stories. TV news, in particular, follows an "if it bleeds, it leads" philosophy. So there is an overreporting of disasters and crime and an underreporting of what is happening at city hall, the county building, and the statehouse. There is generally only one local political or governmental story a night on the television news programs. Anybody trying to make news to bring public support to their cause is competing against the mayor, the governor, and all the other elected officials.

Complicating matters, many in the media have a herd mentality. A feeding frenzy develops especially when scandals occur. There comes a point in a story when almost every TV news reporter seems to be reading from the same teleprompter. This can make some issues seem one-sided or oversimplified. It also means that some stories are neglected, unless all the media are following them.

There are fewer reporters in the newsrooms than there used to be. Sometimes community and small-town newspapers have only a single reporter to cover political, governmental, and community stories. Economic conditions, the role of the Internet, and economic recession have caused many newspapers to fold or declare bankruptcy. Among other things, this has caused newspapers to close newsrooms at the statehouse. The number of statehouse reporters covering state government as a full-time beat declined from 524 in 2003 to 355 in 2009, a 30 percent reduction, with 44 states reporting a net decline.[23]

There have been changes in news coverage generally as well. Now, presidential candidates may get only 20-second sound bites in the middle of a campaign, but other public officials and candidates get a lot less coverage. It is much harder to get news coverage of substantive issues now than it was 30 years ago. In the old days, when Dick Simpson was a

Chicago alderman, he could just walk down the hall to the press room, criticize the mayor, and make news. For community groups, civic organizations, and even most alderman, that strategy won't work anymore.

In addition to all the other problems with the media, news reporters have their own distinctive set of values and codes, and many socialize mostly with other reporters. This is particularly true of reporters at the city or county building press rooms or state capitol press rooms. Political reporters tend to develop a cynicism toward government and a disdain for political officials, which shows up in their stories. As former journalist Mark Magyar has noted: "When [reporters] leave journalism to take a job in public relations because they cannot raise a family on a reporter's salary, they are derided for 'selling out.' . . . When reporters leave for a job in government they are said to be 'going over to the Dark Side.'"[24] These attitudes get conveyed to their readers, listeners, or watchers and lowers the public's view of politics and government.

If state and city politics is hard, news coverage of suburban government is particularly difficult to do well. For instance, the *Miami Herald* and the *Sun Sentential* in the Miami metropolitan area often assign reporters to the suburban beats who aren't experienced enough to provide very good coverage. Cities have hundreds of suburbs, making it hard to cover the news in so many different towns and villages. So citizens don't know much of what is going on in their own communities. If the press in the major cities have trouble covering the news, it is even less likely that the smaller towns and suburbs are going to be well covered.

Libel suits have also become a problem for the press, such as the lawsuit filed against the *Chicago Tribune* by a local county states' attorney. The *Tribune* won the lawsuit, but at a cost of millions of dollars, which most media outlets can't afford and seek to avoid. The previous judicial standard of libel was reckless or premeditated distortion, but lawsuits are beginning to be won without proof of such distortions. The threat of libel lawsuits deters aggressive press coverage of politics and government. Previously, it was almost impossible to libel a public official, but the courts are now beginning to uphold some libel claims by public officials against news media. In response to the threat of such libel suits, the media have become more cautious in criticizing public officials.

The press used to be seen as the fourth estate or the fourth branch of government after the executive, legislative, and judicial branches. That doesn't seem to be as true anymore. The press is not as much of a check on government as it has been in critical periods of our history. Without an effective press, the public doesn't have the information to act and hold public official accountable. Nonetheless, journalists are one set of actors besides public officials who have the potential power to affect politics and government profoundly.

Bloggers

Bloggers, the people who write blogs, or Web logs, have been suggested as an answer to the problems the mainstream media have in covering government and politics. Blogs are online journals or diaries that usually present the opinions of one person. While bloggers sometimes break news stories, they often repeat gossip, and they are not held as accountable for their content as the mainstream news media are. They often do not have an editor or fact checker to determine accuracy and whether the story should run. Nevertheless, there has been a great profusion of blogs, and the mainstream media now frequently get news leads and

In the Pink Texas

According to its Web page in 2005:

In the pink is an independent and insightful commentary on what's going down and who's moving up in the Lege [Legislature]. The trick to generating buzz? Make people laugh while making them think. We plan to post several times a day, bathing in the beauty that is free wireless—from the Texas Capitol to local watering holes (hey—sometimes blogging goes better with beer).

Traditional media outlets can't print the kind of stuff we're talking about—and they can't post as fast as us because of those pesky restrictions like "the editorial process," "fact checking," and "journalistic ethics."

In the pink is rabidly nonpartisan—we spread the love to both sides of the aisle. R's and D's, men and women, rural, urban, boxers, briefs—it's only fair.

Who writes this?

Well, we were going to be anonymous but that's already taken . . .

In the pink is written by Eileen Smith, a professionally trained journalist gone bad. She earned her master's degree from Northwestern University. After working for a daily wire news service, followed by a brief stint at a suburban daily reporting on park district meetings and speeding tickets, Eileen covered techno news for PC World, attending conferences in Las Vegas with a bunch of yahoos. From there, she hitched a ride on the dot-com gravy train, writing content and counting stock options which, sadly, never came to fruition. (Damn you, beanbag chairs and empty promises!)

This landed her in the safe nest of state government—the Texas Legislature, where Eileen worked as an analyst for the House Committee on Appropriations, researching healthcare policy and, more importantly, observing the political animal that we call the Lege. After working in media relations and governmental affairs, Eileen decided to try her hand at political blogging, making next to nothing (OK, nothing).

So there's my story. It's sad but true. What's yours? E-mail me at Eileen@inthepinktexas.com. Let's open up the Lege and see what's inside.

Source: In the Pink Texas Web site, www.inthepinktexas.com.

stories from bloggers who cover the national government, the state capitol, and city hall. Even some legislators, like Texas state legislator Aaron Pena, for example, have begun blogs on their Web sites to keep constituents better informed.[25] More common are official blogs put out during election campaigns by candidates about their experiences "on the road" or their latest policy stands.

Often blogs simply summarize the news from various sources and bloggers put their own slant on it. According to one news story about blogs, "Texas bloggers haven't blown open any big scandals. But they do scoop the statehouse press corps from time to time. A few are winning over a sizable audience with their witty writing and their blunt, opinionated analysis."[26] For an example of these Texas bloggers, see the Insider's View feature about www.inthepinktexas.com.

Bloggers are forcing the mainstream media to adapt. The *Austin-Statesman* in 2005 launched its own blog on the legislature called "Postcards from the Lege," to which five of their reporters contribute. The *Chicago Sun-Times* in 2006 began a new Sunday newspaper section called "Controversy," which in addition to the usual editorials and op-ed pieces included stories from blogs and Web sites around the country on various topics. Some of their regular op-eds are now taken from blogs and e-magazines.

With blogs anyone with writing talent can vent their political views and share important political and governmental information that is undetected or undisclosed by the mainstream media. It is also now common for regular newspaper reporters to blog on their newspaper's Web site, which does have the advantage of letting them cover stories in more depth than they can publish in the paper version of the newspaper. That allows citizens who want more coverage by regular journalists to get more information than in the traditional newspaper format. However, this hardly compensates for all the news organizations that have gone out of business and the fewer investigative and political reports overall.

CITIZEN ACTIVISTS

In a democracy, not only elected officials, appointed bureaucrats, lobbyists, and journalists are important. Citizens matter. But they don't matter as much if they are not organized. And it takes citizen activists to organize them into a formidable group whose demands must be taken seriously.

Many different kinds of citizen-based organizations can have an impact on state and local government. There seem to be at least four basic types: civic groups, interest groups, social movements, and community organizations. Their form of organization, their issues, and their methods differ vastly.

Community activists, who play a very different role from lobbyists and journalists, come in all colors, shapes, and sizes. Very often they become community leaders, involved in politics and government, because of decisions that would negatively affect their neighborhood. One community activist, Florence Scala, was born on the west side of Chicago near the famous Hull House, one of the first settlement houses in America. Scala's father was a humble tailor in this very poor neighborhood. Her mother, who also worked long hours in the tailoring shop, sent her to Hull House for classes and recreation. In her teenage years, Scala became a Hull House volunteer.

In the early 1960s, Mayor Richard J. Daley decided to move the campus of the Chicago branch of the University of Illinois from its old, inadequate site at Navy Pier to Scala's neighborhood just west of the downtown Loop. However, building a new university required destruction of several blocks of small businesses and modest neighborhood homes and apartment buildings. One day in 1961, little kids went door to door asking Scala and her neighbors to attend a community meeting to protest the city's plan to demolish the neighborhood. Soon Scala was chosen as the spokesperson for the neighbors trying to change this decision at city hall. She and the neighborhood fought hard, and when they lost, many longtime residents were displaced to make room for the new college campus. Florence Scala continued to be a fighter and a spokesperson for her neighborhood all her life. She died a community hero in 2007.[27]

Another example of community activism occurred when the citizens of Yuba County, California, organized to keep open Beale Air Force Base, which was the largest employer in the area. Community leaders believed that the base would have closed, had citizens not done what might have seemed to be sophomoric activities, such as send to policy-makers an aerial photo of citizens spelling out "Save Beale!" on a football field. They also sent out over 30,000 letters to the governor, and they won their battle to save the base.[28]

Civic Groups

In a major city, civic groups include such organizations as the League of Women Voters, the Parent Teachers Association, Businesspeople in the Public Interest, the Civic Federation, Common Cause, the Better Government Association, the City Club, Kiwanis, and the

Metropolitan Planning Council. **Civic groups** are organized local groups who see their job as promoting the "public interest." They may specialize in elections, government accountability, lawsuits, public policy recommendations, government budget and tax policies, corruption investigations, or future planning for the metropolitan region. Some clubs like the Lions or Kiwanis raise funds to undertake their own charity work in the community, like providing help for the blind or for sick children.

Most frequently, these civic groups have a membership from several hundred to several thousand, although they are primarily governed by their paid staff and official governing board elected by the members. They have meetings and fund-raising banquets a few times a year. In between these meetings, they pursue their particular strategies to improve government and the lives of individual citizens. Sometimes civic groups will join together in a coalition to press for legislation or government policy change in a particular area of concern. Joining in a coalition gives them greater numbers through their collective membership, shared financing of the project, and shared staff to direct the project. A coalition, however, cannot usually take radical action or champion radical proposals for change.

It is easy to join one of these civic organizations, by paying nominal yearly dues and attending an occasional meeting. They tend not to demand much in the way of time and money from their members. But by the collective effort of their members, guided by a paid staff and board of directors, they can have considerable impact. This is especially true in specific policy arenas like schools, housing, or local government taxes. Civic organizations are viewed positively for the most part by journalists and governmental officials. Their collective opinions are heard and can bring policy changes. While they usually don't have the clout to enact fundamental changes, with a fairly minimal individual membership commitment they are able to help towns, cities, and states make improvements. Since they tend to exist for a very long time, their generally positive effects are cumulative. Some of the leaders also go on to have political and governmental careers, carrying out some of the policies and programs developed from their civic leadership days.

Interest Groups

An **interest group** is an organized group that tries to influence policy affecting the group's particular area of interest. Interest groups thus differ from civic groups in that they are usually concerned about their own "special interest." The term "special interests" has a negative connotation, but most people don't regard the group to which they belong as a special interest. Students who organize to lobby for lower tuition or better food in the cafeteria, for example, would think of themselves as an interest group rather than a special interest.

The major interest groups are organizations such as labor unions (concerned with workers compensation and benefits) and business groups like manufacturing associations or chambers of commerce (concerned with higher profits, less regulation, and lower business taxes). They also include "public interest" groups concerned with the environment (like the Sierra Club), parks and recreation facilities (like Friends of the Parks), and schools (like the local school PTA or School Council). The issues with which they are concerned affect not only them individually but the broader public as well.

Interest groups generally have a large membership base, such as workers in a single union like the United Automotive Workers (UAW), which may organize thousands of workers at all the auto plants in a region. Their power comes from their large membership, organizational strength with paid staff members, and required dues raising relatively large sums of money. Interest groups use their clout to lobby directly or through paid lobbyists and to affect elections by their endorsement of candidates, work on campaigns, and large campaign contributions from their PACs (political action committees).

In recent years, special interests have also borrowed some techniques from community organizing, just as community organizations borrowed many of their techniques from labor union organizing decades ago. For example, Local 880 of the Service Employees International organizes low-income child care and health care workers by going door to door in Chicago's African American South Side neighborhoods, getting these workers to join the union. They then use their larger membership to force changes. They convinced Governor Rod Blagojevich to make it easier to unionize these employees and gain collective bargaining for them; in return, the union gave a $250,000 contribution to the governor's reelection campaign along with promises of phone support and precinct work on his behalf. Local 880 combined community tactics, traditional union organizing, and smart lobbying to gain their goals.[29] Similar projects are going on in other states, such as the Stamford Organizing Project in Connecticut, which has unionized over 4,700 service workers in four affiliated unions. These workers lived near their work, in what is ironically one of the richest areas of the country but in which worker salaries and benefits lagged.[30]

People are participating more than in the past in various interest groups by giving funds to the group's political action committee, joining together in contacting public officials, and participating in marches, demonstrations, and "lobby days" at the state capitol.[31] You probably have a membership card in some organization that represents your collective interests, such as Greenpeace, the Christian Coalition, or People for the American Way. Interest groups provide one of the best means for citizens to amplify their voice and to get government to respond positively to their concerns.

Social Movements

Social movements differ from long-term associations and organizations. A **social movement** is activism by people who share a common concern about a social issue and come together in often ad-hoc organizations to pressure the government or private corporations for change. Examples from America's past are the suffragist movement that began in the 19th century to secure women's right to vote; the civil rights movement of the 1960s, which sought to gain equality and justice for racial minorities and to end segregation; and the antiwar movement in the 1970s, which sought to end the war in Vietnam.

It is not hard to find current examples of movements on both the liberal and conservative side of politics, such as the anti-abortion or pro-life movement, the movement to legalize marijuana, and movements for tort reform, a national health care system, and an end to nuclear weapons. Often action on the part of one group will spark a reaction from another. For instance, in May 2006, over a million people demonstrated nationwide for immigration law reform. A week later, the Minuteman Project launched a much smaller cross-country tour to rally support for tighter border controls and stricter immigration laws.[32]

Social movements arise most forcefully when some great evil is seen and people can be organized in large numbers to take direct action on the problem. Martin Luther King Jr. was the face of the civil rights movement, but many local groups were organized to undertake specific aspects of the civil rights movement, such as protest marches, sit-in demonstrations, and lawsuits. The civil rights movement, like all social movements, had to change public

STRUGGLE FOR POWER

Using the Internet to Mobilize a Movement

MoveOn.org., the online liberal political action group, said that in 72 hours more than 106,000 people signed up for the Wednesday, August 17, 2005, night's vigils against the war in Iraq, in more than 1,600 towns nationwide.

"It was unbelievable," Tom Matzzie, director of MoveOn's headquarters in Washington, said. "We've organized other vigils in the past, but this was the largest MoveOn has ever organized. The response was overwhelming."

And the other side counterorganized using the same tools to bring out counterdemonstrators, letters, and petitions in support of the war.

Source: Adapted from Susan Kuczka, "War Protesters Find Shortcut on the Internet," *Chicago Tribune*, Metro Section, August 29, 2005, 3.

---

opinion about segregation and racial discrimination before laws could be changed. While the tactics of social movements from different eras differ to suit their times, they all seek to enlighten public opinion so as to force changes in government policies and social practices.

Social movements often draw attention to the injustice of certain laws and procedures in society. In cases like the civil rights movement, the movement can eventually lead to changes in those laws—for example, the adoption of the National Voting Rights Acts in 1965. Movements mobilize people and focus media attention on problems. They can, for instance, change public support for existing policies, as when news coverage of the Vietnam War helped to change attitudes about the war, and the decline in public support eventually caused America to withdraw.

Movements today use technology to further their causes, as demonstrated in the Struggle for Power feature on using the Internet to mobilize. When the Vietnam War protests were undertaken in the 1970s, students and others went around campuses and communities pasting up posters to announce the date and time of a protest march or demonstration in order to recruit participants. Now blast e-mails have replaced posters and door-to-door personal contacts. But the power of a social movement still comes from citizens joining together to bring about change.

COMMUNITY ORGANIZATIONS

A **community organization** is a group formed by the people of a community to promote the community's general well-being or provide a social service. There are many different types of community organizations, ranging in size from a single block club to comprehensive neighborhood organizations. Some provide social services such as affordable housing, economic development, job training, pantries for the hungry, or shelters for the homeless. They organize around a neighborhood or geographical community area. They can have considerable political and governmental impact beyond direct service delivery.

Saul Alinsky was one of the great strategists for community organizing in the 20th century. He argued that the "have-nots"—the people in poor, working-class, and even some middle-class communities—could come together and pressure local governments and private corporations like utilities to meet their demands.

STRUGGLE FOR POWER

Saul Alinsky's Tactical Rules for Community Organization

In his book *Rules for Radicals*, community activist Saul Alinsky offered the following advice about tactics.

1. Cut the issue, pick the target, freeze it, personalize it, and polarize it. An issue must be cut as completely black and white as our forefathers characterized the war against England in 1776. They knew the benefits of the British Empire to the colonists would dilute the urgency of the Revolution. An honest "Declaration of Independence" might have weighed the scales 60 percent on our side and 40 percent on the British. To expect people to leave their crops, pick up a gun, and join the Revolution for a 20 percent difference in the balance of human justice defies common sense. So the Declaration of Independence was a 100 percent denunciation of the role of the British government as evil and unjust . . . in no war has the enemy or the cause ever been gray. Beyond the cutting the issue, an organization must select a particular enemy like the energy company, personalize the enemy as the CEO of that company, and portray him or her and the company as the villain for their air pollution which is killing people in the community.

2. Power is not only what you have but what the enemy thinks you have. While the people you organize don't have money or positions of power, the "Have-nots" have large numbers of people on their side. If you have organized a vast, mass-based people's organization, you can parade it visibly before the enemy and openly show your power. If your organization is small in number, then do what Gideon did: conceal your members in the dark but raise a din and clamor that will make the listener believe that your organization numbers many more than it does. If your organization is too tiny, even for noise, stink up the place.

3. Never go outside the experience of your people in planning an action. Members of a community organization may never have participated in a protest demonstration before. If so, you must prepare them to take this action and make it close to union strikes or demonstrations they have seen on television.

4. Whenever possible go outside the experience of the enemy. If the enemy has dealt with protest demonstrations before and has tactics to deal with them, try new actions like tying up their phone lines with complaints or going up to bank windows to open new accounts so as to block regular customers from using the bank and causing a legal disturbance in the business.

5. Make the enemy live up to their own book of rules. It is hard for politicians to meet with hundreds of constituents or company presidents to meet with hundreds of clients who want to complain. But their rules say that they have to do so.

6. Ridicule is a peoples' most potent weapon. If you can make the public laugh at your enemy, you have embarrassed them in a way that may make them willing to negotiate.

7. A good tactic is one that your people enjoy. Doing something different, something fun is better than doing what every other group has always done in applying pressure.

8. Keep the pressure on with different tactics and actions. Rarely are the demands of a community organization met at a single negotiating session with a company or government official. A single demonstration by a group will not usually win the day. Community pressure campaigns should be planned to last for six months to a year.

Source: Adapted from Saul Alinsky, *Rules for Radicals* (New York: Vintage Books, 1971), 27–28, 126–30.

Some groups rely on trained community organizers to get them started. Like lobbyists, some community organizers are paid professionals and some are unpaid volunteers. As a first step, the community organizer talks to the local citizens and existing groups to see what their problems or issues are. The existing groups then join an umbrella organization that represents the entire community. Leaders are elected for the umbrella organization and an issue is chosen, such as blocking a proposed expressway that would destroy the neighborhood, fighting property tax increases that would harm existing home owners, or stopping air pollution from a utility company's coal-fired plant. Fighting against some negative effect on the community

On the Waterfront

Water wars have raged in Florida for decades, but state Senator Paula Dockery brokered a truce. . . . After election to the state Senate in 2002, Dockery became chair of the Environmental Preservation Committee. She began hearing from community developers that Florida's efforts to set aside water for environmental preservation were hindering the state's ability to handle its growing population.

The state needed a water plan. But environmental groups, developers, agricultural interests, and local communities all were locked in a struggle to wrest control of an increasingly overburdened water supply. Dockery spent much of 2003 and 2004 developing legislation to determine how much water the state should devote to conservation, and how much should be available for drinking and irrigation. As the end of the 2004 legislative session approached, however, Dockery realized that not everyone was on board. She could push a bill through, but the legislation wouldn't satisfy each of the shareholders involved.

So Dockery started over, and for the next 12 months, she devoted herself to bringing together anyone with an interest in the state's water future to help craft a workable water plan. Ultimately, more than 120 people—business leaders, agriculture representatives, environmentalists, home builders, city and county officials, tourism workers, public utilities managers—began showing up at monthly meetings.

Throughout the process, Dockery always prized consensus above all else. "She said, 'Look, we either have everybody agree to the language in this bill, or we're not going to have a bill,'" says one admirer. But the work group meetings didn't always go smoothly.

As the meetings progressed, Dockery realized the solution to the problem was deceptively simple: increase the amount of water available.

The measure, which was signed by Governor Jeb Bush, provides $300 million to help local officials develop alternative sources, including desalination, aquifer storage, aboveground reservoirs and sewer reuse. Local water districts must match state investments dollar for dollar, so the plan encourages municipalities to form regional authorities and focus on water-supply problems across community lines.

But the law's most surprising accomplishment is that it pleases the players on every side of the water debate. That feat, says Eric Draper, policy director for Audubon of Florida, was almost surreal. "When we announced the bill, you had environmental lobbyists sitting at a computer working on a press release with representatives from the state Chamber of Commerce. I looked around and said, 'Take a picture of this, because you're never going to see this again.'"

Source: Adapted from Zach Patton, "Paula Dockery: On the Waterfront," *Governing*, November 2005.

---

helps to unite the group. Then, the community organization can promote positive programs that would benefit the neighborhood, such as creating Tax Increment Finance districts to improve the business community.

Alinsky recommended nonviolent direct action by citizens to force reaction by the government or business that would sway opinion in favor of the neighborhood's demands.[33] See the Struggle for Power feature on Alinsky's tactical rules for effective community organizing. Nonviolent protests and the use of pressure tactics are some of the ways that a community organization can force private companies or state and local government to pay attention to their needs and their demands for change.

Another common tactic of community organizations is to force public officials to attend community meetings or a public accountability session. In the friendliest of these sessions, local aldermen, city council members, or state legislators meet with community organizations within their district to report on legislation at city hall or the state capitol that may affect community residents. The official and the organization then agree to work together to support some legislation and to defeat others. Implied in this conversation is that if the

official represents their issues and concerns well, the members of the organization will support the official's reelection. A community organization's power is increased when the group and an elected member of local or state government are united in a goal. See the Insider's View feature for a case in which such cooperation helped to address water issues in Florida.

Better organized community groups plan more structured accountability sessions and invite all the members of the community to attend. After a public official makes a report, the group presents a set of written demands to the official. The group might ask the official to sign a pledge to propose certain legislation, or to vote against certain taxes, or to get state or city agencies to intervene to keep an institution such as a hospital from leaving the community. If the official agrees personally with the demands or if the voting strength of the group is strong enough, he or she signs the pledge and faithfully carries out the promise. This gives the organization prestige in the eyes of the community. Even if the official fails to sign on the dotted line, the group gains prestige from confronting public officials on behalf of the community they serve.

ROLES CITIZENS CAN PLAY

Any group—a political party, civic group, interest group, social movement, or community organization—requires spokespeople, organizers, and foot soldiers to carry out the strategies and plans. In their different ways, based on very different resources and tactics, citizens in these and other groups have a profound impact on state and local government, from getting a minor adjustment to a local administrative rule to affecting major policies such as state funding for stem cell research.

With all of these groups, you have options. You can choose not to join, remaining a nonparticipant. You can choose to be a member, pay your yearly dues, and perhaps cast a vote for the candidates for public office that your group has endorsed. At a higher level of involvement, you can choose to be an organizer, getting others to join, planning strategy for the organization, and directing activities of other members and supporters. If you like a more public role in the spotlight, you can choose to be a spokesperson for the group, its representative to the media, public officials, and others.

Local citizen groups have a more profound impact on local government than on national government. It is easier to get city council members to act the way that a particular local group wants than to get the president of the United States to behave differently. It is easier to get a state legislator to sponsor legislation that a local group wants than to get a member of the U.S. Congress to do so.

Citizen activists can be party members or party officials, paid or volunteer lobbyists, professional journalists or part-time bloggers. They can be active in civic groups, interest groups, social movements, or community organizations. All shape the direction of state and local government. And you can choose to become an activist in any of these organizations and play any of these important roles. The struggle for power and influence is not just the domain of elected and appointed government officials. Local citizens all have important roles to play if we are to have a functioning democracy.

CHAPTER SUMMARY

Local members of political parties, lobbyists, journalists, and citizen activists in community organizations all exercise various degrees of influence or power over state and local government officials. Political parties work to get their own advocates elected to government. Parties play three roles in the political process: the party as an organization, which recruits and supports candidates; the party inside government, or the elected officials; and the party in the electorate, comprising party members and others who identify with or support the party.

The party organization has a pyramid structure: at the bottom are the precinct committeemen, generally elected in primary elections. These active party members in turn elect the county chair or ward committeemen. Above the county chair is the state central committee chair and state party chair. Although many states have nonpartisan elections at the local level, the power and influence of a party lies in its ability to control nominations and get the nominees elected.

Lobbyists work to get a group's interests heard in government. Some lobbyists are unpaid; paid lobbyists have to register with federal, state, or local governments. In the states, there are thousands of paid lobbyists who try to influence the outcome of legislation and executive policy-making. There are five types of lobbyists: (1) contract lobbyists with a number of clients; (2) association lobbyists who represent a single group such as a labor union; (3) single company lobbyists; (4) intergovernmental lobbyists representing states, large cities, or governmental associations; (5) cause lobbyists who work for issues that have no direct benefit to themselves, such as children's issues.

Lobbying includes testifying at hearings, helping to draft legislation, contacting legislators, and mounting grassroots campaigns. A lobbyist's chief task is to personally meet with legislators.

Journalists are watchdogs of our states and communities. Reporters uncover scandals and malfeasance in government, but it takes time to acquire the specialized knowledge needed to cover a beat well. The press was once revered as the fourth estate of government, but what was once considered newsworthy because of its intrinsic value is being replaced in television news coverage by what sells ads. Bloggers, who write online journals, do not have the standards of traditional media; most blogs do not check their accuracy or consider news value before posting an entry. Nonetheless, they can uncover and publicize important political and government information.

Citizen activists, if organized, can have an impact on state and local government. Civic groups, interest groups, social movements, and community organizations have been able to take direct action on serious local problems or issues. Civic groups are organized local groups whose members promote the public interest. An interest group is an organized group that tries to influence policy affecting the group's particular area of interest, its "special interest." A social movement is activism by people who share a common concern about a social issue and come together in often ad-hoc organizations to pressure the government or private corporations for change. Community organizations are formed by the people of a community to promote the general well-being or provide a social service such as food pantries or job training.

This chapter has given you a sample of the various ways you can become involved in the struggle for power and influence. Becoming involved in politics is but one way. You can also become involved through supporting a political party, lobbying, working as a journalist or blogger, or through a civic group, interest group, social movement, or community organization. We all have important roles to play.

KEY TERMS

civic group An organized local group whose members promote the public interest, p. 205
community organization A group formed by the people of a community to promote the community's general well-being or provide a social service, p. 208
grassroots lobbying Lobbying that arises (or appears to arise) from the concerns of many individuals rather than being undertaken by representatives of an organization, p. 198

interest group An organized group that tries to influence policy affecting the group's particular area of interest, p. 206
lobbying An organized effort to influence public policy on behalf of a particular interest area, p. 190
nonpartisan election An election in which candidates run for office without any party label, p. 188
paid lobbyist A full-time professional lobbyist hired by an organization to influence

government legislation, regulations, or decisions, p. 190
precinct The smallest election unit in each state, comprising about 500 registered voters, p. 187
social movement Activism by people who share a common concern about a social issue and come together in often ad-hoc organizations to pressure the government or private corporations for change, p. 207

QUESTIONS FOR REVIEW

1. What are the three roles political parties have in the political process?

2. What are the duties of a precinct committee person?

3. Who are lobbyists? What interests do they represent? Is anyone who contacts a government

official a lobbyist? What prepares someone to be a lobbyist?

4. What are some of the different types of lobbyists, and how does each operate?

5. In what ways do lobbyists at the local government level differ from those who work with policy-makers in Washington?

6. What are the most important tasks of lobbyists in the legislative process? What is the lobbyist's chief task?

7. Why is it difficult for journalists to cover state and local political news?

8. What are the advantages of blogs over mainstream media in covering state and local news? What are the drawbacks?

9. Why are the news media sometimes called the "fourth house" or the fourth branch of government? What gives them power that is equal to the legislative, executive, and judicial branches of government?

10. How can the Internet be used in organizing people in special interest groups and movements? How does this differ and how is it similar to techniques used in the past?

11. How do community organizations different from civic groups, interest groups, and social movements?

12. Why is it easier for a community group to organize first around a threat than a positive program?

DISCUSSION QUESTIONS

1. What are the positive and negative effects political parties can have on the democratic process? Should political parties be abolished? Should they be strengthened? Why?

2. You are on the central committee of a political party in a small county. What are some of the factors you have to consider when slating a candidate for a newly vacant seat in the state legislature?

3. If you wanted to testify before a state legislative committee to request lowering student tuition at the state university and getting more funds for the state's community colleges, how would you prepare your written and oral testimony?

4. In achieving your goals with a legislative body, what is more important than preparing good testimony? How would you attempt to be sure that you got the committee vote you wanted?

5. Why do the media sometimes fail to provide a check and balance to government? What hinders media coverage of certain stories?

6. What do blogs contribute to reporting on local and state government? Do blogs do a better job than the mainstream media in covering local politics and government? Would you want to write a political blog? Why or why not?

7. Which type of citizen activism is most appealing to you? Which type of organized group would you join? What are the different types of roles individuals can play in each of these organizations? Which role seems most appealing to you?

8. If there were no community organization where you live, how would you go about organizing one? What resources would be necessary to succeed?

9. What are some tactics a community organization can use against a company or government agency that is not meeting local needs?

PRACTICAL EXERCISES

1. Who are the chairpersons of your county's Democratic Party and Republican Party? To find out, go to their local political Web sites or to your local newspaper archives.

2. Find out how many registered lobbyists are in your city, county, and state. You can often find the list of registered lobbyists on your state and local government Web sites. If not, contact your city clerk, county clerk, or office of the secretary of state to obtain the list.

3. Who are three or four of the top lobbyists in your city, county, and state? The most important lobbyists are usually written up in newspaper stories like "The Hired Guns of Tallahassee." Do a search of archived news stories on the Web site of the major newspaper in your city or state or a Lexis-Nexis search to find information on lobbyists in your city or state.

4. Write a brief letter to a city council member, county board member, or state legislator on an issue that matters to you. Exchange letters with another student and give each other feedback. Think of how you would react if you were a legislator and received such a letter. Now rewrite your letter and see how you can improve it. If you really believe in what you wrote, send your letter to a public official and see what response you get.

5. Follow one local newspaper, one television newscast, and one blog for a week. Make notes on what stories they cover and how well they cover them. How much space and time did each devote to politics and government stories? Which news source did the best job of providing news about state and local government? Which one did the worst job, and why? What are some stories that you care about that did not get covered?

6. Using various sources, determine the most powerful civic groups, interest groups, social movements, and community organizations in your city or state. For instance, you might do a newspaper search to determine

which organizations get news coverage most often. You might interview two or three knowledgeable people as to which groups they think are most powerful.

7. Once you have determined the most powerful groups in your city or state, find out about at least one major issue they are working on currently. You might call the organizations or their spokespersons and ask them, or you may search for newspaper stories about these groups.

WEB SITES FOR FURTHER RESEARCH

You can find information on your local Democratic Party by following the links at www.democrats.org. For the same information about the Republican Party, begin at www.gop.com.

For information about lobbying, check the local lobbyist registration on your city clerk's, county clerk's, or secretary of state's Web site. For instance, the site for lobbyist information in the state of Illinois is www.cyberdriveillinois .com/departments/index/lobbyist/home.html.

Up-to-date lobbyist data are made available by the Center for Responsive Politics at www.opensecrets.org/lobbyists/ index.asp.

Information on journalism can be found at the *Columbia Journalism Review* site, www.cjr.org, and the *American Journalism Review* site, www.ajr.org.

The informative Center for Public Integrity Web site explains and ranks lobbying disclosure laws for all 50 states. It has a good deal of legal information as well as data on lobbying and campaign disclosure laws. Go to http://www .publicintegrity.org, and to get the information on lobbying disclosure laws for your state, click on the "In your state" link.

A civic group that encourages citizen participation is the League of Women Voters, whose Web site is www.lwv.org. Most state leagues have their own Web sites, such as the Illinois site at www.lwvil.org.

Most major interest groups have Web sites. Two major sites for citizen participation from very different points of view are the liberal http://www.moveon.org and the conservative http://www.grasstopsusa.com.

SUGGESTED READING

Alinsky, Saul. *Rules for Radicals*. New York: Random House, 1969.

Cook, Timothy. *Governing with the News: The News Media as a Political Institution*. Chicago: University of Chicago Press, 1998.

Graber, Doris A. *Mass Media and American Politics*, 7th ed. Washington, DC: CQ Press, 2006.

Hrebenar, Ronald J., and Ruth K. Scott. *Interest Group Politics in America*, 2nd ed. Englewood Cliffs, N.J.: Prentice Hall, 1990.

Mooney, Christopher, and Barbara Van Dyke-Brown, *Lobbying Illinois*. Springfield: University of Illinois at Springfield, 2003.

Nownes, Anthony, and Patricia Freeman, "Interest Group Activity in the States." *Journal of Politics* 60 (February 1998).

Rosenthal, Alan. *The Third House: Lobbyists and Lobbying in the States*. Washington, DC: CQ Press, 1993.

Simpson, Dick, and George Beam. *Political Action*. Chicago: Swallow, 1984.

———. *Strategies for Change*. Chicago: Swallow, 1976.

Simpson, Dick, ed. *Inside Urban Politics: Voices from America's Cities and Suburbs*. New York: Pearson Longman, 2004.

Terkel, Studs. *Division Street: America*. New York: Random House, 1967.

Thomas, Clive S.. and Robert Hrebenar. "Interest Groups in the Fifty States." *Comparative State Politics* 20, no. 4 (1999).

Woodward, Bob, and Carl Bernstein. *All the President's Men*. New York: Simon and Schuster, 1974.

———. *The Final Days*. New York: Simon and Schuster, 1976.

Zeigler, Harmon, and Michael A. Baer. *Lobbying Interaction and Influence in American State Legislatures*. Belmont, CA: Wadsworth, 1969.

People waiting their turn at a Medicaid clinic.

FOLLOW THE MONEY
THE POLITICS OF STATE AND LOCAL BUDGETING

State and local governments across the United States spend more than $8,000 annually per person in the nation. This amounts to more than 17 percent of the gross domestic product of the United States. See Table 9.1 for per capita state and local government expenditures in your state. Citizens, including college students, can play important roles in determining where those monies go—whether to student aid, children's health, environmental

TABLE 9.1 Per Capita State and Local Government Total Expenditures in 2005

National per capita = $8,017[a]

Rank	State	Per Capita
1	Alaska	$14,979
2	New York	11,836
3	Wyoming	11,139
4	California	9,544
5	Massachusetts	9,183
6	Delaware	9,010
7	Washington	8,964
8	New Jersey	8,947
9	Rhode Island	8,794
10	Connecticut	8,585
11	Minnesota	8,515
12	Vermont	8,357
13	Hawaii	8,310
14	Nebraska	8,134
15	New Mexico	8,126
16	Pennsylvania	8,121
17	Ohio	8,022
18	Oregon	8,012
19	Maine	7,789
20	Wisconsin	7,788
21	South Carolina	7,758
22	Illinois	7,706
23	Michigan	7,572
24	North Dakota	7,542
25	Colorado	7,509
26	Maryland	7,406
27	Iowa	7,353
28	Florida	7,350
29	Alabama	7,319
30	Louisiana	7,244
31	Nevada	7,225
32	Tennessee	7,165
33	North Carolina	7,001
34	Mississippi	6,910
35	Utah	6,894

(continued on next page)

TABLE 9.1 Per Capita State and Local Government Total Expenditures in 2005 (*continued*)

Rank	State	Per Capita
36	Kansas	6,893
37	Montana	6,847
38	Virginia	6,827
39	West Virginia	6,794
40	Indiana	6,728
41	Texas	6,652
42	New Hampshire	6,647
43	Arizona	6,596
44	Georgia	6,467
45	Kentucky	6,465
46	Missouri	6,425
47	South Dakota	6,350
48	Idaho	6,248
49	Arkansas	6,224
50	Oklahoma	6,223
District of Columbia		15,634

*Total expenditures includes all money paid other than for retirement of debt and extension of loans. Includes payments from all sources of funds including current revenues and proceeds from borrowing and prior year fund balances. Includes intergovernmental transfers and expenditures for government-owned utilities and other commercial or auxiliary enterprise, and insurance trust expenditures.

Source: CQ Press using data from U.S. Bureau of the Census, Governments Division.

protection, or other programs. Thus, "following the money" is a good way to understand the dynamics of state and local politics.[1]

Budgets are the scorecards of politics and government. These ponderous documents tally in dollars the winners and losers in the struggle for who gets what, who pays, and how much.[2] Elected officials have strong political incentives to do things *for* their voters (e.g., more program spending) and to avoid doing things *to* them (e.g., imposing more taxes). As a result, over time, budgets have a natural tendency to increase faster than revenues to pay the bills, and faster than inflation. In times of recession, it usually takes both increased taxes and serious budget cuts to balance the budget.

The only way to make up the difference in revenues and expenditures is to cut back on spending or increase revenues. If the decision is to increase revenues, especially to increase a major tax like a sales or income tax, then extensive planning, marketing, and trading of favors can be required to win approval of lawmakers. Sometimes the voters in states and localities must also provide their approval in referendums. The governor, mayor, or school board members who initiate revenue increase proposals travel throughout their constituencies, making their case to citizen groups, special interests, and newspaper editorial boards. The state legislators or city council members react coyly, waiting to see how well an executive's appeals play with their voters.

All the major players in state and local politics join in the ceremony. Educators, as major beneficiaries of many tax increases, orchestrate marketing efforts among constituents. Business groups often, but not always, rally the opposition, declaring increased taxes bad for business and thus bad for job growth and general well-being. Economists weigh in with projections of the consequences for the state's business climate. State or local tax "watchdog" organizations, such as Florida Tax Watch and the Civic Federation of Chicago, offer their analyses and recommendations.

When author Jim Nowlan was a Republican freshman member of the Illinois house, the new Republican governor proposed a dramatic combination of tax increases. The governor proposed a new income tax on individuals and corporations and an increase in the gasoline tax and driver's license fees, as well as other fee increases. Immediately after the bombshell became public, the Republican speaker of the house called his members into a private caucus. "The governor has asked me to get a headcount," the speaker told his colleagues, "of how many can support his tax program, and I'd like a show of hands."

As a progressive who felt that education needed more funding and that an income tax could reduce property taxes on farmers in his rural district, Nowlan stuck his hand up in favor. He looked around the 93 members in the caucus and saw only a few other hands. Most Republicans "sat on their hands." Throughout the following four months of the legislative session, Nowlan's seatmates received calls to go individually to the governor's office, a floor below the house chambers. Each received patronage jobs for party supporters, pledges of support for park improvements, and other legitimate favors that only a chief executive can bestow. Ultimately, most of Nowlan's Republican colleagues joined in voting for the tax increases, which were ultimately enacted. Nowlan received nothing for his district because he had already pledged to vote for the increases!

Often this tax raising ritual is unsuccessful. Unless there is a genuine crisis like the Great Depression or the deep recession of 2008–2009, a scaled-back budget is generally contrived, cuts are made in programs, revenue projections are inflated, the payment of bills is delayed, and the tax dance is put on hold for another year or two.

BUDGETS AS ALLOCATIONS OF OUR VALUES AND NEEDS

For state and local governments, the **budget** is the plan for spending financial resources, which include tax revenues, fees, lottery profits, interest income, federal transfers to the states, borrowed money, and monies from other sources. The budget for the state of Virginia shown in Table 9.2 is representative of state budgets. As can be seen in the right-hand column, Virginia spends most on education ($14.2 billion) and health and human services ($9.6 billion).

State and local government budgets operate within a fiscal year (FY), which is generally the 12-month period of effective operation of a budget. Fiscal years often begin in the middle of a year: running from July 1, for example, until June 30 of the following year. (The federal government fiscal year begins October 1.)

Most state and local budgets operate with a number of separate funds. The **general fund** of a budget tends to provide general tax (income and sales) revenues for operation of the general administrative departments of a government, including education, social services, and departments such as corrections and the courts. Nongeneral fund spending often draws on specific revenues such as gasoline taxes for highways or fees from fish and game licenses.

TABLE 9.2 Virginia Department of Planning and Budget Statewide Summary 2007–2008

Secretary	General Fund FY 2008	Nongeneral Fund FY 2008	Total FY 2008
Administration	645,386,540	238,605,923	883,992,463
Agriculture & Forestry	46,300,544	35,701,910	82,002,454
Central Appropriations	1,324,296,745	46,313,093	1,370,609,838
Commerce and Trade	111,445,079	738,170,783	849,615,862
Education	7,859,550,694	6,373,766,518	14,233,317,212
Executive Offices	34,673,613	14,209,782	48,883,395
Finance	773,234,978	30,760,583	803,995,561
Health & Human Res.	4,043,584,452	5,538,868,737	9,582,453,189
Independent Agencies	306,785	337,615,265	337,922,050
Judicial Agencies	363,537,506	26,376,619	389,914,125
Legislative Agencies	63,811,475	3,833,762	67,645,237
Natural Resources	136,862,458	208,186,590	345,049,048
Nonstate Agencies	26,713,850	0	26,713,850
Public Safety	1,748,208,990	787,767,256	2,535,976,246
Technology	4,386,548	58,385,388	62,771,936
Transportation	150,844,067	4,193,230,167	4,344,074,234
Statewide Totals	**$17,333,144,324**	**$18,631,792,376**	**$35,964,936,700**

Source: Department of Planning and Budget, Commonwealth of Virginia, 2007, http://www.dpb.state.va.us.

Most governments also operate with a number of **separate funds,** or **dedicated funds**, to segregate monies for particular purposes and no other. For example, a Fish and Game Fund might use fishing and hunting license fees to support fisheries. A Coal Technology Development Fund might support development of clean-burning coal processes. In difficult economic times, a governor and legislators may agree to allow the chief executive to "sweep" part or most of all separate fund balances and transfer the monies into the general fund, even though the fees were created for special, dedicated purposes.

State budgets like the Virginia budget are determined by the functions of state government. Education, health care, human services, and transportation generally represent the largest expenditures of state government. In contrast, the primary general fund spending categories for municipal governments are for police and fire services, as shown in the Albuquerque, New Mexico, municipal budget in Table 9.3. Police require almost 30 percent of the city budget, while fire protection takes 14.5 percent, far more than any other municipal expenses.

Many state constitutions require that the state and local governments have a **balanced budget**, that is, revenues must equal expenditures. Many budgets that appear on paper to be balanced are, in fact, not. Think of this in terms of your own budget. For example, you may show a $100 balance in your checking account and yet have $1,000 in unpaid bills on your

TABLE 9.3 City of Albuquerque: General Fund Spending by Department (in thousands of dollars)

Expenditures by Department	Approved Budget FY/07	Proposed Budget FY/08	$ Change	% Change	% Share FY/07	% Share FY/08
Chief Administrative Officer	3,971	3,177	-794	-19.99%	0.83%	0.66%
City Support	23,199	20,640	-2,559	-11.03%	4.87%	4.26%
Council Services	2,489	2,847	358	14.38%	0.52%	0.59%
Cultural Services	35,198	36,003	805	2.29%	7.38%	7.43%
Economic Development	3,124	3,140	16	0.51%	0.66%	0.65%
Environmental Health	13,341	14,262	921	6.90%	2.80%	2.94%
Family and Community Services	35,147	37,276	2,129	6.06%	7.37%	7.69%
Finance & Administrative Services	23,506	24,221	715	3.04%	4.93%	5.00%
Fire	66,498	70,328	3,830	5.76%	13.95%	14.51%
Human Resources	2,453	2,677	224	9.13%	0.51%	0.55%
Legal	8,097	10,102	2,005	24.76%	1.70%	2.08%
Mayor	873	904	31	3.55%	0.18%	0.19%
Metropolitan Detention Center	15,429	0	-15,429	-100.00%	3.24%	0.00%
Municipal Development	44,408	45,120	712	1.60%	9.32%	9.31%
Office Internal Audit and Investigations	1,234	1,304	70	5.67%	0.26%	0.27%
Parks & Recreation	24,726	23,484	-1,242	-5.02%	5.19%	4.84%
Planning	14,924	14,824	-100	-0.67%	3.13%	3.06%
Police	130,503	145,257	14,754	11.31%	27.38%	29.96%
Senior Affairs	5,045	5,334	289	5.73%	1.06%	1.10%
Transit (Operating Subsidy)	22,515	23,931	1,416	6.29%	4.72%	4.94%
Total	**476,680**	**484,831**	**8,151**	**1.71%**	**100.00%**	**100.00%**

Source: City of Albuquerque, New Mexico, Fiscal Year 2008 Proposed Budget: Part I: Financial Plan, www.cabq.gov.

credit card. Your personal budget then would be unbalanced. Similarly, governments sometimes incur obligations, such as for services provided by doctors and hospitals, without the money to pay for those obligations in a timely fashion. Delaying payment into the next fiscal year is one device budgeters use to show the fiction of a balanced budget.

THE PRESSURE TO SPEND

The pressure on state and local governments to spend is enormous. In 2004, state and local governments combined expended almost $2.5 trillion. The 50 states and their local governments, respectively, each expended about one-half of the total. State and local expenditures of $430 billion for social services was greater than the $411 billion spent on elementary and secondary education. All other major functions of state and local governments received less than half those amounts each, with transportation spending at about $200 billion and higher education at $150 billion. Administration of state and local governments accounted for only $90 billion in spending.[3]

Scores of interest groups seek increased spending for their special groups, often with compelling needs, such as the mentally retarded, the mentally ill, children without parents, abused spouses, distressed rural areas, drug and alcohol rehabilitation, the arts, and many more. These interest groups include liberals, moderates, and conservatives who believe passionately in their respective causes.

State legislators and city council members never see a person or group come before the appropriations or finance committee to request *less*. Every cause and program makes its case for more spending and greater appropriations. **Appropriations** are the amounts that the legislature has authorized the government to spend on specific budget categories.

There is always an argument for changing spending priorities. When author Dick Simpson was a Chicago alderman, only 7 percent of the city's operating budget was spent on all social services combined, which included human services, homelessness, hunger, the elderly, culture, and the arts. Simpson argued for increasing these services even if other city services and waste had to be cut to do so. The city provided its human and social services almost entirely by grants from the federal and state governments. Because of these fiscal policies, Chicago stayed solvent while New York City, with its more generous human services, went bankrupt in the 1970s. There is always pressure to spend government money for worthy purposes, but it is not always wise to do so. In addition, many government spending programs increase because they are driven by formulas that require increased spending on the basis of individual eligibility and legal requirements.

For example, in 1965 the federal government enacted **Medicaid**, the program that supports health care for low-income people and nursing home care for low-income and even some middle-income elderly. The federal government shares the cost of the program with the states, providing 50–76 percent of total state expenditures through the Medicaid program. The federal share is based on a state's wealth and demographic factors. This determines the percentage of costs for Medicaid that different states must pay. California and Delaware are among a small number of states that pay 50 cents of every Medicaid dollar. Georgia pays 37 cents of every Medicaid dollar; Mississippi pays 24 cents.[4] When states were rocked by deep recession in 2008–2009, the American Recovery and Reinvestment Act of 2009 provided special funds from the federal government to help states pay their share of Medicaid costs, but that funding is only temporary. With health care costs rising, states as well as the federal government will face increasing Medicaid costs, unless health care reforms can be enacted or the formula for the money the federal government provides to the states is increased.

Originally expected to be a relatively small program, Medicaid now serves 62 million Americans and expends more each year than the Medicare program for the elderly. While the national and state governments each have some authority for determining the services provided and eligibility thresholds for participation, the states administer the program. Since the 1980s, the federal and state governments have expanded eligibility to encompass more children and disabled persons. With average life spans increasing, there has been an increase in the number of persons in nursing homes and the number who are paid for by Medicaid, at slightly more than $10,000 per elderly enrollee per year.[5] About two of every three nursing home residents in America are supported wholly or in part by Medicaid. Indeed, health care has become a new function of state governments primarily because of Medicaid.

Prisons and jails also put pressure on budgets. The states spend at least $10,000–$30,000 per inmate per year, more than for each public school or university student. In the 1960s, Illinois paid for housing 8,000 prison inmates; by 2006, the state had more than 42,000 inmates in its prisons. Throughout the 1980s and 1990s, Illinois built about one new prison every year to accommodate the growth. In that same period, the total population of Illinois grew very little.[6] The costs of confining prisoners have increased for all counties that fund jails. They have increased even more for states, which fund most of the prisons in the United States.

Getting tough on crime is costly. Each additional prisoner must be provided basic housing, food, health care, and possibly rehabilitation services. Budgeters have little latitude in funding prisons as inmate numbers increase. Some reduce prisoners' sentences or sentence prisoners to home confinement. Studies have shown that the length of sentences actually served is often affected by budgetary constraints.[7] Filling prisons to overcapacity because of lack of funding for additional prison space has sometimes resulted in early release of inmates, which means getting "less tough" on crime, often the opposite of public wishes. Often budget constraints or reductions have consequences that are not planned.

Constraints on budgetary funding can cause some social service program managers to make tough choices. As one mental health program director lamented, "the state is asking me to play God." He told of looking at children sitting in his waiting room and thinking: "How can the state force me to say you can have services, and you cannot!"[8]

Parts of state and local budgets appear to be sacrosanct and thus almost immune to cuts. Education spending is an illustration. Public opinion surveys show that most citizens want states and local schools to increase spending for education. So local school boards often lobby their state governments to increase state spending for local schools, to reduce pressure on the local property tax. Spending for colleges and universities, however, is often less sacrosanct than spending for kindergarten through high school. Some states have actually cut spending for higher education, which means students have to pay higher tuition charges.

Public safety is another area where budget items cannot be cut. Citizens demand that states, cities, and counties increase funding for public safety, that is, police protection. Transportation and mass transit represent another state and local government function where voters want more, not less, investment and spending.

After all these demands for spending are taken into account, only the cost of government administration remains. In most states and cities, this amounts to one-quarter or less of the total budget. Strong pressure exists to reduce administrative costs in order to free funds for the programs voters want. "Cutting waste in government" in order to free some funding for Medicaid, police and fire protection, and other demands has had universally popular appeal. As a result, budgeters today find relatively little "waste" or administrative costs left to cut.

Expenditures for many state and local governments have increased at rates higher than that of inflation. This causes consternation for most fiscal conservatives, who favor lower rather than higher taxes. Taxpayers prefer small budgets except when it comes to their own special programs, such as special education or the environment, where they often want—indeed, demand—increases in spending. Because of all the demands for increased funding, lawmakers, governors, mayors, and school boards look for ways to increase revenues—without raising taxes, if at all possible. One way to increase current spending for this year's budget is to reduce appropriations for long-term, deferred obligations such as public pensions or infrastructure maintenance.

Pension funding provides an illustration of the competition between current state and local government expenditures and long-term deferred obligations. Each year, state and local governments appropriate monies into one or more retirement systems for state or local government employees, including judges, legislators, administrators, schoolteachers, garbage collectors, and public university faculty and staff. If the pension funds have balances large enough to pay for the current year's obligations to retirees, lawmakers can reduce the annual appropriation without shorting anyone *at the moment*. This often happens when money is tight, as it definitely is during a recession.

As a result of such practices, many government pension funds have "unfunded liabilities," that is, liabilities in the future for which no funds have been put away and invested. In effect, underfunding the pensions shifts the ultimate burden into the future. Many governments are now under pressure to increase their present appropriations to start reducing the unfunded liabilities, which of course will take money away from current expenditures for education, social services, and other functions of governments.

At some point in budgeting, consideration will have to be given to increasing revenues, whether through new or increased taxes, fees, or public gambling like the lottery or licensed casinos. Nearly all state and local officials face the choice of underfunding some critical government functions or having to take unpopular actions like raising taxes, which may cost them dearly when they face reelection.

FINDING THE REVENUE

Elected officials don't like to levy taxes on constituents. But the bills have to be paid and the operating budget balanced. To meet this responsibility, public officials search for revenues that are least painful. For example, an increase of one-quarter of 1 percent in the sales tax adds only a small amount to each purchase. They might levy taxes tied to the use of a product or government service, such as a motor fuel tax and vehicle license fees for the use of the roads. One source of voluntary, nontax revenue is a state lottery. When you buy a state lottery ticket, about 35 cents of each dollar becomes net profit to the state.

The issue of who pays is complicated. Should governments impose taxes on property, income, goods, services, gasoline, utilities, or the "sins" of consuming tobacco and alcohol? If government taxes property, should farmland, residential, and business property be treated the same, or should certain types of property be taxed at a lower rate? Who should bear the greater burden—individuals or corporations; the wealthy, middle class, or poor citizens? These are tough questions for governors, mayors, school boards and lawmakers who want to be reelected.

Governors, mayors, and lawmakers try to be sensitive to the impact that taxes might have on politically important groups, such as senior citizens, and on the state economy, which directly affects citizens and the future ability of the cities, counties, and state to raise money. For instance, in a few states, pension income of all retirees is exempt from the state income tax. Some states provide special exemptions that reduce property taxes for senior citizens. In some Georgia counties, homeowners over age 62 are exempt from property taxes for schools. Military veterans sometimes receive special tax breaks.

In 2006, revenues for all state and local governments in the United States totaled over $2.7 trillion. Taxes from sales, property, and income comprised about half that total. The largest single category of revenue for state and local governments comes not from these traditional taxes but instead from the category of "fees, charges, and interest," which includes tuition payments of public college and university students. Another major set of fees comes from the sale of motor fuel and driver's and automobile licenses. Together, these fees represent almost almost one-fourth of all state and local income. Federal aid is the second largest source of state and local revenue, at about 16.5 percent of the total.[9]

The Big Three Taxes: Income, Property, and Sales

Most state and local revenue comes from taxes on what we earn (income), what we own (property), and what we buy (sales). The income and sales taxes are collected by state governments, and slices of these tax receipts are often shared with various local governments such as school boards, counties, townships, and cities. The property tax, once the primary source of state government revenue, has instead become the primary tax for local governments. (See Table 9.4 for the state and local tax burden in your state as a percentage of income.)

Forty-three states collect a tax on income at an average of $818 per capita in 2006. Connecticut ($1,653) and Massachusetts ($1,629) generated the most revenue on a per capita basis. By contrast, Florida's state constitution prohibits the enactment of a personal income tax; as a result, approximately three-quarters of Florida's general revenue is from the sales tax. In states such as Texas or Florida, a proposal to enact an income tax would likely result in the end of that lawmaker's career in politics. (See the Struggle for Power feature on funding government in Florida.) But most states depend on income taxes for a large portion of their budgets, while most cities and local governments are prohibited by state laws or constitutions from enacting them.

TABLE 9.4 State and Local Tax Burden as a Percentage of Income in 2007

National percentage = 11.0% of Income

Rank	State	Percentage
1	Vermont	14.1
2	Maine	14.0
3	New York	13.8
4	Rhode Island	12.7
5	Hawaii	12.4
5	Ohio	12.4
7	Wisconsin	12.3
8	Connecticut	12.2
9	Nebraska	11.9
10	New Jersey	11.6
11	California	11.5
11	Minnesota	11.5
13	Arkansas	11.3
14	Kansas	11.2
14	Michigan	11.2

(continued on next page)

TABLE 9.4 State and Local Tax Burden as a Percentage of Income in 2007 *(continued)*

Rank	State	Percentage
16	Washington	11.1
17	Iowa	11.0
17	Louisiana	11.0
17	North Carolina	11.0
20	Kentucky	10.9
20	West Virginia	10.9
22	Illinois	10.8
22	Maryland	10.8
22	Pennsylvania	10.8
25	Indiana	10.7
25	South Carolina	10.7
25	Utah	10.7
28	Massachusetts	10.6
29	Mississippi	10.5
30	Colorado	10.4
31	Arizona	10.3
31	Georgia	10.3
33	Virginia	10.2
34	Idaho	10.1
34	Missouri	10.1
34	Nevada	10.1
37	Florida	10.0
37	Oregon	10.0
39	North Dakota	9.9
40	New Mexico	9.8
41	Montana	9.7
42	Wyoming	9.5
43	Texas	9.3
44	Oklahoma	9.0
44	South Dakota	9.0
46	Alabama	8.8
46	Delaware	8.8
48	Tennessee	8.5
49	New Hampshire	8.0
50	Alaska	6.6
District of Columbia		12.5

Source: Tax Foundation, "State and Local Tax Burdens Compared to Other U.S. States, 1970–2007," www.taxfoundation.org.

STRUGGLE FOR POWER

Funding Government in Florida

In legislative lingo it's referred to as the "I-word," and the mere mention of it can send shivers down a Florida politician's spine. It's a personal "income tax," a revenue source tapped by most state governments but outlawed by the Florida constitution since 1924, when the Sunshine State was on the threshold of a historic land boom. The idea behind the ban was to lure Northerners with income and wealth to Florida. It succeeded. It became a perennial given in state politics. Today, Florida remains one of only nine states without a personal income tax, and it would take a statewide vote to alter that.

With homeowners clamoring for tax relief, legislators are casting around for a way to cut taxes on homes and other types of real estate, but so far remain stumped over how to replace the lost revenue. Some legislators privately say a personal income tax would be the fairest and best way because it would tax each Floridian according to his or her ability to pay—but most Republicans and Democrats agree there is neither the political will nor the requisite popular support for such a drastic change in the constitution and public life.

"I believe the citizens of the state of Florida will adopt a personal income tax shortly after we send out for the snow plows," said Senate Democratic Leader Steve Geller, of Cooper City, who doesn't like the idea and laughed at the thought of proposing it to voters.

"If someone proposes it as a bill, I'm sure we will dispose of it appropriately," said House Rules Chairman David Rivera, R-Miami.

Instead, House Republicans and Democrats have advocated an increase in the state's sales tax. A penny increase in the 6 percent sales tax would raise $3.9 billion a year. In comparison, a 1 percent tax on people's income would result in roughly the same revenue for state coffers—$3.1 billion, according to state economists. But opposition is deep and broad among politicians and the public to empowering the state to take a cut of a Floridian's annual earnings, as opposed to tacking pennies onto the price of a purchase. Even suggesting a state income tax has meant political suicide, as Sam Bell found when his Central Florida constituents kicked him out of the legislature in 1988 . . . . Now a Tallahassee-based lawyer and lobbyist, Bell said the legislature is maneuvering itself into a position where it might not have many other options left . . . .

When it comes to taxation, the burden on Floridians is less than on residents of most states. As of March 2006, according to the U.S. Census Bureau, Floridians were paying $1,756.36 apiece to run state government—$258 below the national average, and ranking Florida 36th in the nation based on how much each resident pays to help run schools, prisons, the courts, and other programs, including Everglades cleanup and road construction. For Florida, the largest single source of state revenue is the sales tax, which will raise almost $20.4 billion in the state budget year that begins July 1. Tax specialists and lobbyists for the poor contend that an increase in the sales tax—though food and medicine are exempt—hurts the poor, costing them a disproportionate amount of money in relation to what they earn . . . .

Bell said that state leaders would have to be desperate for new revenue to resolve some as-yet unforeseen crisis to even consider asking voters to repeal the ban on personal income tax. But he warned that the legislature could be moving in that direction if it makes government more reliant on a sales tax that goes up or down with the economy, especially since already this year legislators have committed the state to paying off possibly billions in homeowner insurance claims in the event of a disastrous hurricane.

"Once you have a real need, like a couple of really bad storm years, you'll have to find a revenue source," Bell said.

Source: Adapted from Linda Kleindienst, "Income Tax? Mention It at Your Peril," *South Florida Sun-Sentinel*, April 9, 2007.

The tax on real estate property (homes, farmland, and commercial-industrial properties) is a primary source of revenue for local governments, especially for funding schools. Property is assessed as to its value and then a percentage of that assessed value is captured as the annual property tax. Often the assessed value is a percentage of the actual market value. The property tax is usually paid in just one or two installments, which often means payments of thousands of dollars. So many homeowners pay the property tax along with their monthly home mortgage payments, as a way of spreading out the pain.

New Jersey ($2,217 per capita) and Connecticut ($2,052) generated the greatest amounts of property taxes, which averaged $1,563 per capita nationwide in 2005. Alabama collected the least: only $395 per capita.[10] All but 3 percent of the total of $336 billion in property taxes nationwide went to local governments.

But property taxes are even more of a burden when homes and businesses lose 30 percent or more of their true market value, as occurred during the financial crisis that hit the United States beginning in 2007. Property tax assessment by local government does not quickly change to reflect a new reality, so many properties continue to be assessed at their pre-recession value. In a recession with job cutbacks and reduced income, many property owners will find it harder to pay that tax bill.

Property taxes for homes of similar value may vary dramatically, even within a state. Figures 9.1 and 9.2 show property tax bills for two Georgia homes, one in Cobb County, near Atlanta, and one in Glynn County, along the Atlantic coast. As shown in the figures, the two homes have similar assessed valuations, about $131,000 and $172,000, but the property tax owed for the home in Glynn County is more than four times greater than that for the home in Cobb County, $3,522 versus $764. The reason for the difference is that the home-owner is over age 63. In Cobb County, people over age 63 are exempt from paying the school general and bond levies. In Glynn County, senior citizens are not exempt from the school property taxes, which represent more than three-fourths of the total tax due in that county.

©JOHN S. PRITCHETT

PROPERTY TAX:
A VOTER'S REACTION
TO THEIR TAX BILL
by John S. Pritchett.

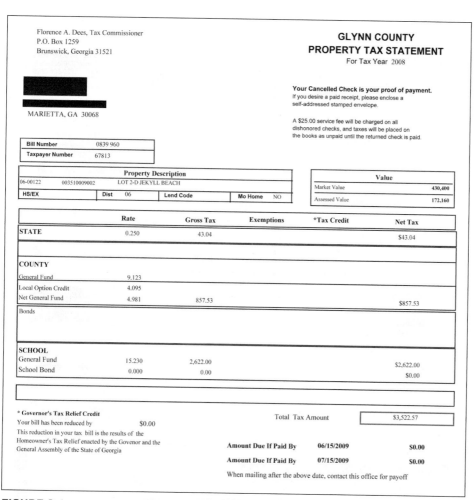

FIGURE 9.1 Property Tax Statement, Glynn County, Georgia

Many states levy taxes on sales but may exempt certain items such as food and medicine. In 2006, state sales taxes generated an average of $758 per capita. Hawaii used the sales tax most aggressively, extracting $1,842 per capita. Alaska, Delaware, Montana, New Hampshire, and Oregon had no general sales tax.[11] One new reality facing state governments, however, is the amount of goods sold via the Internet, where online spending for the first half of 2006 totaled $80.8 billion.[12] While there may be a sales tax on purchases in stores on Main Street or in the mall, sales over the Internet are generally not taxed. As a result of the shift toward greater Internet sales, state and local governments are losing significant sales tax revenue and will lose even more unless Internet sales are taxed just like sales in stores. Part of the problem is that each state that taxes sales has its own tax system, with a unique set of rates and exclusions. There is no uniform tax that could be universally applied to Internet sales. For example, some states tax food for home consumption; some do not. In addition, many local governments within a state tax sales at

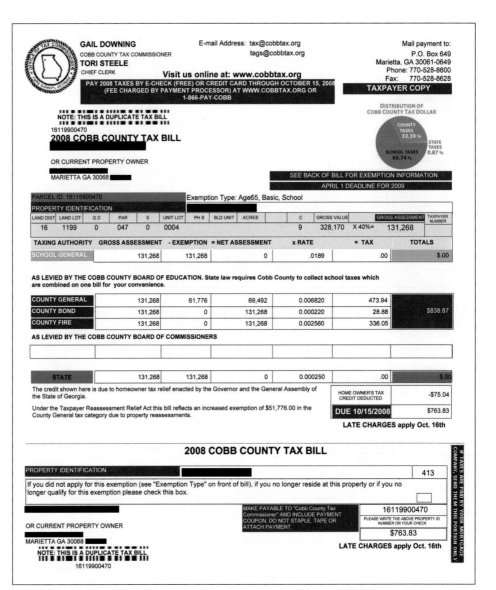

FIGURE 9.2 Property Tax Statement, Cobb County, Georgia

different rates. Another issue is whether the tax should be paid to the state where the customer lives and works or the state that is home to the company that makes the product or provides the service.

The National Conference of State Legislatures has created a Streamlined Sales Tax Project to try to resolve this dilemma. Diane Hardt, director of sales and income taxes for the state of Wisconsin, co-chaired the task force trying to streamline the sales tax and convinced 20 states, as of 2003, to come together to streamline their sales taxes. Since then, more have done so. But Internet sales taxes have yet to be uniformly and effectively enacted. Only if streamlining occurs will businesses selling over the Internet to consumers in 50 different tax systems be able to collect and distribute the appropriate tax.[13]

Lotteries, Gambling, and Sin Taxes

In the quest for less painful revenue sources, policy-makers have been attracted to taxing "sin." **Sin taxes** are taxes levied on goods or services that many consider vices, such as tobacco, alcoholic beverages, or gambling. The purpose of such sin taxes is not only to raise revenue but also to discourage unhealthy activities. In 2007, cigarette taxes per pack ranged from 7 cents in South Carolina to $2.58 in New Jersey, and the average state cigarette tax per pack was about $1.[14] Most state and local governments along with the national government keep raising cigarette taxes.

One form of gambling, lotteries, is a big source of revenue in states, and many states allocate part or all of the lottery profits to funding education. The typical family in Ohio in 2007 spent more than $500 annually on lottery tickets.[15] As of 2008, 42 states operated lotteries, and a number of states had legalized casino and riverboat gambling.[16] With a population of 6.2 million, Indiana generates about $600 million in net revenues from casino gaming, or about $100 per resident.[17] Even with the large amount of gambling done in states where it is legal, gambling usually represents only 1–6 percent of total state revenue, but in Nevada, gambling is the primary revenue source.

Legal gambling does not come without costs to the state. It is widely assumed that poor families spend higher percentages of their income than the wealthy on lotteries and other state-sanctioned games. According to one observer, the biggest negative is that the state legitimizes making new gamblers: "The state's slick television ads sustain hope among many, especially the poor, of a big hit, a jackpot, and then let the good times roll. Forget hard work. Saving is for suckers. Why sacrifice for the future when the future is now? And, like alcohol, widespread gambling reveals more and more people for whom it is an addiction. What a regressive way to serve the public."[18]

Indeed, with state-operated and authorized gambling, citizens must lose for the state to be a winner. On the other hand, gambling revenue provides billions of dollars each year for schools and social services for people in need. Gambling revenue is attractive to many governors and state lawmakers because participation seems voluntary and the revenues from gambling may make it possible to avoid increasing a general tax on all voters. When at the slot machines, gamblers don't think about subsidizing state government, whereas taxpayers instantly recognize a tax increase and, often, vote public officials out of office who pass laws increasing taxes.

Lottery tickets many people buy.

One-Time Revenue Sources and Borrowing

To find extra operating revenue during times of budget stress, governments may use one-time revenue sources or borrowing for operating purposes. Public finance economists frown on these techniques, as they have the effect of borrowing from the future, but often they are used to balance state and local government budgets.

One-time revenue sources used to help balance state and local government budgets include tax amnesty programs and "fund sweeps." Many delinquent taxpayers face large fines and interest for failing or refusing to pay their taxes, such as state income taxes or city parking tickets. Amnesty programs seek to generate cash quickly by offering to waive the interest or fines owed for payment of delinquent taxes during the amnesty period.

Governments may also borrow short-term against future anticipated income in order to provide immediate revenues. Borrowing against the future to fund current operations is a bad idea, as the government may find itself short of revenues when the future arrives.

Rainy Day Funds

In good financial times, some state and local governments put a small percentage of annual revenues into a dedicated fund known as a **rainy day fund**, for use during weak economic times when revenues might not be adequate to meet demands for services. In a recession, for example, when a state's or city's economy weakens and tax revenues decline, officials can dip into the fund to provide money to cover the state's revenue shortfall. Many states but very few cities have created such funds.

Proponents say rainy day funds smooth out state and local government revenues, which tend to be cyclical, and reduce the need to consider increases in taxes when budgets might otherwise be in deficit. However, when budgets are tight, it is hard to put money into these funds for some future crisis. Still, although rainy day funds could not offset all the lost revenues during the recession of 2008–2009, cities and states with rainy day funds had a much easier time balancing their budgets.

THE REGRESSIVE VS. PROGRESSIVE DEBATE

The only good tax is one you pay and I don't! Yet if schools, universities, and myriad other programs and services are to be provided, tax we must. Public finance experts evaluate the efficacy of revenue sources on such factors as *yield or capacity*, that is, will the revenue source generate the amount of revenue needed. Other factors are *ease of administration*, *public acceptability*, and *fairness*.

Yields from a tobacco tax tend to decline over time, as fewer citizens smoke because of the cost and for health reasons. Property taxes produce a good yield but are complicated to administer. Moreover, many homeowners and businesses think their properties are assessed at higher values than similar homes or businesses, making property taxes less acceptable than other taxes. Notwithstanding administration and public acceptance issues, the yield of the property tax is so important to local governments that the tax cannot be abolished.

Fairness in taxation is in the eye of the beholder. For decades, this issue has been framed by the terms regressive, progressive, and proportional. A **regressive tax** is one that extracts a higher proportion or percentage of the income from a poor person than from a wealthy person. For instance, a poor couple with a large family may pay a much higher percentage of their income on a sales tax on food to feed their family than does a wealthy person. Many see this as unfair, because a greater burden is placed on a poor person. So a sales tax on food is seen as regressive, and some governments exempt food from their sales tax.

A **proportional tax** is one that extracts the same percentage of tax from each person's income. For example, many states have "flat rate" income taxes: the same percentage applies to all income.

A **progressive tax** takes a higher percentage as the income level increases. For example, low-income people might pay 1 percent of their income in taxes, whereas people earning more than $100,000 a year might pay 5 percent of their income in taxes. The premise is that higher-income earners have greater capacity to pay taxes and have also benefited more from the government that maintains order, harmony, and other conditions that allow them to be prosperous. Thus higher-income taxpayers should pay a higher percentage of tax on their income.

THE BUDGETING PROCESS

Budgeting is a simple five-step process: (1) find out what you have in the bank; (2) estimate your receipts; (3) decide what you want to have in the bank at the end of the budget period; (4) subtract item 3 from the first two items; and (5) allocate that amount among competing programs.[19] Simple, perhaps, but never easy.

Demands will almost always outstrip resources, and the growth generated primarily by inflation will not be enough to cover new programs. Elected officials, always interested in trying new programs, often declare that their pet program for school reform or alcohol treatment can be paid for from "natural growth" in annual revenues. True, revenues tend to grow with inflation and growth in a state or city's economy. Unfortunately, demands for employee pay increases, increased eligibility for Medicaid coverage, education spending, and unfunded pension liabilities also grow somewhat naturally, using up the increased revenues, leaving nothing for new programs.

Unless some new revenue stream is available, most of the spending for the coming year's budget will mirror that of the previous year. Education, transportation, health care, assistance to the poor and abused, and state agencies consume most of the budget every year, leaving little room for dramatic changes at the state level. At the city level, major expenditures for necessities such as public safety (police and fire protection), sanitation (garbage pickup), street repairs and traffic control, and water and sewer services leave only a small amount of the budget for human and social services.[20]

As former Illinois state budget director Robert Mandeville observed:

> Working on the budget is like spending an afternoon watching *Poltergeist II* and the *Attack of the Killer Bees*. Just as Carol Anne yelled, "They're Baaack!!" so lobbyists and advocates will soon descend upon the state capitol arguing for a bigger piece of the budget pie. . . . Killer bees have not yet reached [this state], but the Killer P's have. Pensions, Public Aid (Medicaid), and Prisons have all made their assaults on the state's programs.
>
> These programs are not killers in themselves. They are legitimate program needs that deserve as high a level of funding as possible. They can be budget killers because they

require large increments to make changes. . . . Full funding of the pension systems is a noble goal. Cutting the funding from the poor to pay for state pensions is not. Restructuring Medicaid to make poor children healthier is a noble goal. Cutting the funding for poor children's education to fund Medicaid restructuring is not. If passing tough sentencing laws to put criminals behind bars is a noble thing to do, providing taxes to build the prisons and pay the guards is a noble thing.[21]

Government budgeting is difficult because state and city budget demands tend to be countercyclical to the economy. When the economy declines, demand for public assistance, health care, and other public services increases. When unemployment rises, so does enrollment in higher education as people turn to community colleges and other campuses to improve their skills and employability. At the local level, when the economy weakens and the tax revenue falls, there is an increase in the number of homeless for whom city governments try to provide shelter and food. The countercyclical nature of budgeting creates serious problems for budgeteers because as demand for state and city dollars goes up, tax revenues decline. That is why forward-thinking cities and states maintain a rainy day fund that can be used during economic recessions to avoid raising taxes.

Budgeting is a tough game. Satisfying everyone is impossible. Satisfying even a majority of voters is almost impossible.

Public Budgeting: A Primer

The government budget process requires a series of interactions between the legislative and executive branches, as depicted in Figure 9.3. A typical budgeting process goes like this:

1. Each *government agency* prepares a request and sends it to the budget office.
2. The *budget office* (Office of Management and Budget) reviews the requests, then prepares a draft budget for chief executive.
3. The *chief executive* approves the budget proposal and sends it to the legislature.
4. The *legislature* reviews the proposal, revises it, votes on it, and sends it to the chief executive to sign.

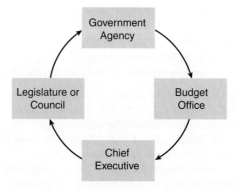

FIGURE 9.3 Budget Process Interactions

5. The *chief executive* either signs the legislation, vetoes it, or possibly uses a line-item veto on some of it. A veto or line-item veto will mean sending it back to the legislature.

6. The *legislature* votes to accept or override the veto which restores items to the budget.

Budget decisions are always difficult. Here are some budgeting realities:

- Zero-base, performance-based, and other budgeting techniques are ideals seldom achieved because of time and other pressures and the difficulty of measuring performance.
- Every spending line in a budget has vocal advocates; opponents are rare.
- It is easy to start a new program and almost impossible to end it.
- It is easy to vote for programs that benefit people now and are paid for later.
- Almost every budget action tends to benefit one sector and hurt another.

The Office of Management and Budget

A government budget office can be called the Bureau of the Budget, Department of Finance, Budget Agency, or **Office of Management and Budget** (OMB). Regardless of its title, the budget office is responsible for developing the budget for the chief executive, or in a few cases, for the legislature or city council, especially in smaller towns and villages.[22] The budget sets out the expenditures and revenue sources and tax rates approved by the government, usually on an annual basis, but some states have two-year, or biennial, budgets.

About six months to a year before a new fiscal year commences, agencies of a government are asked to prepare their requests for the coming year. The OMB may have provided guidelines for the agency submission. For example, OMB might ask how each agency would reduce its programs if it received only 98 percent, 96 percent, or 94 percent, respectively, of the current year's level. Or OMB might ask agencies to prepare budget requests at 90 percent, 100 percent, and 110 percent of the current year's appropriation. And, of course, the agency would have to justify each request for new program expenditures and other proposed increases, ideally on the basis of measurable outcomes.

In states and big cities, major agencies each have an "analyst" at OMB, who is responsible for knowing the agency, its problems and challenges, and its appropriations history. The analyst receives the agency budget request at OMB. All agency requests are typically evaluated within a context of the overall budget picture for the coming fiscal year. Will revenues likely be up or down, and by how much? In addition, does the chief executive have any program initiatives she or he wants to propose, at what cost, and how might that affect the rest of the budget?

From all the agency requests, outside funding estimates, and the chief executive's program priorities, OMB develops a proposed budget for the coming fiscal year, which OMB then submits to the governor, mayor, legislature, or city council.

The bureau analyst is an important player in the budget process, and that role is often filled by a young professional, often with a master's degree in a field such as public policy, political science, economics, public administration, or public management. The typical analyst is neither an advocate nor an opponent of the agency's goals. The agency director, interest groups, and legislators play the advocacy role. Agency directors often attempt to

AN INSIDER'S VIEW

Linking Budgets to Results: Performance-Based Budgeting

In the late 1990s, as an alderman in Somerville, Massachusetts, Joseph Curtatone was perpetually frustrated by the budgets his local legislature was supposed to be helping shape and approve. "Budget time really used to get me," he recalls. "It was a straight line-item budget. . . . [T]here was nothing that told you how much we spent on what—no inputs, outputs or outcomes." It was classic best-guess budgeting, coupled with a typical tactic: Any department that had any money left in its account at year's end was guaranteed to get a budget cut. . . . "So the message to all our departments was, 'Spend down your budget.' "

While this was one of the most significant points of aggravation, Curtatone also was nettled by how the city managed its decisions in general. "Here we were, a multimillion-dollar operation with absolutely no real-time information on even the most basic services. We weren't measuring anything. How many potholes were we filling? How were we filling them?" . . .

[But despite] the continuing state of general ignorance or indifference when it comes to using data to make important decisions, the ranks of governments that are turning to more results-based programming and budgeting seems to be slowly adding up, frequently because one or two people, such as Curtatone, . . . become frustrated enough to try moving government in a new direction. . . .

His early budgets had both line items and performance-related costing to ease them into the program. Now, using data to discuss budgeting has become standard operating procedure within his local legislature.

In Somerville, Curtatone says he ran for mayor on a straight platform of "performance-based budgeting and costing out and measuring activities and results." He adds with a laugh that a certain mayor just across the Charles River in Boston "doesn't believe in all this Activity-Based Costing stuff—yet." But in Somerville, citizens were certainly ready to support a mayor who knows his ABCs.

Source: Jonathan Walters, "Data-Driven Decisions," *Governing*, June 2007, 76–78.

increase the funding, number of employees, and the programs their agency runs. The analyst's job is to sit on the chief executive's side of the table and constructively suggest that perhaps there may be better, more economical approaches for meeting the goals of a particular agency. After all, every dollar saved in one area can be spent elsewhere in a budget that is unlikely to fulfill all the outside demands as well as the chief executive's own program initiatives and promises.

Analysts may also be asked to represent the executive's views to the legislature or its staff. Sometimes analysts have to defend a proposal they may have advised the executive against. For example, even though an analyst may believe that state scholarships should be awarded on the basis of financial need, if the governor decides that scholarships should be awarded on the basis of merit, then the analyst is obliged to support that decision. It is the same for agency heads. They may argue with the chief executive internally that scholarships should be distributed in a particular fashion, or that their agency must have more funding, but publicly and in legislative hearings, they are expected to support the chief executive's policies.

Legislatures and major city councils often have their own budget analysts, who work for budget or appropriations committees. The role of these analysts is similar to that of their counterparts at OMB, although they often represent a political party's membership in the legislative body.

After the annual budget process is complete, the chief executive has the responsibility to match spending with revenue. Budgeting at this stage is basically cash management, and it is challenging. The monthly balance in the state's general fund often falls dangerously low. Payment of bills may have to be postponed, and vendors may be put in a queue for payment. Management of the budget falls to the OMB, which must ensure that budget promises are being met. Budget management often falls as well to a state or city comptroller, who often holds the state or city checkbook and must delay payments of valid claims to avoid overdrawing the accounts.

The checkbook balance of cash on hand can give a misleading picture of the fiscal situation of a state or city because the balance is subject to manipulation. The balance can be built up by delaying payments to nursing homes, hospitals, and school districts and by speeding up collections of certain taxes. If a governor wants to help build a case for a tax increase, a budget office can plunge the available balance into the fiscal warning zone by speeding up the flow of payment requests to the comptroller and by slowing the processing of collections.

In recent decades, government officials have begun to consider performance-based budgeting—linking budget dollars to results, as detailed in the Insider's View feature.

FUNDING CAPITAL EXPENSES

In addition to expenditures from current revenues for operating purposes, state and local governments borrow to make long-term investments in highways, mass transit, new streets, university laboratories, elementary school classrooms, wastewater treatment plants, prisons, jails, airports, domed stadiums, convention centers, sewer and water facilities, and other "capital" or physical infrastructure projects.

Just as individuals pay for a home with a mortgage, state and local governments borrow money and retire the debt by paying off principal and interest of bonds over 10 to 30 years. A state or major city may have one or more agencies authorized to borrow money by issuing bonds, which are sold to investors with the promise to repay the principal amount plus interest on the use of the money. As of 2006, state and local debt in America totaled over $2 trillion, or almost $7,400 for every person in the nation.[23] That is a high but not yet impossible debt level. However, the debt level for some cities and states can get much higher. For instance, both New York City and the Chicago public school system became essentially bankrupt in the 1970s, their financial management taken over by specially created outside agencies. They did not regain full control of their budgets until their debts were brought to a more reasonable level. Hard hit by recession in 2008–2009, a number of smaller cities and counties considered filing for bankruptcy. But this would mean that court officials rather than local government officials would control local government budgets for many years in the future.

The politics of who gets which capital projects are intense because big money, political credits, and local boosterism are involved in the competition. Although capital spending differs from operating expenses, the two budgets are nevertheless in competition. For example, Illinois created a special tax on used cars in the 1990s, and the revenue from that tax was to be dedicated to "Build Illinois" bond indebtedness projects. In the first year, however, the used car tax generated only $41 million of the $83 million needed. As a result, an additional

$42 million in general revenues had to be deposited in the Build Illinois account, money that might otherwise have been available for education or child care services.[24]

Taxpayer-supported sports stadiums and convention centers are capital projects paid for over many years with revenues from varied sources. Among the domed stadiums built in the 1990s with taxpayer support were the Georgia Dome (Atlanta), Metrodome (Minneapolis), Superdome (New Orleans), Kingdome (Seattle), Alamodome (San Antonio), and the RCA Dome (Indianapolis). In some instances, government officials have the authority to issue the bonds, but many states require referendums in which citizens vote on whether to go into long-term debt for such projects. Voters sometimes reject them, especially if general property or sales taxes from all taxpayers will be funding projects that provide events for specialized audiences.[25] Yet city and state pride are at stake, as well as the all-critical tourism dollars for hotels, restaurants, and museums, and the all-important money from hotel, entertainment, and sales taxes to help fund future government budgets.

Elected officials and taxpayers should carefully evaluate these proposals. Are the benefits over time greater than the costs? Is the project multipurpose (generally better than single purpose) and does the proposed location offer adequate hotel rooms, transportation, and other infrastructure support? Would the project, or something similar, be built without taxpayer support?

The tax and revenue sources to pay off bonds should also be evaluated carefully. Multiple sources are considered better than a single revenue source. Will the revenue sources grow with inflation over the course of the debt financing, often 30 years? The sources of revenue for stadium financing have included taxes on property, sales, hotel rooms, liquor, tobacco, car rentals, and airport departures, and revenue from taxes on tickets and refreshments sold at events. Because each of these revenue sources affects different sets of payers, interest-group lobbying for and against each proposed revenue source is spirited.

There is a final important consideration about tourism projects such as stadiums and convention centers. Once the enabling tax and funding legislation is passed, these facilities are often run by independent agencies, not by the general state or city government, and are thus removed from direct control by elected representatives and become unaccountable to the voters. Since they frequently tie up hundreds of millions of dollars in tax and fee revenues to pay off the bonds by which they are built, they have a major impact on state and local government. These projects may turn out to either be great or terrible decisions. They may spur economic development and increase overall tax revenue or they may sap vital resources and return little benefit. So capital projects need careful consideration before they are begun, and mayors and governors must carefully appoint the board members for independent agencies to ensure appropriate monitoring and accountability.

THE POLITICS OF SHIFT AND SHAFT

The politics of federalism can be seen in large part as a tug of war in which each level of government—federal, state, and local—tries to expand its authority over programs while inducing the others to pay for the changes. For example, through legislation in the 1970s, the federal government began to require programs for educating the physically and mentally challenged. Although Congress authorized the federal government to fund about 40 percent of state costs, they actually never covered more than 18 percent of total state costs for special education.[26] Just as this legislation and the later No Child Left Behind Act imposed education requirements on states, state legislatures often impose mandates on local schools, shifting the cost from state taxes to local property taxes. For example, a state legislature might mandate that local schools teach a unit on drug and alcohol abuse, but not provide any funding to implement the program.

Local government leaders are also skilled at playing the game of shift and shaft. In 1993 in Illinois, for example, municipal leaders negotiated an increase from one-twelfth to one-tenth in the share of state income tax revenues automatically distributed to cities and counties, thus reducing the amount available to be spent in the state budget. In 2004, Michigan passed a law requiring real estate owners to pay part of their county property taxes five months sooner the next year, a move that amounted to a tax increase, according to some analysts.[27]

One justification for shifting money from one government to another is that the government level that receives the funds can deliver the service more effectively. These monies can also act as a carrot to induce state or local governments to initiate or change programs. Prominent examples of federally supported but state-administered programs are Temporary Assistance to Needy Families, Medicaid, and highway construction. Being able to play the game of shifting and shafting is important to state and local officials, since so much of their revenue comes from a higher governmental level. As noted earlier, about 16.5 percent of total state and local revenue in 2006 came from the federal government, with most of it going primarily to state governments.

State budgets are even more important to cities and counties than is the federal budget, in part because most federal dollars headed for local governments must first pass through the state. About 60 percent of all state expenditures are distributed in grants and awards to local governments, nongovernmental agencies, and individuals, including public schools, community colleges, roads and highways, and social service agencies. Local governments also often share a portion of income and sales taxes collected by the states.

AN' TAR-BABY DON'T SAY NOTHIN'

STATE MANDATES: ENTRAP
LOCAL GOVERNMENT
by Bill Campbell.

Source: Campbell Cartoon Collection/Carl
Sandburg College.

THE POLITICAL DYNAMICS OF BUDGETING

Years ago, political scientist Thomas Anton saw budgeting as a ritualized game in which each interested party played its part and, if all went well, everyone came out a winner. The agencies would request more than they needed, and the governor and legislature would each make some cuts. The final budget would provide the agencies what they actually wanted—sometimes even more than expected—and the politicos could claim to have cut the budget.[28]

Players in the budget game still anticipate the actions of others. Since the 1960s, however, budgeting at the state and local government levels has generally become more professional. Executives have added Offices of Management and Budget staffed by professionals with strong analytical skills. Legislatures and big-city councils have also created budget staff. In addition to the line-item veto, some executives have a **reduction veto** that enables them to reduce particular appropriations. As a result, state agencies probably have less influence than in an earlier era and the legislature has become more involved in the substance of budget negotiations.

Annual budget proposals are transformed into scores of appropriations bills, often one for each major agency. Sometimes the budget is presented to the legislative body in an "omnibus" budget bill. For instance, the Chicago city budget of nearly $6 billion is about 400 pages long with more than 200,000 line items. Most aldermen don't even pretend to understand the budget. Instead, they analyze only the part of the budget that delivers specific services, like street repair, to their individual ward. To make this easier for them, the Chicago city government gives $1.2 million to each alderman to distribute as she or he chooses among a menu of approved city services, such as street, curb, and sidewalk repair. This means an individual alderman directly controls only $1.2 million of expenditures in a $6 billion budget.

Throughout the process, interest groups and beneficiaries of state and city spending press their claims. If budget recommendations are seen as inadequate in a particular area, there are cries of catastrophe. University presidents cite a brain drain of faculty who are leaving for higher salaries elsewhere. Nonprofit agencies that provide services to the homeless decry too little funding, which will cause the homeless to die on the streets next winter. The rhetoric often heats up. As one state school superintendent wrote in a press release: "School districts are going down the financial tube; [city] school reform is jeopardized before it gets underway. . . . We are in disgraceful condition and it's time for the people of this state to say we've had enough."[29]

Chief executives generally dominate the budget process because they shape the original budget. But the legislative branch has the capacity to impose its own spending priorities. Indeed, in Texas the state's biennial budget is prepared by the Legislative Budget Board, which is chaired by the lieutenant governor and composed solely of legislators.

When policy-makers decide cuts have to be made, the tools for making the cuts are less than ideal. Information is always imperfect and inadequate, and decisions must be made quickly and under pressure. One technique is "across the board cuts," in which every agency receives the same percentage cut in its appropriation. This blunt instrument fails to take into account that the needs of some agencies and programs are greater in a particular year than those of others, or that some functions of government should have higher priority or value to the state and local government than more mundane and less essential services.

In the 1990s, new Illinois Governor Jim Edgar had a budget badly out of balance and an economy in recession. The required costs of Medicaid and pensions were increasing

STRUGGLE FOR POWER

Ohio Cities Feel Budget Crunch

Toledo's not the only city in Ohio digging through its pockets for spare change. While there is much handwringing and gnashing of teeth at One Government Center over the city's budget woes, other big city governments in this state are suffering similar pains.

Ohio law requires all city governments to pass balanced budgets, so it's the process of making revenues match up with expenditures that hurts the most. "You do it by tightening your belt," Columbus Communications Director Mike Brown said. "But tightening your belt can be difficult."

Toledo, which is facing an $11.9 million deficit in 2007 and a projected $17 million in 2008, has laid off 23 city employees and has proposed to shut swimming pools, not hire police and firefighters classes this year, and charge residents $6 a month for garbage pickup.

In Akron, Mayor Don Plusquellic is asking residents to vote for the first income tax increase for city expenses in 26 years. Officials there say the tax hike isn't to combat a deficit, but additional funds are needed to hire more police and boost economic development efforts.

And in Cincinnati, government leaders made about $15 million in cuts this year, with the possibility of more cuts in 2008. "You can want all the stuff you want, but if you can't pay for it, you can't pay for it," Cincinnati City Councilman Jeff Berding said. "But at the same time, cities have to grow their way out of these problems, not cut their way out. If you continue to cut services the middle class wants, they will continue to move to the suburbs."

Mr. Berding's point illustrates the basic conundrum facing all those trapped in a budget crisis: How do city leaders increase revenues and decrease spending? In Columbus, Mr. Brown said the city had flat income tax revenues and increasing budgets during the first part of this decade, causing Mayor Michael Coleman to take action. He laid off 30 city employees in 2003, sparing police and fire, made a habit of dipping into the city's reserve funds to pay off debts, and insisted the city spend less than it budgeted.

Mr. Brown said Columbus hasn't faced a budget deficit since those layoffs, but the city has had some good luck to help change its fortunes.

Joel Taylor, Columbus' director of finance and management, said the city received a one-time debt payment of $55 million in 2004 from the Solid Waste Authority of Central Ohio that went directly into the city's reserve fund. Mr. Taylor said the waste authority sold bonds to reduce its debt on a lease with Columbus on a trash-burning power plant. "Without that payment, that fund would've been depleted long ago," Mr. Taylor said.

Cincinnati's Mr. Berding said that to grow a tax base and increase revenue, a city must make sure its residents are safe and its economic climate is attractive for businesses. This year and next, Cincinnati will hire 60 new police officers. About $500,000 also was budgeted for a program to tear down dilapidated buildings.

Akron officials said Mr. Plusquellic's proposed tax increase is meant for those same principles - boosting safety and economic development. Mr. Plusquellic's proposal calls for an income tax boost of .33 percent that would generate about $17 million annually. The revenues would, in part, pay for 25 additional police officers. The money also would go to economic development and assorted capital projects.

Communications Director Mark Williamson said Akron has about 1,600 fewer city employees than it did 25 years ago, though a mayor has not laid off a city employee in 20 years. "We've created neighborhoods, built homes, and done so much with so little," Mr. Williamson said. "We can no longer provide the things our citizens have grown accustomed to without [the tax increase]."

In 2005, officials in Dayton identified a series of troubling trends they said could lead to a budget shortfall of $90 million by 2010. To counter the looming crisis, a committee of government staff, local union leaders, corporate representatives, and residents drafted a comprehensive plan to curtail spending and boost development.

Barbara LaBrier, Dayton's budget director, said the comprehensive effort has worked so far but included some painful side effects. Ms. LaBrier said the city had a projected budget deficit of $9 million in 2006, yet actually ended the year with a $1.6 million surplus. She said Dayton's strategy is to stem the tide of deficits by a systematic reduction in work force. Mostly through attrition, Ms. LaBrier said Dayton has cut 510 jobs since 2001, including 78 police positions. She said the city would reduce its work force by 2 percent each year through 2010.

(continued)

(continued from p. 241)

Ms. LaBrier also said no full-time employees have been laid off since 2001 and only a handful of part-time staff members have received pink slips. "Whether or not we can continue to do it this way into the future is the question," she said.

Ms. LaBrier said the good news for Dayton is the city would pour $3.5 million into an economic development fund this year and already has $21.3 million in a reserve fund. The bad news, she said, is a projected shortfall of $11.6 million for this year. Dayton must also continue to deal with the fallout from Delphi Corp.'s financial troubles, which caused the auto-parts supplier to announce several plant closings in the Dayton area.

The foresight that saved Dayton last year accomplished similar wonders for Cleveland, according to Mayor Frank Jackson. During his state of the city address on March 1, Mr. Jackson said rising costs and flat revenues were projected to put Cleveland's 2007 budget in a $20 million to $30 million deficit. So the mayor mandated a 3 percent reduction in costs citywide, which he said saved Cleveland $16.4 million. "This savings, along with other savings and cost reductions, and a 1 percent increase over budgeted income tax, allowed the 2007 general fund budget to be balanced, averting $20 million worth of layoffs and preserving the jobs of hundreds of employees," Mayor Jackson said during his address.

Like most other heads of cities, Mayor Jackson said he valued public safety and said he would be hiring 30 additional police officers.

Source: Joe Vardon, "Other Ohio Cities Face Balancing Act," *Toledo Blade*, March 18, 2007.

substantially. Edgar had to make deep cuts in his first budget. In an effort to guide the governor, analysts for a business trade organization categorized each budget program and item as either essential (must be funded), important but not essential (should be funded), or "nice but not necessary." Although several billion dollars could have been eliminated by striking the "nice but not necessary" items, Edgar rejected this approach, saying that voters did not really want to eliminate long-standing programs like fairs or support for the arts or support for rural development. Edgar did have one bottom-line criterion: "Will anyone die if I fail to provide the funding?"

For several years into his first term, Edgar delayed paying bills (in effect, running up the overdue balance on the state's "credit card") and reduced funding for many programs, while meeting the requirements for increased funding for Medicaid. As he had promised in his campaign for election, he did not propose raising taxes. After about four years, the Illinois economy revived, unpaid bills could be paid, and modest increases in most programs could again be accommodated.

In his second term in office, Edgar proposed a major tax swap of income tax increases for property tax reductions and increased funding for education. His proposal appeared to have the votes to pass both houses of the legislature. The tax swap never became law, however, because the president of the Illinois senate, like Governor Edgar a Republican, refused to call the bill for a vote. Coming from the prosperous suburbs, the senate president believed, correctly, that his voters would end up paying more in taxes while taxpayers downstate, where Edgar resided, would generally pay less. A single player in the budget game can block the will of a governor and a majority of the members of the legislature.

Since cities must obey often strict balanced-budget guidelines imposed by their states, local government flexibility is often less than that of states, as seen in the Struggle for Power feature about Ohio cities.

In summary, the budget represents accommodation. Seldom does any participant—agency head, state budget analyst, governor, mayor, lawmaker, legislative staffer, lobbyist—see her or his budget agenda fully satisfied. Each uses influence and power, whether it is expert information, a key vote in committee, veto action, campaign support, or editorial comment to pursue budget goals. Finally, after the legislature or city council responds to the governor's or mayor's vetoes, the final budget roughly approximates the will and values of the society.

BUDGET TRENDS

State and local budget fortunes ebb and flow with the economy. A meltdown of the stock market at the turn of the 21st century and the terrorist attacks of September 11, 2001, brought a recession that hit state and local governments particularly hard. It was a period of major cutbacks in government spending and a reallocation of more spending for public safety. By 2006, most major cities had recovered, as had all but 10 of the 50 states. This led many governments in 2006 to approve tax cuts or invest more heavily in services like roads and education.

In 2008–2009, the country was again gripped by a recession. Income and sales tax receipts fell dramatically, putting most state and local governments into deficit budget positions, that is, spending more than was being generating in tax and fee revenue. Adjustments were thus needed to achieve the balanced budgets required by most state constitutions. For example, California had to adjust its budget continually throughout 2008–2009. However, making major changes in the California budget is difficult because the state constitution requires a two-thirds legislative majority to adopt budgets and increase taxes; this requires support from some of the minority party Republicans. In addition, elected officials must often go to the voters in special elections for approval of budget changes.

Facing a $40+ billion deficit across two years in a $130 billion annual budget, California lawmakers agreed in February 2009 to a series of budget cuts and tax increases, as well as some borrowing, to close the gap. The sales tax was increased by one cent on the dollar; the income tax and vehicle fees were increased. As a result of the increases, it was estimated that a family of four earning $75,000 a year would pay $963 more a year in taxes. At the same time, $7 billion was slashed from public school funding of $58 billion and $2.5 from health and human services. The California State University system planned to cut enrollment, and tuition charges at the state's universities were expected to go up 10 percent. Cuts at the state prison system had officials there contemplating shorter parole terms and sentences, with more credit given for good behavior.

The history of the last two decades illustrates the continued adjustments that governments must make as conditions outside their direct control change. These battles of the budget are not only economic but also political. Generally speaking, Democrats, Republicans, and political factions within both parties have different priorities and philosophies about how government money should be raised and spent. And the political struggle over power—especially in hard times—is intense. The recovery plan adopted by the national government under President Barack Obama's urging has been a great help to cities and states as they recover from the recession: $79 billion was provided to help states forestall cutbacks to local schools, public universities, and police protection; $83 billion was provided for extending unemployment benefits, job training, and health care for the unemployed. However, states are still facing a budget shortfall for 2009 and 2010 of at least $312 billion. This comes at a time when state and city tax revenues are declining.

CHAPTER SUMMARY

Whether they are passed by states, cities, or smaller governments, budgets are the scorecard of politics and the best estimator of who gets what, who pays, and how much. They are the best determiner of who is ahead in the struggle for power and influence in government. Budgets are the plans for spending government resources and determining what services demand the most expenditures. Sources of funds include taxes, fees, federal money, and borrowed money. The big three are income, property, and sales taxes.

Most governments operate with fund accounting. States may have hundreds of separate accounts to segregate money for particular purposes, such as education, health, police, fire, and economic development. Some state constitutions require a balanced budget, that is, revenues must equal expenditures. However, this may not be the case, especially during a recession.

The pressure on most states to spend is enormous. Each government is pressured by interest groups with compelling needs. Yet it is becoming increasingly difficult for states and local governments to find the revenue needed to provide services without raising taxes. Many try to raise revenue in the least painful way possible, such as small increases in sales taxes or imposing user fees.

To determine taxes, officials must consider the *yield or capacity* of a revenue source, *ease of administration*, and *fairness*. Some voters prefer regressive taxes, which take a higher proportion from a low-income person than from a wealthy person; others prefer progressive taxes that do the opposite. A proportional tax extracts the same proportion of tax from each person's income.

States and cities have an Office of Management and Budget (OMB) or some other budget office which is responsible for determining the budget for the chief executive. Budget analysts in these offices develop ways to maximize revenues and economize where possible.

One way of increasing expenditures is borrowing, especially by issuing bonds to repay at a later date with interest. Some cities and states take a chance by approving huge capital expenditures for convention centers and stadiums in the hope that they will bring in income over time.

The budget represents accommodation. Seldom is any participant fully satisfied. Each uses influence and power, whether it is expert information, a key vote in committee, veto action, or public support to pursue budget goals.

KEY TERMS

appropriation The amount that the legislature has authorized the government to spend on a specific budget category, p. 222

balanced budget A budget in which revenues equal expenditures, p. 220

budget Plan for spending financial resources, which include tax revenues, fees, lottery profits, interest income, federal transfers to the states, borrowed money, and monies from other sources, p. 219

general fund The fund that supports general state government operations and major functions such as education, p. 219

Medicaid The federal program that supports health care for low-income people and nursing home care for low-income and even some middle-income elderly; the federal government shares the cost with the states, p. 222

Office of Management and Budget (OMB) The government office responsible for developing the budget for the chief executive or sometimes for the legislature, p. 235

progressive tax A tax set at a higher percentage for higher-level incomes than for lower-level incomes, p. 233

proportional tax A tax set at the same percentage for all income levels, p. 233

rainy day fund Monies set aside in good times for use during weak economic times when revenues might not be adequate to meet demands for services, p. 232

reduction veto A veto that reduces a particular appropriation, p. 240

regressive tax A tax that extracts a higher percentage of the income of a poor person than of a wealthy person, p. 232

separate funds or dedicated funds Funds dedicated to specific purposes and often supported by special fees, p. 220

sin tax A tax levied on goods or services that many consider vices, such as tobacco, alcoholic beverages, or gambling, p. 231

QUESTIONS FOR REVIEW

1. Why is it important for state and local governments to have a balanced budget?

2. What are some policy areas in which state and local government find tremendous pressure to spend?

3. How do sales, income, and property taxes raise revenue for state and local government. Which tends to be least popular among voters?

4. What are the costs and benefits to states of lotteries and other forms of gambling?

5. What is the difference between a progressive tax, a proportional tax, and a regressive tax?

6. What are the steps in a typical state or local government budget process?

7. How can the balance in the general fund be manipulated? Why would a state official try to give a misleading picture of a state's "cash on hand"?

8. How much of state and local revenues come from the federal government? Which programs are most dependent on this revenue?

9. What are some ways that state and local governments deal with budget deficits?

DISCUSSION QUESTIONS

1. How many prisoners does your state have at present? Does that represent a major increase from, say, 1970? How much is spent per inmate annually?

2. How many fees and licenses can you list that you or your parents pay to state and local governments? Are fees for a specific service fairer than general taxes on everyone?

3. What items that you buy are taxed by a sales tax? Are any items you buy exempt from the sales tax? How many cents per dollar is the sales tax in your locality?

4. Does your state impose a tax on individual income? If so, at what rate? If not, do you think your state should impose an income tax? Why or why not?

5. Should sales over the Internet be taxed just like sales in a local store? How would a state benefit from a tax on sales made through the Internet? What are the drawbacks?

6. Does your state authorize gambling? How many different types of gambling? How much of the total state budget comes from the gambling revenue? Do you favor or oppose legal gambling?

7. Which do you think is fairer: a regressive, proportional, or progressive tax? Why?

8. Are there major sports stadiums in the major cities in your state? Did they receive taxpayer support? Do you favor taxpayer support to build such stadiums? Why or why not? What about taxpayer support for convention centers? Should taxpayer approval be required?

PRACTICAL EXERCISES

1. Imagine that you are a governor who will be facing a strong opponent in your reelection campaign. But your next budget proposal has to close a $2 billion hole in the state budget, basically through reductions and/or tax increases. Working with the draft budget shown in the chart on the next page, how would you close the gap? Could you close the gap?

2. Categorize the following programs as either essential, important but not essential, or nice but not necessary. If you had to choose, would you eliminate the nice but not necessary programs, or would you make reductions across most or all of the following programs?

   ■ state arts council (grants to local arts and music groups)
   ■ state humanities council (grants to local groups regarding history, culture, literature, traditions of the state)
   ■ rural affairs economic development center
   ■ public universities and colleges
   ■ tuition aid grants to students at private colleges and universities
   ■ job training programs
   ■ department of veterans affairs
   ■ department of historic preservation
   ■ the state library
   ■ the state fair
   ■ gambling revenue to support horse racing
   ■ convention center bond financing
   ■ office of lieutenant governor
   ■ individual projects sponsored by legislators for their districts: e.g., grants for rural fire trucks, parks, renovations of historic sites, community centers
   ■ state math and science academy, a residential school for talented students, at $25,000 a year per student
   ■ grants to local parks

Budget item	Reduce Expenditures	Add Revenue	Add Spending
Cut Medicaid for poor	$500m		
Cut state aid to schools	$300m		
Reduce hours at parks	$ 25m		
Increase income tax from 3% to 3.5%, or		$1.2b	
Increase income tax from 3% to 4%		$2.5b	
Extend sales tax to 22 services, e.g., barbers, accountants, lawyers		$500m	
Borrow, or		$1.0b	
Borrow		$500m	
Add fees at state parks		$ 25m	
Cut optional services for Medicaid, e.g., dental, eye	$100m		
Legalize video poker in bars		$300m	
Increase college tuition $1,000		$220m	
Increase university spending			$500m
Eliminate aid to private colleges	$300m		
Increase aid to public schools			$500m
Budget deficit reduction	$	$	$+
Net budget deficit reduction			$

WEB SITES FOR FURTHER RESEARCH

For data about state and local government finances, go to the Census Bureau site, www.census.gov.

Information about tax policies is available from the Center for State Tax Policy and Data, at the Tax Foundation site www.taxfoundation.org.

See the Center for Budget and Policy Priorities, www.cbpp.org, for information about budgets.

Other information about the states is available from the Council of State Governments, www.csg.org.

SUGGESTED READINGS

Anton, Thomas. *The Politics of State Expenditure in Illinois.* Urbana: University of Illinois Press, 1966.

Chen, Greg, Dall Forsythe, Lynne Weikart, and Daniel Williams. *Budget Tools: Financial Methods in the Public Sector.* Washington, DC: CQ Press, 2008.

Gove, Samuel K., and James D. Nowlan. *Illinois Politics and Government: The Expanding Metropolitan Frontier.* Lincoln: University of Nebraska Press, 1996.

Rubin, Irene S. *The Politics of Budgeting: Getting and Spending, Borrowing and Balancing,* 5th ed. Washington, DC: CQ Press, 2005.

New York, an example of a global city.

METROPOLITANIZATION AND GLOBALIZATION

Two worldwide developments of great importance affecting our cities and states are **urbanization**, the growth of cities and urban areas, and **globalization**, the growth of connections such as international trade, immigration, and communications throughout the world. Continuing a trend that began a century ago, more people are moving to cities. At the same time, our local and state economies are becoming even more dependent on the sale of goods and services beyond America's borders. For a long time, most people in the United States have lived in metropolitan regions. Now, for the first time in global history, most

FIGURE 10.1
Los Angeles: Urban sprawl.

people in the world live in cities. Through globalization, the connections between cities and the rest of the globe continue to multiply.[1]

In the United States, the replacement of small family farms with huge agribusiness operations has meant that fewer families live on farms or even in rural areas. Many move to cities. In this they are joined by scores of immigrants from abroad. Neither group necessarily moves to inner-city neighborhoods, as migrants and immigrants in previous decades tended to do. They are equally likely to move directly to the suburbs. Since the 1960s, America's suburbs have exploded and now dwarf central cities in population. As of 2000, New York City had 8.1 million people, but the New York greater metropolitan region had over 21 million. Los Angeles's population in 2000 was 3.7 million, but its region had around 16.4 million. Chicago had 2.9 million and its region over 9 million.[2] It is like this for most big cities in the United States. Figure 10.1 reflects this urban sprawl as seen from satellite.

With immigration and rural migration, the United States by 1920 had shifted from being majority rural to majority urban. By the time of the 2000 U.S. Census, it had become majority suburban. In a few short centuries, we shifted from a frontier nation, to a farming nation, to a nation now characterized by large metropolitan regions composed of cities and their many suburbs. By 2000, 8 out of 10 Americans lived in metropolitan areas. Western cities are growing fastest, while northeastern and some midwestern cities are declining. At the same time, suburban growth outpaced cities in both growing and declining urban areas. Although cities have tended to be more interracial and multicultural, the demographic differences between cities and suburbs are narrowing. This is shifting the nature of politics, the local political and governmental coalitions, and the struggle for political power at the most local level.[3] In such a metropolitanized and globalized setting, the effort to achieve genuine representative democracy is becoming more challenging.

Immigration is a major factor in changing racial and ethnic divisions in cities and suburbs. In the Chicago metropolitan region, whites are the majority population at 55 percent; Latinos are the next largest group at 20 percent, followed by African Americans at 19 percent, and Asians at 5 percent. In the city of Chicago itself, blacks and Latinos have for two decades outnumbered whites. Yet median earnings are not equal among racial and ethnic groups. For instance, in 2003, non-Latino whites earned an average of $36,620, blacks $26,456, and Latinos $21,495 a year.[4] These racial and economic differences, sometimes called the "color gap," are one of the most vexing problems of urbanization and suburbanization in America.[5] While racial and economic patterns differ in American cities, Latinos are reshaping both urban and suburban growth in many cities and states and will continue to do so in the years ahead. National immigration policies directly affect cities, metropolitan regions, and states. Acculturating and accommodating new residents from abroad is a major task and a divisive issue for state and local governments.

Sometimes the struggle for power is not just the fight between individuals seeking to win public office or to get their preferred policies adopted. Sometimes the struggle is against larger trends and forces such as globalization, immigration, and metropolitanization. State and local government officials find their cities and states are all affected by these larger social and economic trends, but they respond differently. How can democracy, economic and social justice, and a good life for most citizens be achieved in face of these powerful forces? State and local government officials struggle with that question.

A problem emerges because governmental institutions and processes, which were invented in the 19th century, are inadequate to cope with 21st-century changes. The institutions established early in our nation's history were not created to control transnational corporations, the present levels of immigration, or suburban sprawl.[6] It is difficult at either the state or local government level to provide for representative democratic decision-making in metropolitan regions of 10–20 million people, economies larger than most nations of the world, and hundreds of separate governments with overlapping powers, taxes, and authority.

GOVERNING METROPOLITAN REGIONS

Unlike the traditional units of government for village, town, city, county, and state, there are few legally recognized single authorities to govern the multiple cities, towns, suburbs, and counties that make up a **metropolitan region**. Metropolitan regions now also include areas known as **exurbs**, developed areas beyond a central city and its suburbs. In these regions, characterized by a plethora of governments with often overlapping jurisdictions, the struggle today is how to achieve effective metropolitan governance and promote cooperation and coordination among very different local governments.

To take a single example, there are 540 taxing units of government in Illinois's Cook County, including the City of Chicago. In the larger Chicago metropolitan region, there are more than 1,200 units of government. A taxpayer in Chicago pays property taxes to 7 separate units of government. Many suburbanites pay property taxes to 13. This fragmentation of government makes coordination, planning for growth in the region, and the delivery of government services more difficult. It undermines government accountability and transparency. It creates waste and duplication. If we were starting over, no rational person would design metropolitan governance like this.

Our current system grew organically because of restrictions on local governments' indebtedness. Most governments rapidly reached their debt level, and so when new functions like building a library or recreational facilities were to be undertaken, a separate government was created to raise the necessary funds through taxes and bonds. The result was the myriad governments in metropolitan regions.

So now we are faced with many governments providing a multitude of different services in various overlapping jurisdictions. There are, of course, advantages to small general suburban governments like towns and villages. Their elected trustees, city council members, and mayors live in the same communities as the people they govern. As local residents, these officials fully understand most local problems and concerns. Citizens in smaller, often residential communities know the police officers and fire fighters who protect them as neighbors. School districts are smaller, and parents can fairly easily affect schools through their local elected school boards. Serving in local governments teaches the arts of politics and government to local leaders, who can then go on to hold elected positions in state and national government. Thus, local governments are often the training grounds of our representative democracy.

But fragmentation and waste make governing larger metropolitan regions a major challenge in the 21st century. Misgoverned metropolitan regions are also less able to deal with the pressures of globalization, for it is metropolitan regions that compete among the new globalized cities of the 21st century.

Regional Governance

To solve the problems of governance in our large metropolitan regions, some theorists have proposed a unigovernment structure, while others favor city/county mergers or other forms of regional governance. Although support for regional government has been growing, actual practice has not developed. This is because local interest in government autonomy is widespread, especially in the suburbs. While there were 83 ballot referendums to create various city/county/metro governments from 1921 to 1979, only 17 succeeded, and only 3 of those were in major metropolitan regions of over 50,000 people. Mergers have not increased much since then.[7]

It is obvious, therefore, that a simple merger into a single county or regional government is not likely to happen in most metropolitan regions. One reason is while there are citizens who identify strongly with their home city or state, there are very few metropolitan citizens who identify with the region as a whole. Instead, they identify with their suburban town or big city and are mostly unconcerned about the fate of those in other communities in the region. At the most, they identify with specific areas within the metropolis, such as Brooklyn, New York; Los Angeles County, California; or the city of St. Louis, Missouri.

In a study of the St. Louis metropolitan region, only 7 percent of the people living outside the city believed that there was a very close connection between the quality of life in the city of St. Louis and the quality of life in their community. The same study found some support for centralized decision-making in limited policy areas and even support for some limited redistribution of local tax revenue to help poor areas of the region. But urban politics has failed to produce the type of metropolitan citizens necessary for any major moves toward broader metropolitan governments.[8]

Although the St. Louis region—which is made up of 12 counties in Illinois and Missouri and almost 800 local governments—created a regional agency responsible for community needs including sewers, community colleges, parks, and a medical center, it served only the city of St. Louis and St. Louis County. The seven-county Bi-State Development Agency, now called Metro, only focuses on transportation issues and has not been able to get ballot measures passed to link the regional public transportation system. In the last two decades, voters have rejected all ballot measures to create a metropolitan economic development commission. In 2000, voters finally approved the creation of a regional parks system for five counties.[9]

While there is little support for merged metropolitan regional government, there is increased support for creating regional planning agencies, intergovernmental compacts for better service, elimination of unincorporated areas, ending older structures like townships, and some sharing of tax revenues. However, for these more modest reforms to come about in metropolitan regions, state government must pass new laws that encourage and mandate these changes. For this to occur, according to some analysts, the poorer sections of the inner city and the poorer suburbs must join forces with enlightened residents of the fortunate quarter of the metropolitan region to gather the necessary votes in the state legislature and to convince state governors to place metropolitan reform on their agenda. This has happened in some areas like the Minneapolis–St. Paul region in Minnesota.[10] But creating a broad-based political movement for reform across racial/ethnic, geographical, and local government divisions is hard work. In addition, city officials face the dilemma of the **commuter tax**, a tax levied on the people who work in the city but do not reside there. While they might believe that the suburbanites who enter cities to work each day and otherwise enjoy a city's benefits cost more than they contribute, suburbanites generally disagree and many have succeeded in getting their state legislators to block such taxes.[11]

Even areas such as Houston, Texas, that have a strong culture of personal and community independence, are beginning to foster greater regional cooperation and some regional governance. With liberal annexation power, the city of Houston has long annexed new suburbs. When it could no longer prevail by annexation, it used its home-rule powers to implement a regional water plan, build an international airport far from the city, and gain virtual control of regional boards such as the Port Authority and Metropolitan Transit Agency.

There is some progress in regional governance. The Ohio state legislature has allowed localities to form joint economic development districts and cooperative economic development agreements, which provide suburban revenue in exchange for city services. Texas legislation passed in 1999 allows for "limited purpose annexation," which is used by Houston to raise revenue from suburbs in return for the promise it won't annex for a set number of years.[12] But these efforts are no longer sufficient. Many metropolitan regions, like Houston, face new challenges that only regional solutions can answer. The civic community and business leaders have begun to recognize this.[13] However, recognizing the challenges facing a region and enacting state legislation to force hundreds of local units of government to share or relinquish power and authority are very different matters. It is difficult to get the votes in the state legislature and the support of the governor to achieve better metropolitan regional governance without a strong movement to demand it. Economic pressures, a desire to better protect the environment, and specific problems like transportation gridlock are slowly bringing the political pressure to enact some changes in metropolitan governance, but change is very incremental.

One of the key challenges of the 21st century will be to create better, more coordinated, and more cooperative metropolitan governance. Metropolitan government by a single unit of government is not currently possible in most of the United States. Better metropolitan governance will require direct involvement of state governments, because they are the only unit of government with the power and legal authority to mandate such change. States are likely to act only when they realize that they will lose out in global economic competition if they don't.

County governments play a dramatic role in metropolitan governance in some parts of the country, particularly in the South and West, which are experiencing rapid population growth. County governments are more significant in highly urban counties. In many of these, they serve the same general government function as the cities. In other places, such as Chicago's Cook County, they play an important but limited and specific role. They provide health care and run the criminal justice system. There are currently more than 3,000 county governments in the United States. They must be considered if we are to understand metropolitan governance, even though there is a lot of variance in their powers.

In fiscal year (FY) 2002, county governments raised $256.7 billion in general revenue, almost the same as city municipal governments, which raised $285.5 billion. While counties in 1962 accounted for only 39.3 percent of municipal revenue, they now account for 48.4 percent. By 2002, they employed nearly as many people as the cities: 2,340,200 county employees in comparison to 2,474,400 city employees. They provide government services to about 252 million Americans.[14]

Counties are governed by a board of supervisors or a county commission of three to seven members, usually elected by districts. They often have an elected county board president or county executive. In about 800 counties, the county board appoints a chief administrator much like a city manager.[15] As counties take on more functions and become better organized, some observers note, "urban counties serve to further the regionalizing of metropolitan America."[16] They have the potential to become a middle stage between the inner city and the suburban metropolitan regions. However, often they don't have the power or the ability to force coordination of the multiple governments that are within their boundaries. Most metropolitan regions span a half-dozen or more counties, and intercounty coordination is sadly missing in most places.

Urban Sprawl

Debates are ongoing about whether **urban sprawl**, the unplanned, uncontrolled development of land at the boundary of an urban area, is positive or negative. Those who argue that sprawl is good point out that many Americans, and others throughout the world, have opted to move to the suburbs to achieve "the American Dream," or their version of it. They want to own their own home with a lawn, trees, and flowers; good schools for their children; good neighbors; and clean, honest government run by people they trust. And it is not just middle-class whites who move to the suburbs. Blacks have been moving out of the city centers even when they have to move to racially segregated suburbs. East Asians, Middle Easterners, and Latinos are often settling in suburbs directly, rather than locating first in a city, as many earlier immigrants did.

Urban sprawl is not necessarily universal. European metropolitan regions tend to have greater density and less sprawl than American cities. Wealthier Parisians prefer to live within the city while the poor are confined to the less desirable suburbs, for instance. Sprawl is not automatic or a law of human nature. But in the United States and in many countries, it has been the dominant pattern since the 20th century.

The arguments against sprawl in U.S. metropolitan regions today are that sprawl makes it much more expensive to provide the necessary governmental infrastructure and is ecologically damaging. Critics also argue that many of the new suburbs are ugly and tasteless. Sprawl, with its separate taxing units, deprives cities of the resources necessary to provide services to the poorest citizens who need them most. Moreover, the very problems that many suburbanites fled the city to escape have followed them to the suburbs. Suburbs can be segregated and have troublesome levels of poverty, unemployment, crime, and drug use. Many suburbs lack the racial and ethnic diversity as well as the cultural amenities of the city. And long commuting times for those living in suburban communities can be almost unbearable. In many metropolitan regions, commuting to work takes longer than one hour in each direction, resulting in hundreds of lost hours of productive work and relaxation for each commuter each year. Moreover, in an economic recession, there may not be sufficient funds to provide or maintain the public infrastructure of roads, schools, utilities, and other services that urban sprawl demands.

Political author Fred Siegel has argued that we can only curb sprawl locally rather than by national legislation, because doing so involves difficult tradeoffs that are best worked out in each metropolitan region. We can ameliorate sprawl's side effects by supporting the consensus that has developed in recent years around preserving open space and recreational areas. We must end exurban growth subsidies for transportation, water, and sewers. Like cities in Minnesota, we must share a part of sales taxes to support cultural institutions, pay for affordable housing, and provide subsidies for communities serving the poorest citizens whose tax base is inadequate.[17] If sprawl is to be contained and made manageable, it will have to be done by local and state governments conscious of the specific costs and benefits of different forms of development. Local and state political leaders must also create local voter support for specific restrictions.

Coping with urban sprawl and creating better metropolitan governance is not a matter of altruism. The competitors in the new global economy are metropolitan regions, not central cities or individual suburbs. Ugly regions without dynamism or vision are unlikely to succeed in global competition. Regions will lose out if they cannot produce a highly educated workforce because of inadequate education, or if they cannot lure global corporations to their area because they lack cultural amenities, diverse peoples with international languages and skills, and world-class hotels and attractions.

Gated Communities

One response to the problems of cities and modern public life has been **gated communities**, private communities with limited access points and security guards to protect the property. This arrangement is also known as a **common-interest development (CIDs)**, a private community whose land and common facilities are owned by the members of the housing association and governed by a homeowners association board of directors. CIDs take primarily two forms. The city form is most often a single building of multiple apartments or condominium units in which each unit is owned by its occupant. The common service areas are owned and paid for by all. The entire building is governed by a condominium or homeowners' association. Larger buildings usually have a doorman or security guard who ensures that only owners and their approved guests can enter the building.[18]

Entrance to a gated community in California.

In more suburban and exurban areas, CIDs take the form of housing developments comprising numerous private homes and often common areas. They may be gated communities with security personnel, or they may have more open access. Such developments are governed by a homeowners' association board similar to the condominium boards that govern city CIDs. These associations can regulate many aspects of the community, from the color of the houses, the care of lawns, and the display of American flags or political posters. Some of these restrictions are seen as excessive, but gated communities and other CIDs are only beginning to be regulated by law.[19]

These private enclaves provide both positive and negative aspects for cities, suburbs, and states. They provide desired residences to attract wealthy homeowners, which raise greater tax revenue for local and state government. And many otherwise public services such as garbage collection and police protection are provided and paid for privately by the individual homeowners' association, lessening the demand for expensive government services. These private enclaves can provide wealthier citizens with a retreat from involvement in public life. Many homeowners' associations prevent full exercise of First Amendment rights, restricting lawn signs or other public indications of support for political candidates. These enclaves frequently become a strong conservative political force opposing tax increases to pay for public schools or opposing the building of affordable housing because it might lower their property values. When rallied to vote in large numbers for what they perceive to be in their self-interest, they can have a significant impact in local elections and policy-making.

Gated communities and other CIDs are a form of local government. They now perform many of the functions that were previously done by village and town government. Their populations range from 100 to 1,000 families or more. Because they also have an impact on the state, city, or metropolitan region of which they are a part, their governance is of great importance to millions of Americans. The Insider's View feature depicts a meeting of the board of the Summit Towers condominiums in Hollywood, Florida, a private community of two 22-story towers of apartments with several hundred family owners.

The board of directors for a homeowners association is normally small, and the officers are elected by all of the member homeowners in annual elections. These are unpaid positions by law, although these boards have the responsibility of hiring employees and overseeing the work of often multimillion dollar corporations. The board enters into contracts for the association, sets salaries of employees, and determines regulations—even those which may limit First Amendment rights—that govern all homeowners in the association. Unless there is a controversial issue on the agenda, most board meetings of most associations have few or no residents other than board members or staff at their meetings. On some issues, emotions may run high because there are real consequences to the decisions these bodies legally make. They are in effect "private governments" with broad and important powers.

These homeowner boards are now a part of metropolitan governance even if they are seldom recognized as governments because they are private not public agencies. They have many of the powers and authority—and sometimes greater power—than other local governments. States are only beginning to pass laws to regulate these new governments and to prevent possible abuses of their power. Any attempt to understand urban and metropolitan governance must include an understanding of these private governments within the broader urban landscape.

AN INSIDER'S VIEW

A Board Meeting of the Summit Towers in Hollywood, Florida

All nine directors, the building manager, and building secretary were present along with more than 100 residents. This was the first meeting of the new board elected in 2005. After the roll was called, President John LaMarche made an opening statement: "We have a lot that has to be done. Working together as a team we can get a lot done. There is a new head of maintenance and we have a lot of projects. Cement repairs to the building especially will be expensive. The pool elevators will be fixed, new equipment placed in the gyms, and worn carpet will be replaced. Accountability will be the buzz word around the Summit. We will be looking over people's shoulders to make sure that you are getting what you are expecting and paying for."

The treasurer reported that four units are delinquent for January assessment payments. For two, the "checks are in the mail." Two are being sent to the condo's attorney for appropriate legal action.

The Budget Committee reported that staff salaries were reviewed, and they added one extra mechanic and deleted one housekeeper. The budget for salaries for all staff, including security guards and maintenance personnel was recommended to go from $499,000–$515,000.

Board Member Harold Levine objected that the manager's contract needs certain conditions. Furthermore, the association is offering a raise, but the job contract is not yet signed. Sam Fox moved to take out the manager's salary from the budget approval until contract is signed. His motion was carried by a vote of 5–4. At this point in the meeting, the condominium manager picked up his papers and walked out. The condominium manager met with the condominium secretary in the hall and was persuaded to return to the meeting. Board Member John Munroe moved that pending the signing of the manager's contract, they give the manager the salary recommended in the committee report. The board will have approval over his final contract. This time the motion on manager's salary passed by a voice vote.

The Legal Committee reported that a person in the building has received permission, contrary to the general condo rules, to have a dog in his apartment. The person is on disability and is covered by the Americans for Disabilities Act. The Florida Commission on Human Relations has ruled that the building must allow the dog (there is a doctor's testimony that the dog is needed) but the commission has agreed now that the condo owner must abide by legitimate rules for the animal. The dog is to be transported in and out of the building in a carrier, can only use the freight elevator, and may not be taken to the fourth floor pool area. There is a letter from the commission stating that these regulations are reasonable.

The Restaurant Committee reported that the owners of the restaurant within the condominium are allowed to take a two-week vacation, but they aren't allowed to take the two weeks consecutively, as they did last summer. There is a need to replace the carpeting and molding in the restaurant. Degreasing needs to be done in the kitchen. The restaurant's bathrooms need to be expanded in future construction. All contracts with the restaurant are to be approved by the building attorneys.

The board then discussed the report of the parking garage engineer. There is concrete spalding in the garage and there is an expansion of the joints between the concrete, which causes leaks. The cooling towers of the building also need concrete repair. The building manager estimated the cost of repairs at $610,000. The president asked if we should get a second engineering opinion or go to competitive bidding on the contract for repairs. Eventually all the motions were withdrawn to send out for bids and the item was tabled until the next board meeting in three weeks. Three engineers among the owners would meet as a special committee to provide a recommendation.

There was also discussion of partial replacement of gym equipment. The proposal from the committee was to spend $15,000–$20,000. The motion was made to upgrade equipment up to $20,000. Hall carpets are to be replaced at a cost of $58,000 to cover 13 floors. There has been no new carpeting for the lasts two years. The survey of the beach elevator will cost $800. The motion to pay for the survey so the elevator could be properly fixed passed unanimously. The fire alarm maintenance contract was reduced $1,100–$900. The motion to approve the new contract passed unanimously.

As the meeting was breaking up, the president remembered there needed to be an open part of the meeting for the audience. A moment of silence was taken for Paul Finkelstein who had recently died. One audience member complained about the South Tower Stop Sign in the garage. People don't see the sign. The sign needs a light by it because many people don't stop. The board ended the meeting requesting that e-mail addresses of owners be collected so that the association could communicate by e-mail.

Special Districts

There are many different kinds of special districts, and there can be hundreds or thousands of them in a single metropolitan region. Among the most common types are school systems, parks, libraries, and transportation, water, and sanitation districts. Special districts are still being created very rapidly: over 13,000 have been created since 1967. By 2002, 23 percent of all special districts covered either a single county or a larger metropolitan area, so they play a particularly important role in metropolitan governance.[20]

Local school boards are a good example of special districts. They govern nearly all of the more than 15,000 school districts in America, making up about one-sixth of the local governments in the country. Most school boards have 7–9 members. In many districts, they are elected. In some, they are appointed by the mayor with the approval of the city council. In some of the larger cities like Boston, Chicago, New York, and Philadelphia, the mayors have taken over the schools more directly, appointed school superintendents of their choice, and enacted policies that they favor. In general, studies seem to confirm that mayoral control has a positive effect on student achievement as long as there is some oversight on the mayor's choices of school board members, such as a nominating committee to recommend appointees.[21] Two-thirds of the states have passed legislation allowing either the mayor or the state to manage school districts that are underperforming. But in most of the country, the school board still governs.

In general, school boards retain broad popular support although nearly half of the population now favor having mayors appoint them. Nonetheless, most school boards remain elected, particularly in suburban towns and rural areas. They must cope with difficult problems with no special training.[22] The thousands of other boards that run our libraries, public transportation systems, parks, and water and sewer systems are much the same. In some places, it makes sense to merge districts for efficiency. In other areas, it makes sense to bring them more directly into the city or town or county government under the control of a mayor and city council. But most of these districts will continue to provide key governing services under a small governing board elected by the voters, with a professional staff that does the day-to-day service provision.

GLOBALIZATION

Cities, metropolitan regions, and states in modern times confront not only the forces of urbanization but also the forces of globalization. In the 20th century, America's towns, hamlets, cities, and states were increasingly connected through a national economy and a national culture. This was brought about by better transportation (evolving from horse and buggy to trains, automobiles, trucks, a national interstate highway system, and air travel) and communication (from telegraph to telephone and from newspapers to radio and television networks). With the 21st century's innovations of the Internet, cell phones, and easy intercontinental travel and shipping, local economies and societies are now connected globally. Given these changes, some observers say the world is now "flat," made more homogenous and uniform by new technologies of transportation, communication, and production.

Globalization has dramatically changed the economy in many cities and states. Older factories are being replaced by computer-automated assembly lines, with parts manufactured around the world, and by "just in time" production and distribution rather than warehoused inventory. The simple manufactured goods that cities used to produce are now being partially replaced by financial services and producer services as the chief moneymakers and employers in the United States.

The place of a city, metropolitan region, state, or nation in this new world economic order is determined by the number of transactions within its borders. First-tier global cities are at the nodes of the financial and producer services and communication networks, They are banking, financial, and market centers like New York, London, Paris, and Tokyo. Second-tier cities, like Chicago and Los Angeles, are regional capitals of the global economy. Other cities like Miami, Houston, and Atlanta are attractive global markets. But cities like Detroit, Indianapolis, and Newark get only the crumbs from the global table. Small towns and less populated western and southern states like Wyoming, North Dakota, and Alabama are affected by the global economy and global society but are not central to either. Table 10.1 shows the cities in the United States that have at least five Fortune 500 headquarters, which are the top revenue-producing companies in the United States. However, as Table 10.2 indicates, only four cities in North America—New York, Toronto, Houston, and Atlanta—qualify as top cities at the global level.

Even though many cities aspire to be a "global city" economically, there are major problems for cities, regions, and states caught up in the whirl of globalization. Globalization, while it provides new opportunities for wealth and cultural enrichment, has its dark side.

TABLE 10.1 Cities in the United States with Five or More Fortune 500 Headquarters, 2007

City	Number of Fortune 500 Companies
New York	45
Houston	22
Atlanta	12
Chicago	11
Dallas	11
Minneapolis	8
Philadelphia	8
Pittsburgh	7
St. Louis	7
Cincinnati	6
Milwaukee	6
Seattle	6
Cleveland	5
Columbus	5
Fort Worth	5
Los Angeles	5
Omaha	5
Richmond	5
San Antonio	5
San Francisco	5

Source: *Fortune*, April 30, 2007, http://money.cnn.com.

TABLE 10.2 Global Cities Ranked by Number of Global 500 Companies and Revenue, 2006

Rank	City	Country	No. of Global 500 Companies	Global 500 Revenues ($ millions)
1	Tokyo	Japan	52	1,662,496
2	Paris	France	27	1,188,819
3	New York	U.S.	24	1,040,959
4	London	Britain	23	1,054,734
5	Beijing	China	15	520,490
6	Seoul	South Korea	9	344,894
7	Toronto	Canada	8	154,836
8	Madrid	Spain	7	232,714
8	Zürich	Switzerland	7	308,466
9	Houston	U.S.	6	326,700
9	Osaka	Japan	6	180,588
9	Munich	Germany	6	375,860
9	Atlanta	U.S.	6	202,706
10	Rome	Italy	5	210,303
10	Düsseldorf	Germany	5	225,803

Source: *Fortune*, July 24, 2006, http://money.cnn.com.

It accentuates the widening the gap between the rich and the poor. The wealthy chief executives of global corporations, law firms, and banks become fabulously wealthy with salaries in the millions and golden parachutes to ease their exit if they should fall from power. But fewer and fewer workers are needed to run factories and to do other jobs in global companies. So there are more poor people, and in this two-tiered global economy, people such as janitors, salespersons, paralegals, and office staff no longer make a middle-class wage. They receive no health care benefits or pensions. In short, the rich get richer and the poor get poorer faster in the global economy.

On the one hand, cities and states want global corporations' headquarters, research centers, and factories to shore up their tax base. Government can, after all, tax the property, factories, and business income of these companies. And a global company's local purchases "trickle down" to other firms and service companies. But the two-tiered nature of this service economy leaves too many people without jobs and those with jobs with low salaries and benefits. In addition, while a second-tier city thrives in the global knowledge economy, often the smaller surrounding urban communities do not share in its prosperity. According to one study, while the greater Boston area has enjoyed a 51 percent gain in the number of jobs between 1970 and 2005, the formerly thriving smaller manufacturing towns of Massachusetts known as its "gateway cities" lost 11,000, or 3 percent of their job base. Over 70 percent of the global-economy-friendly knowledge-based jobs in the state were in the greater Boston area, while only 23 percent were in the gateway cities.[23] Two other effects of globalization on our second-tier cities and their surrounding areas are artificially higher housing prices and increased traffic congestion.[24]

Despite the drawbacks, many believe it is better to be part of the global economy than to be left out. So cities and states offer lower taxes, free infrastructure improvements, and cash incentives to get key companies to locate in their area. Some cities open development offices overseas to attract companies to their city and to encourage business. In February 2007, for instance, Chicago opened such an office in Shanghai, China.[25] But being part of the global economy means also being tied to its ups and downs. So all U.S. global cities suffered in the recession after the terrorist attacks of 2001 and have experienced high rates of unemployment and poverty from the 2008–2009 recession. Moreover, global cities like New York, Tokyo, and London have become terrorist targets. After the terrorists attacks of September 11, 2001, other U.S. cities not directly attacked lost tourism, conventions, and jobs while experiencing an economic recession. Many cities, metropolitan regions, and states are at risk from events like terrorist attacks and global forces like economic recessions over which they have no control.

Likewise, as many cities and states have learned, transnational corporations make business decisions strictly for their own reasons. Without consultation, global companies move factories, corporate headquarters, and jobs around the nation or overseas. There is nothing that governors, mayors, workers, or communities can do to stop them from doing so, no matter how devastating such decisions may be to individuals, communities, regions, or states.

Globalization is more than blind economic forces at work. It is about people, especially those caught up in migration and immigration. Having a multiethnic, multicultural population is often a benefit for a city, metropolitan region, or state. People from other countries bring language skills, knowledge of other cultures, and personal contacts that make buying, selling, trading, and communicating overseas much easier. They bring a willingness to work hard in their new setting. They bring gifts of art, culture, and wisdom, which are invaluable. Of course, immigrants from other cultures also create clashes with older established groups in the metropolitan areas, which can lead to discrimination and violence.

Accommodating millions of immigrants who may not speak English well or understand our laws and customs places great stress on local and state governments. These new immigrants have not yet become **acculturated**, or accustomed to their new country's norms, culture, and ideals. As they adapt to our society, they may need support services, such as classes in English. Government offices may have to translate documents into other languages. Some citizens and politicians want to slow or end immigration, even by building fences to block the nation's borders. Yet others want to make naturalization and citizenship easier to obtain. Many people, including probably your universities, want to make travel and study in the United States easier for foreigners, while others want to make it much more difficult in order to thwart what they fear will be future terrorist attacks.

It is up to our state and local governments to manage these aspects of global society and public safety. How well they perform these functions affects their future economic and social success. But often the responses to the pressures of globalization are uniquely local. They depend on the public-private regime of political and business leaders in control of local government and in the support they receive from the public—voters, consumers, and customers.[26] Each city or state must find its own niche in the global economy. Those cities and states that have more resources—a higher level of economic development, greater social capital, and better leadership—have more latitude in their choices. Poorer cities, states, and nations have fewer choices and end up with less desirable roles in the global economy.[27]

Thus, the struggle for power occurs not merely among politicians seeking elected office but between cities and states fighting for a better place in the global pecking order. In this, as in all the other struggles, having leaders with a positive vision and public support affects the outcome. Individuals, cities, metropolitan regions, and states are caught up in a new world order that, according to one observer, "combines global deregulation of business with high mobility of goods, services, and capital . . . fueled by expanding credit and speculation."[28] One consequence of this new world order is that there are winners and losers among states, regions, and cities. The task of local leaders is twofold: ensuring that their communities and states are important nodes in the global economy while at the same time easing the negative effects, such as a widening gap between rich and poor. Leaders at the state and local levels must directly confront the demands and stresses of globalization. Good leaders in both the public and private sectors will have to shape the local responses to global forces. The health and well-being of our communities and states depend on their responses.

Economic Development in a Global Economy

The principal problem facing states, metropolitan regions, towns, and other communities is the competition for economic development. American farms and corporations compete not only with rivals in other parts of the United States but also with producers in the global marketplace. In this competition, farms, factories, and service companies in other parts of the world have some major advantages. For example, given the lower wages paid to farm workers in Mexico or South America, it is cheaper to grow some crops there and then ship them to U.S. supermarkets than for American farmers to produce the same crops in the United States. Lower wages in China mean that many of your clothes were made there. If you call for service when your computer doesn't work, you may be talking to a technical assistant in India. And technological advantages are not always ours. Car companies in Japan and other countries have made the smaller, more fuel-efficient automobiles that many Americans prefer, driving down the sales of American automakers who were slow to respond to market changes and finally faced bankruptcy in 2009.

Towns, cities, and states must promote economic development if they want to create solid local economies, money for charity, a healthy middle class, and a good tax base. They must help local firms grow and succeed in the global economy, and they need to attract new companies to their communities, metropolitan regions, and states.

To attract global companies, they must create the right infrastructure and amenities. Global firms require good transportation and communication systems, a highly educated workforce, and attractions worthy of a global city. A highway, rail, and water transportation system has to be supplemented by a major international airport. Cities are constructing fiber optic access to the Internet, and smaller towns are acquiring wireless access for their area in order to be competitive. Cell phones have to work well in exurban areas far from downtown. Modern business communications are impossible without a massive communications infrastructure, which no one firm wants to pay to create.

A bigger problem in making our cities competitive in the global marketplace is our education system. The No Child Left Behind legislation was a response to the perception that many schools were failing to adequately educate their students. Local school boards and state policy-makers are working hard to improve our educational system from preschool to university. If grade scores are any indication, education in at least the basics—reading, writing, and math—is improving. But future economic development depends on a well-educated workforce that can compete at the highest levels for the best jobs.

American cities have long attracted immigrants from around the world who come to the United States for jobs, education, and freedom. We also have a massive influx of immigration to suburbs and even rural towns. Racial, ethnic, and cultural differences can stress societies, so we have to adjust to the diversity that comes with large numbers of immigrants. In the global economy, cities and states promote immigration for the skills, languages, cultural understandings, and connections abroad that immigrants bring, but at the same time, we have to create tolerance and understanding. The movements to make English the official language and to impose strict immigration laws suggest that tensions still remain, and managing those tensions occurs primarily at the state and local levels.

In addition to infrastructure, good education, and being immigrant-friendly, states and cities must attract tourists and conventions with world-class museums, parks, cultural attractions, convention centers, and sports teams. Thousands of visitors a year are a major source of revenue for cities and states. Supporting these cultural amenities for residents and tourists can cost hundreds of millions of dollars a year, yet they are what make living in global cities and metropolitan regions so special. They also draw people from the suburbs and exurbs, not just downtown residents.

Each city, region, and state has to develop a strategy for economic development and provide public subsidies to encourage specific developments. Public officials at state and local levels of government have to figure out how their communities can fit into a particular niche in the global economy. For instance, an oil boomtown like Houston, Texas, may focus on energy and energy-related firms. Given the location of Wall Street, it makes sense for New York City to specialize in stock, bank, and financial transactions.

In general, U.S. cities have a competitive advantage in high-end production such as the creation of machine tools; producer (as opposed to consumer) services such as accounting, management consulting, and financing; and development of new products such as computer software, microchips, and medical machinery and treatments. This does not mean that cities and states can ignore producing products for the local, regional, and national markets, but they have to fit profitably within the global hierarchy of trade as well if they are to flourish in the 21st century.

The competition for economic development, especially luring corporate headquarters and major new factories, is intense. For instance, when Boeing Aircraft announced that it was moving its corporate headquarters from Seattle a few years ago, many states and cities offered major tax incentives and subsidies to have the company locate in their city. The final three contenders were Chicago, Dallas, and Denver. Chicago won the contest based on its cultural amenities and business connections in addition to the tax breaks Illinois offered. A number of other firms soon followed Boeing to Chicago. Despite this coup, Chicago, like most major cities, has had a net loss in corporate headquarters over the last decade and a half. And bigger banks in New York and Europe have bought up Chicago's major local banks.

Because of the fierce global competition, governors now lead trade missions around the world to promote their state's economic enterprises. States and cities have strong and well-funded departments of economic development promoting everything from the location of new factories to encouraging film-making in their state. Mayors promote "sister city" programs around the globe, pairing up with similar cities in other countries. Every local community is hoping to become the next Silicon Valley, developing unique new products that the world just must have.

Local Responses to Globalization

Cities can face significant problems if they are left behind in the global competition. What happens to economically stressed cities like Detroit, Buffalo, and St. Louis, or cities like New Orleans that take a decade to recover from national disasters? Building a new downtown convention center or sports stadium—even if hundreds of millions of dollars are spent by state and city government—is an insufficient economic development strategy by itself. And smaller towns like Peoria, Illinois, can't compete with convention cities like Orlando or Las Vegas for most conventions. These cities and states will have to find a different strategy to fit into the global economy.

Even in successful global cities, there is a growing gap between rich and poor. A region might attract **transnational corporations**, firms whose facilities and employees span more than one nation, but as the professionals who manage them or provide the legal, financial, and consultant services become very wealthy, the region can develop a two-tier economy. At the lower level are secretaries, paralegals, janitors, food service workers, and salespeople who do not earn a sufficient salary to bring them into the middle class. The middle class shrinks and the working class slips further into debt. At the very bottom of the hierarchy in the global city are the poor—the permanent underclass—who increase at an alarming rate.

So cities and states not only have to promote economic development and find their niche in the global economy. They have to do so in a way that makes their communities livable and desirable. This means that they must create better-paying jobs without driving away business firms; promote affordable housing for their working class; provide job training to make those now unemployed employable; provide affordable health care, which the private sector does not supply for everyone; allow for quality recreation and parks; and make communities safe by curbing crime and drug abuse. Our local communities cannot survive with a small rich class and more than half of their population struggling economically.

The task of making our cities globally competitive, livable, and desirable is daunting. In addition to providing strategies for economic development, state and local governments also offer distributive services and redistributive services.[29] **Distributive services** are ordinary city services such as police protection, parks, garbage pick-up, and street repair, which can be distributed to all citizens. **Redistributive services** are services like public housing or welfare payments that wealthier taxpayers fund and poorer taxpayers receive, which provides a limited redistribution of wealth. Providing programs and subsidies for economic development is noncontroversial for the most part. Every public official wants to promote economic development, and the public applauds such efforts as long as they don't cost too much in tax subsidies or produce negative consequences for a neighborhood such as garbage dumps and polluting factories.

Distributive services, however, which include the usual meat and potatoes of local and state politics such as police protection, roads, public schools, libraries, and parks, are an arena in which many local political struggles for power are fought. These services can be distributed unevenly depending on whether you live in a rich suburb or a poorer section of the metropolitan region. If state or local taxes are lowered for the highest income levels, people at the lower income levels probably pay more taxes in proportion to their income. So local political battles are often fought over who gets which services and who pays. There are major arguments about what is fair. Should those who are most in need of public services get greater services or should those who pay the greatest taxes to fund the services get better ones? These are fighting words for citizens, interest groups, communities, and the politicians they elect.

Redistributive services pose other challenges. These government programs are paid for by the wealthier taxpayers and property owners to help the economically less well off. To help the poorest individuals and families with welfare and medical payments, subsidized public housing, and special education programs is beyond the financial ability of most cities. No city or state today wants to be in the position of having provided too much in the way of redistributive services from their own funds. They are perfectly happy to distribute federal funds for all of these programs. However, even when the federal government shares part of the burden of Medicaid programs, it nearly swamps state and local governments to pay for health care for the poor. This was true even when additional federal funds were made available in the 2009 economic recovery plan of the Obama administration. What happens when the extra federal funds are withdrawn and states are still committed to paying the health care costs of the poor?

If cities or states raise taxes to support their service programs, then the very businesses and wealthy individuals they need in order to provide jobs, wealth, investments, and tax revenues may move to communities with lower tax rates. Most cities don't have the ability to levy an income tax. Although states sometimes share part of the state income tax with them, local governments don't have the billions of dollars necessary to provide a significant redistribution of wealth from the rich to the poor. Only the federal government can do that on a grand scale. So most redistributive services have to be funded by the federal government. Cities and states face the task of ameliorating the negative effects of global economic competition with limited resources.

Boarded-up homes and shops found in many rustbelt cities.

A New Metropolitan Agenda

On February 19, 2009, about a month after he was inaugurated, President Barack Obama signed an executive order creating the White House Office of Urban Affairs. As the order noted, "About 80 percent of Americans live in urban areas, and the economic health and social vitality of our urban communities are critically important to the prosperity and quality of life for Americans. . . . In the past, insufficient attention has been paid to the problems faced by urban areas."[30]

The White House staff members charged with implementing new programs for our metropolitan areas include Valerie Jarrett, a longtime friend of the president, who was named his senior advisor on intergovernmental affairs and public liaison; former Bronx borough president Adolfo Carrion, director of urban affairs; Derek Douglas, special assistant to the president for urban affairs; Cecilia Munoz, director of intergovernmental affairs; and David Agnew, deputy director of intergovernmental affairs. It is not clear what these new programs will be or how well they will work, but as one observer commented: "Despite the many obstacles that might undermine Obama's approach to urban policy, he's still got one big advantage—he's serious about pursuing it. The creation of the new urban affairs office and the full engagement of one of his closest advisors mean that local governments have a say again in Washington."[31]

From a metropolitan perspective, the $787 billion American Recovery and Reinvestment Act was a mixed bag of "business as usual though punctuated with surprising new opportunities" because the huge flow of funds was channeled mostly through existing federal, state, and local mechanisms. "Because current federal policy is generally neutral or hostile towards action at a metropolitan scale, [the Recovery Act] is also." Yet, a majority of the funds are delivered to state and local decision-makers. This provides the opportunity for metropolitan coordination, but it is not mandated nor guaranteed with this new infusion of funds.[32]

According to Vice President Joe Biden, in a speech at the University of Illinois in Chicago on April 27, 2009, the administration recognizes that metropolitan areas are key to economic recovery. Cities provide two-thirds of the nation's jobs and three-fourths of our gross national product. Biden reaffirmed that a new metropolitan agenda will be formulated to take advantage of the metropolitan regions which house so much of the population and so much of the economy—particularly the knowledge economy—which will be the basis of economic growth. The American Recovery and Reinvestment Act is pouring billions of dollars into health care research, transportation projects, energy improvements, and higher education in our nation's cities.[33]

Laboratories of Innovation

One advantage of cities and states is that they can be laboratories for new public policies. Among their experiments are charter schools, welfare to work programs, rebuilding the railway system between cities, and living wage ordinances. Cities and states are also laboratories of democracy, testing the webcasting of city council meetings, voting by mail, applying for licenses online, and a host of other experiments in representative democracy and government services.

For example, in May 2007, dissatisfied that the federal government was not doing enough to prevent global warming or to curb the emission of greenhouse gases, 31 states passed a Climate Registry law "to measure, track, verify and publicly report greenhouse gas emissions by major industries." This legislation was pioneered by California and was significant because it required third-party verification of carbon dioxide pollution levels. Then Arizona Governor Janet Napolitano declared this to be an example "of how states are taking the lead in the absence of federal action."[34]

At the same time, New York City held a "Large Cities Climate Summit" at which mayors for 30 world cities discussed strategies to fight global warming, including biofuels, high energy efficiency standards for buildings, new parks, planting millions of trees, congestion pricing of automobile traffic in the central city, hybrid taxis, and pollution-cutting "greenprints" for cities. All of these ecology measures are currently being undertaken by large cities around the world. As Toronto Mayor David Miller pronounced: "Where national governments can't or won't lead, cities will."[35]

OUR NEIGHBORS: CANADA AND MEXICO

One clear example of the pull of globalization is the intertwining of the United States with our neighbors Canada and Mexico. Especially since the implementation of NAFTA (North American Free Trade Agreement) in the 1990s, our economies have become much closer. Products are created and designed in the United States but manufactured in Mexico. U.S. citizens drive to Mexico or Canada to buy medicines that are vastly cheaper than in the United States. Citizens from Mexico move to the United States to make money but send money back to their families and to support projects in their home towns, and they hope one day to retire in Mexico.

Like other forms of globalization, the relationship between the United States and its neighbors is not always benign. As described in the Struggle for Power feature, there is a dispute between American and Mexican truckers as to who should have the right to haul cargoes on U.S. highways and which laws and regulations must be observed. We also have issues with Canada about which documents are required for Canadians to enter the United States. Americans opposed to NAFTA believe that environmental, wage, and workplace safety standards are too lax in Mexico. Canadians have been unhappy about the acid rain created by U.S. coal-fired power plants that falls on their land and U.S. reluctance during the administration of George W. Bush to address global warming.

Views differ among Mexicans, Canadians, and Americans. The Chicago Council of Global Relations conducted a survey assessing the diverging opinions of globalization in Mexico and the United States. The survey found that while 69 percent of American respondents believed that globalization had led to stronger job creation for Mexicans, less than half (49 percent) of Mexicans held the same opinion. Likewise, 78 percent of Mexicans thought that globalization was good for the American economy, but only 42 percent of Americans felt the same.[36] These differing views trickle down to policy-making at all levels. As a result, American politicians may be less likely to endorse a "guest worker" proposal, while Mexican citizens may protest against American foreign direct investment in their country.

Despite our differences, all three countries are mutually influenced by what happens inside the borders of its neighbors. And globalization has resulted in cities being more

Note: Emily Renwick provided much of the information for the section on Canada and Mexico.

The U.S.-Mexico border crossing.

alike throughout the world than they used to be. U.S. states, Mexican states, and Canadian provinces share many similarities even though there are differences in circumstances and culture. According to Mexican political scholar Jesús Reyes Heroles, however, "Mexicans are unquestionably more aware of the United States than the other way around."[37] Similar ignorance has plagued American relations with Canada. In the words of former U.S. Ambassador James Blanchard, Canada is "the invisible world next door." Or as Canadian political satirist Jack Downey says: "Those who seem to understand us the least live the closest."[38]

A public opinion poll conducted in 2004 found that Canada and Mexico ranked among the most critical countries of U.S. foreign policy. At the same time, 78 percent of respondents in Mexico and 72 percent of respondents in Canada said that U.S. foreign policy under the Bush administration made them "feel worse" about the United States.[39] It is still too soon to know if Mexicans and Canadians will feel better about U.S. foreign policy under the Obama administration.

With globalization, increased interaction among nation states demands an improved respect for neighboring countries. Not only are we linked to Canada and Mexico geographically, but we share a dynamic economic relationship, cultural traditions, and similarities within our governments. For instance, Canada and Mexico are two of the United States' largest oil suppliers, and Canadian reserves are second only to those of Saudi Arabia. Moreover, in signing the North American Free Trade Agreement in 1994, Canada, Mexico, and the United States negotiated one of the largest regional trade agreements in the world. Encouraged by NAFTA, the United States is Canada's largest trading partner, and Americans import more food products from Mexico than all European products combined.

Our interactions extend to the local level. For example, towns along the border between Mexico and Texas cooperate to end drug trafficking, while the governments of Vancouver and Seattle work together in monitoring coastguard services and cargo transportation to prevent terrorist attacks. In many border communities, basic political functioning depends on daily interaction with their foreign neighbor.

STRUGGLE FOR POWER

American and Mexican Truckers

Dozens of truckers rallied at Mexican border crossings in California and Texas in September 2007 to protest a pilot program to allow up to 100 Mexican trucking companies to haul their cargo anywhere in the United States. Carrying signs reading "NAFTA Kills" and "Unsafe Mexican Trucks," a few dozen protesters circled in the heat at Laredo's port of entry on the U.S.-Mexican border.

"What do we want? Safe highways. When do we want them? Now!" they chanted.

Government lawyers said the program is a necessary part of the North American Free Trade Agreement (NAFTA) and the trucks would meet U.S. regulations. Critics such as Teamsters organizer Hugo Flores doubt that Mexican drivers will be held to the same rules, such as the length of work shifts and drug testing.

"There are no means to regulate these guys. [President] Bush has opened up highways to unsafe trucks," Flores said. "I don't want them sharing the roads with my family."

NAFTA requires that all roads in the United States, Mexico, and Canada be opened to carriers. The government says it has imposed rigorous safety protocols in the program, including drug and alcohol testing for drivers done by U.S. companies. Additionally, officials have stepped up nationwide enforcement of a law requiring interstate truck and bus drivers to have a basic understanding of written and spoken English.

But besides the safety issues, Flores said there are also concerns about job security and pollution from emissions. "Now they're trying to export all our driving jobs to Mexico," Flores said.

Source: "'I Don't Want Them Sharing the Roads': Protests Target Mexican Trucks in U.S.," *Chicago Sun-Times*, September 7, 2007, 34.

As Table 10.3 indicates, each of these three nations has adopted a federal system of government. As in the United States, Mexican states and Canadian provinces—and local governments—struggle for power and influence within their respective federal systems, with varying degrees of representative and participatory democracy, especially when dealing with minority groups such as Mexico's Chiapas Indians and the Canadian Quebecers. Like the United States, the policies adopted on the national level often have unexpected outcomes when implemented locally. These three nations also have differences in income, as indicated in Table 10.4, which can further affect the struggle for power at all levels of government.

TABLE 10.3 Sizing Up the Neighbors—A Regional Comparison

	Canada	Mexico	United States
Population	33,212,696	109,955,400	303,824,646
Area sq. km	9,984,670	1,972,550	9,631,418
Per capita GDP	$38,200	$12,500	$46,000
Political structure	Federal: 10 provinces, 3 federal territories	Federal: 32 states	Federal: 50 states
Locus of political power	Provinces	National	National
Level of corruption	Low	Pervasive	Uneven
Separatists	Quebec	Chiapas	None
Political party organizations	Strong	Strong	Weak

Source: *CIA World Fact Book*, 2008.

TABLE 10.4 Gross National Income (GNI) per Capita, Atlas Method and Purchasing Power Parity (PPP) in American Dollars

	GNI	Rank	PPP	Rank
United States	44,970	10	44,260	4
Canada	36,170	23	34,610	19
Mexico	7,870	73	11,330	79
World	7,439		10,180	

Source: World Bank, 2006, http://siteresources.worldbank.org/.

Note: GNI takes into account all production in the domestic economy plus the net flows of factor income (such as rents, profits, and labor income) from abroad. The Atlas method smoothes exchange rate fluctuations by using a three-year moving average, price-adjusted conversion factor.

   Purchasing power parity (PPP) conversion factors take into account differences in the relative prices of goods and services—particularly nontradables—and therefore provide a better overall measure of the real value of output produced by an economy compared to other economies. Because PPPs provide a better measure of the standard of living of residents of an economy, they are the basis for the World Bank's calculations of poverty rates at $1 and $2 a day.

Mexico

The sprawl of Mexico City has no peer anywhere else in North America. Mexico's states are divided into *municipios*, which can roughly be equated with county governments in the United States. Although on paper each *municipio* is governed by a council, headed by a *presidente municipal*, or mayor, city governments are controlled politically and economically by the central government.[40] Since municipalities by law in Mexico have little power to raise their own funds, most monies are transferred to them from the central government, and services are poor, due to lack of trained and well-paid public officials. Many Mexican presidents tried to decentralize central governmental functions, but they did this by setting up federal offices in the states and municipalities rather than delegating more power to them.[41]

   As in the United States and Canada, public safety and police patrol are the responsibilities of the local governments. However, corruption is much more a problem in Mexico. And as Figure 10.2 shows, corruption is also more of a problem in Mexico than in the Central American nations to its south. The starting salary for a police officer in Mexico is less than $350 a month, and there is little room for expense money—police officers even have to buy their own bullets. One police officer related how he would drive across the Texas border to Wal-Mart so that he could smuggle back cheaper bullets. One time he was caught, but he was so embarrassed, he did not mention he was a police officer to the border patrol. As a result of these low wages, taking a bribe is an appealing temptation for any officer trying to make ends meet.[42]

   Mexican police officials fight a daunting battle against drug trafficking and are often caught between the deadly clashes of gangs fighting for control of the illegal drug trade. Although police officers and police chiefs have been killed in drug-related murders, some members of the police force have been involved as lookouts and hit men for drug gangs. Two 1999 surveys found that 90 percent of the citizens in Mexico City had "little" or "no" trust in the police, and only 12 percent of the national population expressed confidence in the police.[43]

   In a 2004 survey by Transparency International, a nongovernmental organization devoted to fighting corruption, Mexico was number 64 in a list of the least to most corrupt

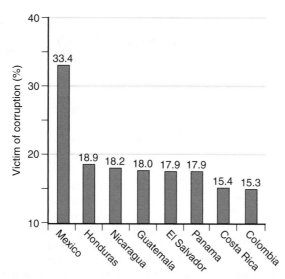

FIGURE 10.2 Corruption in Comparative Perspective Source: Jorge Buendía, Alejandro Moreno, and Mitchell Seligson, "The Political Culture of Democracy in Mexico, 2004," *Mexico in Times of Electoral Competition* (Nashville: Vanderbilt University, 2004), 59.

nations. Canada, at 12, was near the top of the list of least corrupt, with the United States at 17.[44] Transparency International also estimated that Mexicans spend on average 8 percent of their income on ***mordidas***, or "bites," the term for bribes. An estimated 4 out of 10 Mexicans have admitted to paying a *mordida*.

Citizens of Mexico struggle with a much lower standard of living than those in Canada or the United States. Many are poorly paid and have no health care, for example, or must wait in long lines for care. Thus millions of Mexicans have migrated to the United States in search of better jobs and the dream of American opportunity. The CIA estimates that for every 1,000 Mexican citizens, roughly four are leaving as migrants. On the positive side, these immigrants send a reported $10 billion annually to their families back in Mexico.[45] Simply crossing the border into the United States can result in a pay rate 12 times higher than in Mexico.[46]

American and Mexican border towns are as different as night and day. Mexican border towns have poor water and sewer systems, limited toxic waste management, poor infrastructure, and few schools or emergency facilities. Franco Bareno at the Border Environment Cooperation Commission estimates that as much as $3 billion is needed to provide Mexican border towns with adequate water and sewer services. To improve all public services, Bareno estimates that Mexico would need to invest at least $18 billion. Given these circumstances, it is not surprising that northern Mexican states such as Zacatecas have almost half of their population migrating to the United States.

Many Mexicans residing in the United States regularly send money to their families in Mexico, and some have set up hometown associations to help their former communities. In a policy dubbed "Tres per Uno" (or three for one), Mexicans living abroad who are part of a hometown association send money to local authorities back home in Mexico, which is then matched by state, local, and federal funding. Every $1 sent home is matched by $3. All of these

STRUGGLE FOR POWER

Toronto's Plan for More Autonomy

As Toronto became one of the fastest-growing cities in Canada, Toronto officials fought to control urban sprawl. City officials devised a plan for branding all of Toronto proper as a "metropolitan federation." Created in 1953, the Toronto Metro Council brought together 25 members selected to represent each city ward or suburb. Each suburb's mayor participated along with the city of Toronto's mayor, two city controllers, and an alderman from each of the nine Toronto wards. The Metro Council also elected a presiding officer who did not serve as the Metro mayor but instead played an organizational role such as chairing the meetings.

Transportation became the Council's primary responsibility. The traditional responsibilities such as the police, libraries, and licensing all stayed the responsibility of the lower municipal governments. The Metro Council and the municipalities also share the responsibilities for certain public policy measures such as education. A Metro School Board was set up to contribute to funding while the main policy-making responsibilities were set by local school boards.

The Metro Council also had primary responsibility for assessing property in the Toronto area. In 2005, the Toronto Metro Council was reformed to include further expansion. The mayor and municipal government became responsible for over 2.5 million citizens and 44 wards and suburbs.

In 2005, the city of Toronto organized a new plan to seek more autonomous power from the provincial government of Ontario. Toronto officials would like to have a closer relationship with the federal government, granting Toronto similar powers as Canadian provinces are granted. Toronto is frustrated by its limited means of generating funding, which is strictly based on revenue from property taxes. . . . As a part of the empowerment campaign, city officials are working on a charter "which could give Toronto's government powers that go far beyond anything a Canadian city has ever seen." With dynamic municipalities and cities, such as Toronto, the provinces will no longer control local governments at whim.

Source: Adapted from Toronto Government Online, "Strong Toronto, Strong Ontario," June 22, 2005, www.city.toronto.on.ca.

funds are set aside for promoting hometown enterprises, renovating schools, and building infrastructure. The program was created by former tax consultant Guadalupe Gómez, who has gone back and forth across the border to solicit money from the hometown associations.

Canada

Canadian politicians represent a population of 33 million in comparison to the United States' roughly 300 million people. While 59.3 percent of Canadians are English speakers, or **anglophones**, 23.2 percent are French speakers, or **francophones**. They are dispersed across the 10 Canadian provinces and the three arctic territories of the Yukon, the Northwest Territories, and Nunavut. The Canadian provinces are similar to Mexican and U.S. states, but the territories are under direct control by the federal government.

According to the Canadian constitution, the provinces have three main responsibilities: health care, education, and welfare. The health care system, as a responsibility of the provinces, is financed and controlled by the provincial legislature and the premier, who is comparable to a U.S. state governor. However, provinces must follow the guidelines of the Canadian Health Act, a federal act that applies to all provinces. In contrast, the Mexican federal government, not the states, is responsible for these issues. In the United States, states and the federal government share these responsibilities. Provincial governments in Canada therefore have autonomous power themselves, and they also maintain closer control of municipal governments than in the United States.

STRUGGLE FOR POWER

Quebec's Quiet Revolution

The 1960s marked the start of what is called the "quiet revolution" during which some nationalist leaders of Quebec slowly started to pull away from federal politics. Jean Lesage, a formal federal Liberal cabinet minister, pushed for a Québécois identity for provincial French speakers. The quiet revolution surged with René Lévesque, who campaigned for Quebec independence in the 1970s. The movement was marked by violence. In 1970, British diplomat James Cross and Quebec provincial Liberal minister Pierre Laporte were abducted by members of the Front de Libération du Québec (FLQ). Cross survived, but Laporte died.

The FLQ had killed previously for political reasons. In 1963, the group launched a series of bombings that killed and injured innocent citizens. The prime minister, Pierre Trudeau, immediately outlawed the Front as a political organization, and called for an arrest of all its connected "terrorists." He even invoked the War Measure Act, suspending civil liberties. Thousands of soldiers marched through the streets of Montreal to discourage further violence. This became known as the "October crisis."

Following a decade or more of violence, Quebec separatism was proposed to the people of Quebec in a 1980 provincial referendum, which failed to win approval for Quebec autonomy. Nonetheless, as the only province not to approve of the Canadian constitution, Quebec and national politicians met in 1987 to form the Meech Lake Accord. This pact proposed recognition of Quebec as a "distinct society," giving the Quebec government the right and responsibility to "preserve and promote the distinct identity of Quebec." The Accord, however, never received national approval in Canada. Liberal Premier Clyde Wells of Newfoundland led the anti-Accord movement, arguing that "at worst, [the accord] would result in the destruction of the nation." He argued that if special distinction is given to Quebec, then immigrants, aboriginal tribes, and women could follow suit.

A similar attempt with a 1992 Charlottetown Accord also failed. Ironically many English Canadians thought these accords gave too much power to the French, while the French citizens did not feel they gave them enough power. In 1995, a referendum for Quebec autonomy was proposed once again but failed by a razor thin margin of 50.6 percent opposed and 49.4 percent in favor. After such a close vote, an attempt was made to resolve the conflict in 1997 in Calgary, where 10 premiers met to put an agreement together to resolve the issue. However, Quebec leaders felt that the resolution offered Quebec even less autonomy than the Charlottetown or the Meech Lake Accord.

The Quebec Liberal Party does not have separatism as a declared objective, although it stands for a strong Quebec and safeguards for the French language and culture. Formed in 1991, the Bloc Québécois, on the other hand, specifically calls for national recognition of Quebec's distinct identity as its own nation. The party's Plan d'Action underlines the desire for Quebec sovereignty, declaring that "Quebec has always been and will remain profoundly different from Canada." With 54 members of parliament, the Bloc Québécois has the potential to make a major impact in national politics as it already has in helping to install Stephen Harper as the Canadian prime minister.

Sources: Based on "Constitutional Discord: Meech Lake," *CBC Archives*, September 6, 2005, http://archives.cbc.ca; Lucien Bouchard, "Statement Following the Meeting of the First Ministers in Calgary, September 16, 1997"; Gouvernement du Québec, September 6, 2005, www.saic.gouv.qc.ca; Plan D'Action, Bloc Quebecois, www.bloc-quebecois.org.

Canadian municipal governments have historically had less power than the U.S. counterparts. Canadian mayors are generally weaker than mayors of large U.S. cities. As Toronto political scientist Harold Kaplan explains, "Municipal government has been kept close to the people through legislative supremacy and annual elections. These communities have what Americans would call 'weak mayor' systems. At the same time, and perhaps paradoxically, important local programs have been vested in independent administrative agencies, in order to protect these programs from politics in the person of the municipal councilor." According to Kaplan, mayors are elected at large, and they speak for the municipality at ceremonial occasions, but they have "little formal control over the administrative branch and little voice in setting municipal policy" beyond their one vote in the council.[47]

Due to recent reforms, municipal governments have been gaining more autonomy from the provinces since a redistribution of power in 2000. These reforms are largely influenced by the desire to create more transparent, responsive, effective, and accountable government. Toronto, a global leader, illustrates the growth and power of urban Canada, as seen in the Struggle for Power feature.

With the legacy of both French and British colonization, an ongoing cultural battle has divided the French- and English-speaking communities in Canada to a much greater extent than historical language and cultural divisions in the United States. In one recurring struggle for power, the province of Quebec has sought more autonomy and even separation from Canada. The province of Quebec, which is three times the size of France and rich in natural resources, has the largest French-speaking population outside of France, with 5.5 million francophones and half a million anglophones.[48] In Quebec, the primary language is French rather than English. For more information about the separatist movement, see the Struggle for Power feature on Quebec's quiet revolution.

The Canadian constitution specifically gives the provinces the responsibility for health care, and health care system is struggling with modern-day challenges. Patients frequently have to face long waits for common medical procedures. In 2000, the average wait for medical treatment for breast cancer was 27 days.[49] The increased waiting times for treatment are attributed to decreased health care funding from the federal government in the 1990s.

Wealthier Canadians can afford to go to a private clinic and thus avoid waiting for treatment. However, the great advantage to the Canadian system is that in 1995, the average cost for a major surgery was less than $300. Privatization of health care in Canada would increase these costs. Originally, provinces funded around a third of the medical costs, but in 2002 the finance minister, Paul Martin, drastically cut federal funding to 16 percent, resulting in provinces being crushed by the growing cost of health care. As prime minister in 2004, Martin then readdressed this issue with large infusions of new federal money.

GETTING ALONG

Mexican workers migrate to the United States. Hollywood movies are made in Vancouver. Canadians travel to the United States for operations. As Mexicans, Canadians, and Americans increasingly interact, cultural differences blur to some extent, but we need an improved understanding of each other's politics and political structures.

We can learn from one another. For example, Mexico and the United States can encourage more women to hold high political office as Canada does. Perhaps problems of health care in all three countries can be solved by adopting the best parts of the Canadian and American health care systems. Neither two-party nor multiparty systems automatically command the level of citizen participation required in a democracy. By understanding the struggles each country faces, we can establish a more complete understanding of each other and create a more respectful interaction. Improving relations with our neighbors to the north and south will depend on U.S. cities and states, not simply the national government.

The struggle for power in cities and states in the 21st century is made much more complicated by the forces of metropolitanization and globalization. Our connections of trade, immigration, and communications with the rest of the world have provided new opportunities of social and economic development while vastly complicating the governing process. Although metropolitan regions now compete globally as if they were almost separate nations, we have yet to devise adequate institutions, political processes, and governmental structures to govern these regions. We are unable to tame multinational corporations or to easily assimilate the new immigrants needed to make us competitive in the global economy. We are far too ignorant about our neighbors to the north and south. Cities and states along our northern and southern borders in particular must struggle with how best to cooperate and learn from our neighbors while at the same time finding ways to control unlawful immigration, drugs, or potential terrorist attacks over too porous borders.

State and local politics is becoming fast-paced and demanding. This is one struggle for power, control, democracy, and justice in which we all share.

CHAPTER SUMMARY

Two world-wide developments that are greatly affecting our cities and states are metropolitanization and globalization. Cities have sprawled into suburbs and exurbs, creating large metropolitan regions. New immigrants may move directly to suburbs or small towns instead of the largest cities. Acculturating and accommodating the new residents is a major task for state and local governments. These new developments have aggravated the struggle for power. How are power and influence fairly divided among longtime residents and newcomers? How can government be truly representative and participatory with such a mix of cultures? How can governmental structures function as they are supposed to while making sure that political processes are still operative? How can we guarantee that government institutions and processes work properly when people do not know how to use them?

One problem of metropolitanization has been the overlap of the dozens of governments in any one metropolitan region. Although a consolidated regional government might be a solution, this is unlikely to happen because people identify with one suburb or city neighborhood rather than their region. Although there is a movement for establishing regional governmental authorities, there is a problem getting the residents of cities, suburbs, and exurbs to agree to pass state laws initiating such a far-reaching reform.

Some Americans deal with fragmentation by moving into gated communities or other private housing developments. Residents of these common-interest developments (CIDs) can become a strong conservative political force opposing tax increases to pay for public schools or opposing the building of affordable housing because it might lower their property values. Since residents of gated communities can often be rallied to vote in large numbers for what they perceive to be in their self-interest, they can have a significant impact in local elections and policy making.

This fragmentation makes governing larger metropolitan regions a major challenge and can be a setback when dealing with globalization. Regions that cannot produce a highly educated workforce, or lack diverse peoples with language skills and broader cultural understandings, or do not have the resources or political leadership to build world-class hotels and attractions will lose out in this competition.

Globalization holds many problems for American states and cities. Every city aspires to be a "global city" economically, but globalization accentuates the widening gap between rich and poor. Chief executives of global corporations become fabulously wealthy, but fewer workers are needed to run factories and to do jobs in global companies. The struggle for power also occurs between cities and states fighting for status in the global economy. The task of local leaders is to ensure that their communities are important in the global economy and to try to negate globalization's widening gap between rich and poor.

To deal with the problems of globalization and metropolitanization, states and cities must promote economic development, simultaneously providing the infrastructure and amenities necessary for transnational corporations to function well in a community. We must improve our school systems so that our young people will be part of a well-educated workforce. We need to be more cosmopolitan, welcoming immigrants and learning from them, and must market our regions through promoting tourism. In addition, we must help cities and states continue to be laboratories of innovation in providing services to their residents. The expense of distributive and redistributive services necessitates funding from the federal government to help with programs for those who need them.

One clear example of the effects of globalization is the increased intertwining of the United States, Canada, and Mexico, especially since the passage of the North American Free Trade Agreement (NAFTA). All three countries are influenced by what happens inside the other two.

KEY TERMS

acculturated Socialized into another country's norms, culture, and ideals, p. 259

anglophones People whose primary language is English, p. 270

common-interest development (CID) A private community whose land and common facilities are owned by the members of the housing association and governed by a separate condominium or homeowners association board of directors, p. 253

commuter tax A tax levied on the people who work in a city or other taxing jurisdiction but do not reside there, p. 251

distributive services Ordinary city services such as police protection, parks, garbage pick-up, and street

repair which can be distributed to all citizens, p. 262

exurbs Developed areas beyond a central city and its suburbs, p. 249

francophones People whose primary language is French, p. 270

gated community A private community with limited access points and security guards to protect the property, p. 253

globalization The growth of connections such as international trade, immigration, and communications throughout the world, p. 247

metropolitan region The large area of suburbs, smaller towns, and counties around a central city, p. 249

mordida A Mexican colloquial term for bribe, from the word meaning "bite," p. 269

redistributive services Services like public housing or welfare payments that wealthier taxpayers fund and poorer taxpayers receive, which provide a limited redistribution of wealth, p. 262

transnational corporation A firm whose facilities and employees span more than one nation, p. 262

urban sprawl Unplanned, uncontrolled development of land at the boundary of an urban area, p. 252

urbanization The growth of cities and urban areas, p. 247

QUESTIONS FOR REVIEW

1. How does metropolitanization differ from the urbanization of the 19th century?

2. In which region of the United States is urban growth occurring most rapidly? Why?

3. Why is the governance of metropolitan regions an issue for state government?

4. Why must regulations and curbs on urban sprawl be decided at the state and local level rather than by national legislation?

5. What are gated communities? What are common-interest developments (CIDs)? What are their advantages and disadvantages?

6. How are gated communities and CIDs governed? To what degree are these privatized governments

integrated into higher units of local and state government?

7. What do homeowners associations do? What role do they play in metropolitan governance? How well are officers and board members prepared, trained, and supported to play these roles?

8. What are the three types of governmental services provided by local and state governments? Which is the least controversial? What is the most difficult for local and state governments to provide? Why?

9. What are the similarities and differences among Mexico, Canada, and the United States?

DISCUSSION QUESTIONS

1. How much fragmentation exists in the largest metropolitan region in your state? What level of cooperation exists between these multiple units of government? What form of governance exists and what level of metropolitan government do you think should exist?

2. Why are there few "metropolitan citizens"? How does that affect the prospects for metropolitan regional government?

3. What forms of improved metropolitan governance exist in the largest metropolitan region in your state? What degree and form of metropolitan governance do you think are best, and which forms do you think are achievable?

4. Which political groups would have to band together in your state to achieve better metropolitan regional governance?

5. Is urban sprawl good or bad? What are the arguments for and against sprawl? With which do you agree?

6. What distinguishes first-tier global cities from cities and states that play minor roles in the global economy? What are the advantages and disadvantages of being a global city, region, or state?

7. What role does immigration play in a global city? Are high levels of immigration good or bad for a city or state?

8. How should your city and state respond to globalization? Do they have any official policies related to globalization?

9. Should the United States adopt the Canadian style of national health care? Or should Canadians turn to a mixed public and private health care system? Why or why not?

10. Many small towns lose their young people to cities or regions with more job openings and cultural amenities.

If you were mayor of a small town, how could you encourage your young people to remain?

11. In many cities and towns, the number of immigrants from Mexico has been increasing. If you were the mayor, what steps would you take to integrate the increasing numbers of Mexican immigrants in your community?

PRACTICAL EXERCISES

1. Using census data for 1990, 2000, and 2005, determine whether the largest city, the suburban area around that city, and your state are gaining or losing population. Most census data are available at the Web site of the U.S. Census Bureau, www.census.gov. You can complete a table like the following:

	1990 Population	2000 Population	2005 Population	Percent Gain or Loss
City				
Suburbs				
State				

2. For the largest city and metropolitan region, what is the racial and ethnic background? Create two pie charts like those below to compare percentages of the white, African American, Latino, Asian, and Other populations for each location. What are the differences in the populations of the city and the region?

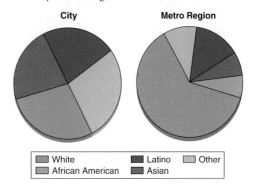

3. How many separate taxing units of government are in the largest county and metropolitan regions? This information is usually available from the county assessor's or treasurer's Web site, from newspaper stories obtained through a search engine such as Lexis-Nexis, or from governmental watchdog groups like the League of Women Voters.

4. Is there a metropolitan regional planning agency in your state? If so, what is its budget and what are its legal powers?

5. Are there any metropolitan regional government units? What are they, and how well do they work? Which specific services do they provide?

6. Is the largest city in your state a major global city or unimportant in the global economy? How many transnational corporations are headquartered in your city? How many global financial transactions like international bank loans, stock purchases, or futures trades occur? What is your city's special niche in the global economy? Sometimes such data are collected by a group such as World Business Chicago or may be published in newspaper stories or collected by city or state economic development agencies.

7. How great is the gap in median income between the richest and the poorest community in your metropolitan region? You can find this information on the Web site of the U.S. Census Bureau, www.census.gov. Is this disparity caused by globalization? What are the effects of globalization on these communities?

8. What laws govern homeowners associations in your state? For this information, try your state legislature's Web site or state law books.

WEB SITES FOR FURTHER RESEARCH

Regional governors associations address issues specific to their geographical location. Two are the Council of Great Lakes Governors, www.cglg.org (8 Great Lakes States with the provinces of Ontario and Québec), and the Western Governors Association, www.westgov.org (19 states as well as Guam, the North Marianas Islands, and American Samoa).

States also work together through the Council of State Governments, www.csg.org.

Some regions have set up limited regional government. The Portland, Oregon, regional government Web site is www.oregonmetro.gov.

The city of Miami and Dade County, Florida, have consolidated government. See http://miamidade.gov.

The CIA *World Factbook* provides regularly updated and detailed information for all countries, such as the government, economy, history, and geography. Go to www.cia.gov/library/publications/the-world-factbook.

The World Bank provides information on technical and financial assistance provided to all countries and gives updates on international issues, including energy, climate change, corruption, and the food crisis. Go to www.worldbank.org.

Information on how to run a homeowners association is available from the Community Associations Institute, www.caionline.org. To get an idea of the regulations in a community governed by a homeowners association, see http://deerwoodhomeowners.com.

SUGGESTED READING

Clarke, Susan E., and Gary L. Gaile. *The Work of Cities*. Minneapolis: University of Minnesota Press, 1998.

Cornelius, Wayne A., Todd A. Eisenstadt, and Jane Hindley, eds. *Subnational Politics and Democratization in Mexico*. San Diego: University of California, 1999.

Dreier, Peter, John Mollenkopf, and Todd Swanstrom. *Place Matters: Metropolitics in the Twenty-first Century*. Lawrence: University Press of Kansas, 2004.

Hambelton, Robin, and Jill Gross, eds. *Governing in a Global Era: Urban Innovation, Competition, and Democratic Reform*. New York: Palgrave, 2007.

Judd, Dennis, and Todd Swanson. *City Politics*. New York: Longman, 2009.

McKenzie, Evan. *Privatopia: Homeowner Associations and the Rise of Residential Private Government*. New Haven, CT: Yale University Press, 1994.

Orfield, Myron. *Metropolitics: A Regional Agenda for Community and Stability*. Washington, DC: Brookings Institution, 1997.

Peterson, Paul E. *City Limits*. Chicago: University of Chicago Press, 1981.

Ponce de León, Beatriz. *A Shared Future: Economic Engagement of Greater Chicago and Its Mexican Community*. Chicago: Chicago Commission on Global Affairs, 2006.

Ranney, David. *Global Decisions, Local Collisions: Urban Life in the New World Order*. Philadelphia: Temple University Press, 2003.

Savitch, Hank, and Paul Kantor. *Cities in the International Marketplace: The Political Economy of Urban Development in North America and Western Europe*. Princeton, NJ: Princeton University Press, 2002.

Simpson, Dick, ed. *Inside Urban Politics*. New York: Longman, 2003.

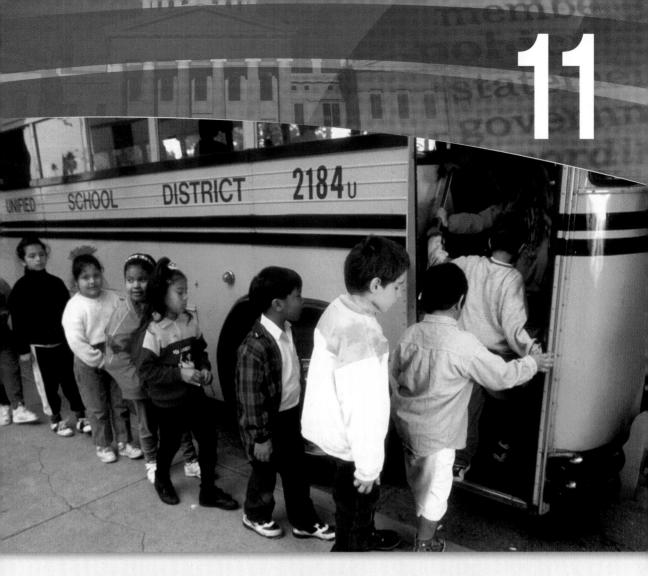

Are these children on their way to a better education?

ISSUES OF THE SCHOOLHOUSE AND CAMPUS

Public education is a part of the soul of our nation. Teachers in America transmit values about democracy. Knowledge and the capacity to solve problems that we learn in school are the determinants of individual success as well as national economic productivity.

Almost one in four Americans is enrolled in education, from the preschool level through our colleges and universities. As of 2007, there were 49 million in the K–12 schools and 17.5 million in higher education. One in every 25 employed Americans is a public school employee.[1] Total state and local education spending reached $689 billion in 2005, or about

$2,330 per person in the United States, and represented the largest expenditure of state and local governments, at about 30 percent of state and local budgets.[2] Total education revenues have represented about 4 percent of America's gross domestic product in recent years.

Given the fundamental importance of education, there are intense struggles over who governs these institutions and who pays. The stakes are also immense—the competitiveness, wealth, and global standing of America depend on the skills and intellectual prowess of our citizens. Yet international comparisons of reading, math and science scores consistently find American 15-year-olds ranking 15th to 20th among developed nations, while American expenditures per pupil rank third among nations.[3]

In this chapter, we examine the struggle among governments and interest groups for power and influence over America's schoolhouses and campuses. School finance is examined, since the struggle in education usually revolves around expenditures for teachers, extra support for children at risk, and the cost of reform initiatives. While citizens are quite vocal in their demands for better education in return for their tax dollars, it is not clear which policies are most effective in improving education.

The study of how cities and states deal with education is illustrative of other public policy conflicts such as health care, economic development, and public safety. In education as well as these areas, the struggle is not just about power and influence over policy but also about controlling billions of dollars, thousands of jobs, and profitable contracts. These decisions illustrate the pull of representative and participatory democracy. Some decisions are made at the most local level by parents, teachers, and community members through PTA and school council meetings open to the public. As students get older, they have a voice through student government, campus organizations, and a student newspaper. But most critical decisions are made by a school bureaucracy headed by school superintendents, elected school boards, and elected and appointed officials in state and federal government.

The institutional structures and political processes determine how education policies get made. And in education, the outcomes matter in a very important and personal way. If a large percentage of students fail to finish high school, or if college students do not have the right skills and knowledge to get good jobs upon graduation, these are factors that affect individual's life choices and the capacity of our communities and nation to compete effectively in the global economy.

THE HISTORICAL DEVELOPMENT OF AMERICAN PUBLIC EDUCATION

In 1642, the colonial legislature of Massachusetts directed "certain chosen men" to see that "children were being trained in learning and labor and other employments—profitable to the state." For the first time in the English-speaking world, a public body enacted legislation to require that children be taught to read. Later, the Land Ordinance of 1785, enacted by the Congress under the Articles of Confederation prior to adoption of the U.S. Constitution, declared: "Religion, morality, and knowledge being necessary to good government and the happiness of mankind, schools, and the means of education shall forever be encouraged." To back up this language, the Congress set aside a portion of every township for the support of schools.[4]

In the Northeast, "district" or "common" public primary schools became a model for the nation. Supported by district taxes, state funds, and tuition paid by parents, these schools were

A one-room country schoolhouse.

open to all. Reformers led by Horace Mann preached that common schools could, if properly run, protect "society against the giant vices . . . against intemperance, avarice, war, slavery, bigotry, and the woes of want." According to historian David Nasaw, "The campaign for the common schools—through the later 1830s and 1840s—was no more and no less than a campaign for public taxation." Without taxes specifically earmarked for schools, there could be no common schools. Opponents of taxation for free schooling fought back, yet ultimately the nation became blanketed by public schools. Public education, however, was not extended to everyone. "Schooling for blacks was, from the 1830s on, proscribed by law in the South and by 'custom and popular prejudice' elsewhere," Nasaw states. The fines for teaching African Americans, whether free or slave, ranged from $100 in Virginia to six months' imprisonment in South Carolina.[5]

In 1862, during the tumult of the Civil War, President Abraham Lincoln signed into law the Morrill Act, which provided for the creation of "land grant" universities. Named for U.S. Representative Justin Smith Morrill of Vermont, the act provided each state 30,000 acres of public land for each senator and representative in that state, with proceeds from the sale of that land to be invested in a perpetual endowment fund to support colleges of agricultural and mechanical arts in the state.[6] Yet even with this strong federal government support for public education, by 1900 only 8 percent of the population 14–17 years of age was enrolled in grades 9 through 12 in public high schools, although some students attended private academies and post-grammar schools.[7]

Ever since Horace Mann and other educators advocated for common schools in America, educational reform has been a continuing theme. In the Progressive Era of the early 20th century, business and civic groups fought to take control of local schools away from urban political machines they believed to be corrupt, inefficient, and ineffective. According to Nasaw, the civic reformers, muckraking newspaper reporters, and business leaders agreed that neighborhood school board members were at fault for overcrowded, understaffed, patronage-rife schools that failed to keep adolescents in school long enough to learn "how to work." Coming primarily from the elite class of society, the reformers and business leaders focused on state legislative change that would encourage a model of a centralized school board with a superintendent and professional staff. In most major cities, centralization was soon the norm.[8]

After World War II, a wave of consolidation closed thousands of one-room schools, whose students were usually then enrolled in township or smaller school districts. Thus, while there were approximately 40,000 public school districts serving the nation in 1960, by 2004 this was reduced to 14,000.[9] At the same time as the number of districts declined, a dramatic growth in spending for public education began. This increased after the Soviet Union beat the United States into space by launching its *Sputnik* satellite in 1957, startling Americans out of complacency. Consequently, while America had spent on average $1,897 per pupil in 1960, the figure was $3,273 by 1970, and by 2000 it was $7,814. All figures are adjusted for inflation to dollars from the year 2000.[10]

WHO GOVERNS OUR SCHOOLS?

Although public schools are primarily creatures of state government, responsibility for the day-to-day operation of public schools rests with local school boards. Nevertheless, as with other aspects of American society, the federal government has considerable influence on American education. Both chambers of the U.S. Congress have committees and subcommittees dealing with education, and there is a Department of Education in our nation's executive branch. Moreover, the U.S. Supreme Court and lower federal courts have handed down many court decisions dealing with education. One of the most important court decision of the 20th century was *Brown v. The Board of Education of Topeka, Kansas* (1954), which struck down "separate but equal" as a justification for racial segregation of public schools.

Fourteen states elect a statewide superintendent of education or public instruction. Boards and commissions select a state schools superintendent or commissioner in 25 states. In 11 states the governor appoints the chief state schools officer, generally with the advice and consent of the state senate.[11] These state offices oversee public schools and, to a much lesser extent, private education at the local level.

In many cases, state policy is carried out through **school mandates**, which are laws or regulations that local school districts must follow. What goes into mandates is determined by diverse government officials and interests, including state legislators, governors, state education departments, teachers unions, and school board associations. With the potential to control almost any aspect of school governance, mandates can set course requirements such as consumer or physical education, the number of days in a school year, or what holidays must be honored.

School superintendents complain that these mandates—usually described in the hundreds of pages of a state's **school code**—are often unfunded, which requires local school districts to pay for them. While a few states have developed procedures for requesting waivers from such regulations, some interested parties, including the federal government (which has its own indirect mandates) and teachers unions, are vigilant to see that their interests are protected. Thus, while most of the 14,567 school districts in the United States are governed locally by elected school boards and managed by local school superintendents, school districts do not have the flexibility of cities' home rule powers.

Local control—the concept that school affairs should be governed primarily by the local school district rather than by state and federal government laws and regulations—continues to be the mantra of public secondary and elementary education in many localities. Yet states and the federal government have been seizing more authority over the public education enterprise. Teachers unions have become more aggressive in recent decades. The business community has reasserted its interest in education policy. And charitable donors such as the Bill and Melinda Gates Foundation are trying to reshape schools. Most of the recent policy-making has been spurred by anxiety about the condition of American education.

Reform Efforts

In 1975, Americans learned that scores on the Scholastic Aptitude Test (SAT) had been declining for 10 years. In 1983, a national commission issued *A Nation at Risk: The Imperative for Educational Reform*, which documented poor performance and low expectations in the schools.[12] The report aroused significant anxiety among America's leaders and helped launch a wave of school reform that continues today.

Prior to the 1970s, education-related groups at the state and local level were the primary initiators of education policy.[13] However, from the 1970s on, new groups outside the field of education—including politicians, governors, and prominent business leaders—began to exert influence on the shaping of educational policy that made its way onto the policy agenda in the states.[14]

Thus the local control model of school governance and finance evolved into a state system of finance, standards, and assessment. In 1960, education professor Myron Lieberman predicted that local control of schools would have to give way to a system of educational controls in which the roles that local communities played would be ceremonial rather than policy-making. According to Lieberman, local control had become "a corpse."[15] State policy activism grew as state funding of local schools increased. In 1971, states provided 38 percent of local school funding; by 2003, that figure had reached 49 percent of total funding, with the local districts providing 43 percent and the federal government 8 percent.[16]

Since local schools are creatures of their states, the latter have powerful statutory tools as well as money to implement state standards, testing, and "report card" comparisons of school districts, individual schools, and even students. Education policy analyst David Conley notes that, given current state policies in accountability and assessment, principals have become as responsible to the state government as they are to the local superintendent and board of education. School superintendents have been called on to support and facilitate improvements on a school-by-school basis, and in many states, local principals have been freed to make major decisions about how they are to meet measurable goals, creating a kind of "bottoms up meets top down" partnership style of educational management.[17]

In the recent wave of reform, Kentucky overhauled its public education system with echoes of the Progressive Era. In the 1980s, a coalition of business leaders and social reformers criticized the Kentucky system of education for politicized hirings and firings by locally powerful yet poorly trained school boards. The Prichard Committee on Academic Excellence (named after a crusading education reformer) culminated in a dramatic decision by the Kentucky Supreme Court that rejected the existing education finance system, and in 1990 the legislature passed the Kentucky Education Reform Act.[18]

The work of the blue-ribbon Prichard Committee, which included four former governors, generated widespread media attention and had strong financial support from Kentucky's leading companies. The 1990 reform legislation increased funding for poor school districts, replaced an elective with an appointed chief state school officer, required extensive statewide testing, created local school councils that took power away from traditional school boards, and created controversial family centers at school sites to provide day care as well as health and social services for new parents.

Major foundations, such as Rockefeller, Ford, MacArthur, and Spencer, have played a big role in education reform. With more than $60 billion in assets, the Bill and Melinda Gates Foundation is richer than many small nations. "Gates has so much money to spend," says education professor Diane Ravitch, "they have the power to set the agenda" for education. As Gates gave grants for small schools, "almost every district lined up to seek their money and to split up their high schools into small schools."[19] Other foundations have included education reform among their funding program areas.

Problems occur, however, when such foundation grants expire. For instance, the Gates Foundation gave grants of $100,000 to 50 new, small New York City public schools. After these grants expired in 2006, one school executive asked, "What happens to that success when there is no more private money?"[20] Often, school districts must either take money from other programs to sustain an initiative or discontinue it. Another option is for schools to be run by private education corporations like the Edison Schools, a for-profit company that in 2006 managed 20 schools in Philadelphia and others across the country.

Teachers Unions

The **National Education Association (NEA)** is the largest teachers union in the United States, with 3.2 million members. The **American Federation of Teachers (AFT)** is the second largest teachers union, with over 1.4 million members, and is affiliated with the AFL-CIO. These unions and their affiliates at the state and school district levels are by far the biggest and most powerful players inside public education.[21] Started in 1857 as a professional organization, the NEA transformed itself by the 1960s into a political juggernaut that has become a proponent of more money for education and an opponent of efforts to weaken the traditional public school system. NEA units contribute big money to political campaigns and energize their teachers to work to elect or defeat gubernatorial and legislative candidates.

Teachers at an NEA rally.

In 1976, for example, the Illinois Education Association contributed $80,000 (about $300,000 in 2009 dollars) to Republican Jim Thompson's campaign for governor, the largest contribution to his campaign. After his election, Thompson asked the association to review every piece of legislation that could affect teachers; he also signed a collective bargaining bill for teachers. Many state legislators also make a point of checking all education legislation with teacher union lobbyists. Lawmakers want to stay on the good side of the teachers unions, to keep campaign contributions flowing and to forestall a union-backed opponent at the next election.

NEA affiliates also take their cases to court. In 2005, the Kansas Education

Association helped convince that state's supreme court that the legislature and governor must increase funding for the public schools.[22]

Opposition to outside efforts to run the public schools comes primarily from these powerful teachers unions. The NEA argues that a number of programs and practices are detrimental to public education: privatization, performance contracting, tax credits for tuition to private and parochial schools, voucher plans, planned program budgeting systems, and evaluations by private, profit-making groups.[23] Although various groups have advocated vigorously for some of these changes, because of the equally vigorous resistance from teachers unions, they continue to be the exception rather than the rule.

Managing the Schools

A century ago, Progressives were able to take control of local schools away from mayors, city councils, and political machines, but by the last decade of the 20th century, mayors in big cities like New York, Chicago, Los Angeles, and Philadelphia began to reassert their control. If they were going to be blamed for bad schools, they wanted to affect how they were run. Mayors wanted better management and top-down control of policies. They were especially concerned with improving student retention rates and test scores. They also wanted to curb the power of teachers unions. So some big-city mayors began installing, with state legislative agreement, their own school boards and handpicked superintendents to achieve these outcomes.

In California, the California School Boards Association, the principals union, and the League of Women Voters fought state legislative efforts to give the mayor of Los Angeles control over the city's schools. The law that emerged from the California controversy created a multiheaded organizational structure that spreads authority across the school board and mayor and gives the mayor direct control of schools with only about 10 percent of the district's students. Yet, according to an official in the California governor's office, "A culture of passive resistance exists among faculty and central office officials who believe that, if you don't like the reform agenda, all you had to do was lie low for two years and the agenda would change."[24]

However, big-city mayors are intent on having a say in their city school systems, and it seems likely that they will prevail. Some mayors bring in nontraditional executives to run their schools, as Chicago mayor Richard M. Daley did when he made Paul Vallas head of that city's schools in the mid-1990s. Vallas went on to head the Philadelphia school system and the Recovery School District of New Orleans. A state and local budget expert, he is of a new breed of education outsiders brought in to attempt the daunting challenge of making big improvements fast in generally low-performing big-city school systems. Similarly, former Colorado governor Roy Romer headed Los Angeles schools from 2001 to 2006, and former U.S. Justice Department lawyer and executive Joel Klein holds the reins in the New York City school system.

The State of Pennsylvania took over Philadelphia schools in 2001. Lawmakers brought in Vallas, who had generated national recognition for improving test scores as head of the Chicago public schools. Intense, with a no-nonsense style, Vallas developed a good working relationship with the teachers union head and state politicos.[25] He then embarked on a multifaceted reform program including a standardized curriculum, competition through privatization, and smaller high schools. In addition, Vallas increased the number of children receiving early childhood instruction by 50 percent, reduced the shelf-life of textbooks to three years, and took unruly students out of regular schools. A government budget expert, Vallas also made changes in budgeting, human resources, and capital planning. In Philadelphia, Vallas started the nation's largest privatization program. The for-profit Edison Schools, for example, operated more than 20 Philadelphia schools; several universities operated others. Kaplan, the test training company, designed a citywide uniform high school curriculum for Philadelphia that is unique in the country.

Vallas also wheedled more money from the legislature and generated large foundation grants. After five years, achievement was up among many of the district's lowest-performing students, enrollment in early childhood programs had doubled, and two dozen theme-based small high schools had been created. But Vallas ended his stint in Philadelphia with hard feelings. According to Vallas, an outside school reformer is given about five years in any top schools post: "The first two years you literally get to do just about anything you want. You're a demolition expert. By year four, there's a lot of people walking around pissed off because you're getting so much credit for it. And by year five, you're chopped liver."[26]

In contrast to the outsiders, former Boston schools superintendent Thomas Payzant was a career school executive, having served as superintendent in four districts in four states, including San Diego, prior to his decade-long stint at the helm in Boston. Named a Public Official of the Year in 2005 by *Governing* magazine, the mild-mannered Payzant focused on basic subjects like literacy and math and on teacher training rather than attempting to revamp Boston's school management or structure.

As in Pennsylvania, Massachusetts had considered a takeover of the city's schools. Instead, state leaders gave control of the Boston school board to the mayor, who staunchly supported Payzant in initiatives such as reducing the pupil-teacher ratio, breaking high schools into smaller campuses, and increasing 10-fold the number of students taking Advanced Placement courses. Between 1998–2004, the percentage of students who passed the state's rigorous 10th-grade math test increased from 25 percent to 74 percent, while those passing the English test jumped from 43 to 77 percent. "I don't know another urban system that has made more progress than the Boston public schools," declared Harvard professor Richard Elmore.[27] Payzant retired in 2006 and is now a senior lecturer at the Harvard School of Education.

Clearly, different styles can make improvements in urban schools. Nevertheless, according to one observer, "The grim reality is that urban school districts are still a mess everywhere."[28] Unwilling to accept that poor and minority students are incapable of achieving at levels similar to those from higher socioeconomic classes, the federal government has become much more deeply involved in local public education in the 21st century.

THE FEDERAL ROLE IN EDUCATION

Education had been primarily a state and local function prior to the *Sputnik* launch in 1957. Since then, however, the role of the federal government in education has grown steadily. According to education policy professor David Conley, "The cumulative effect has been to insert the federal government into local schools in an ever widening circle." A number of these initiatives illustrate the political game of federalism, in which each level of government tries to expand its authority, and get the other levels to pay for it. Even though the federal government strongly affects how schools operate today, the national government provides only 8 percent of all funds spent on elementary and secondary education.[29] This percentage will increase for a couple of years because of the federal stimulus funding aimed at combating the 2008–2009 recession.

Early Federal Involvement

A year after the Soviet Union launched *Sputnik*, President Dwight D. Eisenhower signed the National Defense Education Act (NDEA), which provided funding for science and math education initiatives and loans for college students. In 1965, Congress enacted the Elementary and Secondary Education Act (ESEA), the basic purpose of which was to provide financial assistance to schools for children from low-income families, found in Title I of the act. Both laws provided financial support for focused rather than general education programs.

Not until the Individuals with Disabilities Education Act of 1975 (IDEA) did the federal government begin to leverage its funding to require that local schools change their pedagogical approaches to teaching children. Originally known as the Education for All Handicapped Children Act, this program required participating states to provide a "free and appropriate education" for all handicapped children from ages 3 to 18. Education is to be provided in the least restrictive environment and in regular classrooms wherever possible, an approach called "mainstreaming." In addition, school districts are required to provide an individualized educational program for each child with disabilities.[30]

In return for participating, Congress authorized payment to the states to cover 40 percent of their costs for implementing the IDEA requirements. However, the federal government has never funded IDEA at this level. Estimates are that federal payments to the states are less than 20 percent of the total expenditures for the handicapped.[31] This shortage of federal funding created significant conflict at the local school level between advocates for children with disabilities and school officials who are faced with reducing expenditures elsewhere to meet the high costs mandated by IDEA.

No Child Left Behind

After President George W. Bush signed into law the 2002 No Child Left Behind Act (NCLB), school reform advocate Fred Hess said its passage amounted to a roar of frustration from national policy-makers "demanding that state and local officials stop making excuses and do something about low-performing schools."[32] NCLB requires states to improve performance of students so that by 2013, 100 percent of all students are scoring at least at the "proficient" level on required annual state-developed exams. If schools fail to make "adequate yearly progress," sanctions are imposed. For example, students in underperforming schools are accorded the right to transfer to a better public school within the district and to receive special after-school tutoring. Schools that fail to reach adequate progress after five consecutive years are to be restructured and new staff hired.

As might be expected from such sweeping requirements, NCLB created a storm of reaction from state and local educators. One proponent lauded NCLB as "the most comprehensive federal education law ever written."[33] Supporters of NCLB said that, for the first time, the focus of accountability is on the school's capacity to address the education of each student as an individual rather than on the average performance of a school. Critics declared that, once again, the federal government failed to provide adequate funding to implement the act. In addition, the critics contended that the goal of 100 percent proficiency by 2013 is unrealistic and that teachers are, more than ever, required to teach to the test. (Table 11.1 shows the percentage of eighth graders who were proficient or better in math in 2007.)

TABLE 11.1 Percentage of Public School Eighth Graders Proficient or Better in Mathematics in 2007, by State Rank

National Percentage = 31%

Rank	State	Percentage
1	Massachusetts	51
2	Minnesota	43
3	North Dakota	41
3	Vermont	41
5	Kansas	40
5	New Jersey	40
7	South Dakota	39
8	Montana	38
8	New Hampshire	38
8	Pennsylvania	38
11	Colorado	37
11	Maryland	37
11	Virginia	37
11	Wisconsin	37
15	Washington	36
15	Wyoming	36
17	Connecticut	35
17	Indiana	35
17	Iowa	35
17	Nebraska	35
17	Ohio	35
17	Oregon	35
17	Texas	35
24	Idaho	34
24	Maine	34
24	North Carolina	34
27	Alaska	32
27	South Carolina	32
27	Utah	32
30	Delaware	31
30	Illinois	31
32	Missouri	30
32	New York	30
34	Michigan	29
35	Rhode Island	28

National Percentage = 31%

Rank	State	Percentage
36	Florida	27
36	Kentucky	27
38	Arizona	26
39	Georgia	25
40	Arkansas	24
40	California	24
42	Nevada	23
42	Tennessee	23
44	Hawaii	21
44	Oklahoma	21
46	Louisiana	19
46	West Virginia	19
48	Alabama	18
49	New Mexico	17
50	Mississippi	14
District of Columbia		8

Sources: *State Rankings* 2008; U.S. Department of Education, National Center for Education Statistics, "The Nation's Report Card, Mathematics 2007," NCES 2007-494. There are four achievement levels: below basic, basic, proficient, and advanced. Proficient represents solid academic mastery for eighth graders. Students reaching this level have demonstrated competency over challenging subject matter.

At the end of 2006, many schools were unable or unwilling to implement requirements that all students in underperforming schools be allowed to transfer to better-performing schools. In Los Angeles, for example, 300,000 students attended schools in 2006 that had failed to make NCLB goals, yet only about 600 students transferred to better schools. This is in part because of resistance from the school district and in part because of the practical impossibility of providing spaces in better schools. As the Los Angeles school superintendent declared: "Think about it. We're 160,000 seats short. Where do you transfer to?"[34] States find themselves in a dilemma: they will lose federal funding should they fail to enforce federal education mandates, but they are strapped for the funding necessary to carry them out.

State and local education officials are hoping that the Obama administration will make education funding a priority. Obama proposed a $1.3 billion increase in the 2010 federal budget for education, to $46.7 billion, which does not include the up to $100 billion provided for education programs through the American Recovery and Reinvestment Act. The president's budget request, which raises funding to states for special education and school improvement grants to local education agencies, will be of great relief to state and local school boards, whose revenues are down because of the recession of 2008–2009.[35]

WHO PAYS FOR PUBLIC EDUCATION?

Americans spend more than $600 billion a year on public elementary and secondary schools, or as Table 11.2 indicates, well over $9,000 per pupil on average nationwide. There is, however, great variation in spending among the states. In 2007, per pupil expenditures ranged from an average of 14,675 in New Jersey to $5, 551in Utah. Within many states, the ranges of spending among school districts are even greater. And according to U.S. Census Bureau figures, 81 percent of all public school expenditures in 2003 were for salaries, wages, and benefits of school employees.

Because of the large dollar amounts involved, there are intense struggles over who gets the money, and where it comes from. The groups contending for a larger share of the funding pie are often different from those at the forefront of education policy reforms. In addition, because of large variances in local support of education between property rich and property poor districts, rural, suburban, and urban school districts are often competing for the same state education dollars.

A root cause of conflict over spending for public schools is that the main source of funding for public schools is often the local property tax, which is also the basis on which additional state support has been largely calibrated. The **local property tax**—primarily a tax on real estate such as residential, commercial, and industrial property and farmland—provides a large, reliable source of government funding, yet it is arguably the most unpopular of all taxes. Critics cite a number of problems with the property tax. Because it is usually paid in one or two lump sums a year, the payment can seem onerous. It is also regressive: poor people pay a higher percentage of their income in property taxes than do wealthy people. It is paid on unrealized "paper profits" of homes that have appreciated in value, and it is difficult to administer in a manner that reassures taxpayers they are being taxed fairly. Another criticism is that while a high proportion of this tax goes to fund public schools, it is paid by people who have no children in school, whereas a tax like a gasoline tax is paid only by those who drive cars.[36]

While the state income, sales, and other forms of state revenue are also unpopular, these taxes are collected statewide and generally disbursed across the state, which can remove unevenness caused by local wealth variation. In theory, federal general aid to schools would reduce unevenness among states, but policy-makers at the federal level have never seriously contemplated providing enough aid to reduce inequities significantly because of the huge cost involved.

Prior to the Progressive Era, local school districts were largely autonomous from state government involvement. In the early 20th century, progressive educators and academics began to advocate state involvement in funding schools. In 1923, George Strayer and Robert Haig of Columbia University propounded a state school aid formula based on providing every school district a minimum per pupil **foundation level** of combined state and local support. The Strayer-Haig formula helped prompt arguments that continue to this day over adequacy, equity (fairness), and equalization (funding based on equal shares per pupil rather than fairness) in school funding.

The Strayer-Haig formula established a minimally adequate level of funding. If the per pupil local tax support generates less than that foundation level, the state makes up the difference, so that every district receives the foundation level. If local tax support from a wealthy district is greater than the foundation level, that district has more to spend than the foundation level but receives little or no state support.

TABLE 11.2 Estimated per Pupil Public Elementary and Secondary School Expenditures in 2007, by State Rank

National per Pupil = $9,557

Rank	State	Per Pupil
1	New Jersey	$14,675
2	New York	14,206
3	Vermont	13,385
4	Wyoming	13,328
5	Massachusetts	13,294
6	Connecticut	13,005
7	Delaware	12,565
8	Maine	12,063
9	Rhode Island	11,503
10	Pennsylvania	11,304
11	New Hamoshire	10,792
12	Ohio	10,563
13	Wisconsin	10,432
14	Hawaii	10,431
15	Ilinois	10,404
16	Alaska	10,392
17	Maryland	10,298
18	Michigan	10,209
19	Minnesoia	10,143
20	West Virginia	10,071
21	Virginia	9,785
22	Indiana	9,330
23	South Carolina	9,274
24	New Mexico	9,036
25	Oregon	8,989
26	Arkansas	8,905
27	Colorado	8,895
28	California	8,834
29	Kansas	8,804
30	Georgia	8,799
31	Washington	8,730
32	Montana	8,682
33	Louisiana	8,657
34	Florida	8,493

(continued on next page)

TABLE 11.2 Estimated per Pupil Public Elementary and Secondary School Expenditures in 2007, by State Rank *(continued)*

National per Pupil = $9,557

Rank	State	Per Pupil
35	Kentucky	8,459
36	Nebranka	8,309
37	South Dakota	8,237
38	North Dakota	8,228
39	Missouri	8,170
40	Iowa	8,141
41	Texas	8,048
42	North Carolina	8,000
43	Alabama	7,672
44	Tennessee	7,255
45	Idaho	7,176
46	Okilahoma	7,084
47	Nevada	6,963
48	Mississippi	6,866
49	Arizona	5,696
50	Utah	5,551
District of Columbia		16,540

Sources: *State Rankings 2008*; National Education Association, "Rankings and Estimates, December 2007" (Copyright 2007, NEA, used with permission).

School aid formulas have become complex because schools have different mixes of pupils, and each group has different needs and costs. High school students cost more to educate than elementary students; special education students generally cost more, often much more, than high school students. Students "at risk" of failure often require more spending than other students the same ages. As a result of these factors, there is an old joke that "only five people in the state understand the state school aid formula, and they are not allowed to fly on the same plane."

Even with school aid formulas operating in most states, significant inequities continue. Yet, as Alexander and Salmon observe, "Inequality in income follows inequality in education and vice versa."[37] As political analyst Alan Ehrenhalt puts the conundrum: "Leveling the poorer districts up to the spending level of the rich ones is fiscally impossible. Leveling the rich ones down to make them even is politically unacceptable."[38]

Enter the courts. In 1954, in *Brown v. Board of Education*, the U.S. Supreme Court overturned the separate but equal doctrine of *Plessy v. Ferguson* (1896).[39] Part of the rationale for the *Brown* decision lay in the fact that segregated schools were indeed separate but rarely equal in funding and facilities. But in localities where there was strong resistance to racial integration, many whites who could afford alternatives withdrew from the public schools. Thus, implementation of *Brown* did not significantly alter the funding inequities that lower-income

AN INSIDER'S VIEW

Seeking Equity in Texas School Funding

In 1997 in Texas, then governor George W. Bush introduced a bold initiative to reduce funding inequities across school districts, reduce the local property tax dramatically, and raise state taxes. Bush spent his political capital and staked the prestige of his office on passage of the school finance proposal. He devoted months to lobbying members of the state legislature on behalf of the measure. "I've never seen a governor make it this personal, make the whole tax vote a personal issue on George Bush's personality," recalled a Bush ally, State Representative Paul Sadler.

The bill passed the Texas house but stalled in the senate. Powerful lieutenant governor and senate presiding officer Bob Bullock, a Democrat, told Bush bluntly that the governor needed to put Republican votes on the bill. Bush then called in GOP senators one by one, offering enticements, appealing to their better natures. One important vote was that of Senate GOP caucus leader Florence Shapiro. When she met with Bush in his office, he greeted her warmly and, sitting beside her, appealed for her crucial vote.

"Governor, I'm not drinking that Kool-Aid," she told him. Bush had lost his own party on the issue.

Bush, shortly to be president of the United States, had joined that long line of Texas governors who failed to solve the school funding issue. According to Texas journalists James Moore and Wayne Slater, Bush "put his head in his hands and he began to cry."

Source: Based on James Moore and Wayne Slater, *Bush's Brain* (Hoboken, NJ: John Wiley, 2003).

students faced in many school districts. In 1968 in *Rich Schools, Poor Schools*, Arthur Wise argued that school funding should comport with the concept of equal protection under the law as found in the 14th Amendment to the U.S. Constitution as well as in many state constitutions.[40] In 1971, the California high court declared in *Serrano v. Priest* that all school districts regardless of their property wealth should be funded equally.[41]

In 1973, however, in a 5–4 decision in *Rodriguez v. San Antonio Independent School District*, the U.S. Supreme Court refused to declare differences in per pupil expenditures unconstitutional under the 14th Amendment.[42] This forced school funding reform advocates to focus on state courts and state constitutional language regarding equal protection and adequacy or equity of funding for schools. Subsequent to *Serrano* and *Rodriguez*, advocates in most states launched school funding lawsuits. Between 1971 and 2003, 26 state courts found the school funding systems unconstitutional. In other states, governors and legislatures acted to reduce funding inequities in order to forestall court action.[43] The Insider's View feature describes actions in Texas to reduce funding inequities.

School funding reform is rough sledding in most state legislatures. Lawmakers often weigh the benefits of equalized funding and property tax reductions against the costs of increased state taxes, and they find the costs of increased taxes more politically dangerous.

During recent decades, opponents of increased state and local taxation have mounted vigorous campaigns to limit revenues. Best known of these is a California action in 1978 in which voters approved a ballot initiative known as Proposition 13, which reduced local government revenues by 23 percent and local school funding by almost 30 percent. Only a part of this tax loss was made up by increased state funding. Two years later, Massachusetts enacted a proposition that had similar effects in that state.[44] Notwithstanding all the court victories on school funding, since the late 1960s overall school funding in America has remained at a remarkably constant 4 percent share of the gross domestic product.[45]

Because of all the attention given nationally to school funding, skeptics and academics began to ask if all the apparent increases in spending for schools made a difference in student performance. The heavily cited Coleman Report, authored by education professor James Coleman and others, concluded in 1961 that spending and performance did not appear to be strongly linked.[46] This sparked a debate that continues to today. Eric Hanushek, another school funding analyst, declared in 1991 that, "There is no systematic relationship between school expenditures and student performance."[47]

Many disagree with Coleman and Hanushek. Anita Summers and Barbara Wolfe, for example, found in 1977 that at the level of individual student data, money does matter. Others have found "strong evidence . . . that resources did often matter, though perhaps not always and in all circumstances."[48] In a 2006 article, Paul Tough reviewed the literature on the psychology of learning and concluded that the education gap between poor and middle-class children can be closed, but that it will require spending significantly more per poor pupil than for other students.[49] The money would be needed to provide longer school days and school years for students from lower-income families, as well as for paying teachers who would be involved much more intensively in their students' lives.

ALTERNATIVE SCHOOLS

America has had a strong tradition of private schools, and many are church sponsored. Private schools account for 10.4 percent of students in kindergarten through 12th grade. Schools sponsored by the Catholic Church represent 30 percent of all private schools; other religious schools enroll half of all those in private schools. The remainder are in nonsectarian schools. The percentage of students enrolled in private schools varies by state and region. One-fifth of all K–12 pupils in Delaware and 18 percent of students in New Jersey attend private schools, while only 8 percent of pupils in Wyoming, West Virginia, and Utah attend private schools.[50]

School Choice

In the years after World War II, public schools basically told parents: "Give us your children. We will educate them, and you stay out of the way." Today, many dissatisfied parents have been telling public schools and elected officials: "Give back our children. We have better ways to educate them than in the public schools." Some critics of the public schools advocate **school choice,** which lets parents choose among the schools, generally within a district, including magnet and charter schools.

Magnet schools are specialized public schools that focus on one area of study such as music, performing arts, or science. They draw students with strong interests in these areas. **Charter schools** offer education based on a charter or philosophy of education that may differ from that of the mainstream public schools. They focus on certain aspects of learning, concentrate on a specific school population (such as learning disabled), or take different pedagogical approaches to teaching. Although these schools receive public funding from the local school districts, they are allowed variations from the conventional schools in curriculum and are freed from many of the rules that govern the public schools.[51]

School choice advocates also push for tax-supported **school vouchers** and tax credits to assist parents in paying for their children's education in private schools. Many conservative academics, activists, and foundations are strong supporters of vouchers and other school choice options. For example, Stanford professor Terry Moe and John Chubb at the Brookings Institution in Washington, D.C., have written extensively about competition, choice, and vouchers.[52] The Reason and Heritage foundations and organizations such as the Separation of School and State Alliance also strongly support school choice efforts.

The Milwaukee Parental Choice Program (MPCP) is the oldest and largest public school voucher program in America. Envisioned by Milwaukee state representative Annette "Polly" Williams for her poor black constituents, the program received state authorization in 1990 and was upheld by the state supreme court in 1998. Originally a pilot project for low-income families in Milwaukee, the voucher program provided up to $6,607 per voucher for the 2008–2009 school year.. These vouchers support over 19,500 full-time equivalent students who attend alternative schools in Milwaukee, and 30 percent of all Milwaukee schoolchildren in 2006 attended schools other than the city's public schools. This includes students in the MPCP voucher program, as well as private-pay students in private schools, and city students attending public schools in the suburbs under a voluntary racial integration program.[53]

Many parents have decided to home school their children. **Home schooling** is schooling that is carried out in the home, usually by a parent, rather than in an established school. Often the decision to home school is religiously motivated. An estimated 1.1 million children in K–12 years are schooled at home by their parents, 2.2 percent of the school-age population. Home education is legal in every state, and the only common requirement is that students meet compulsory attendance rules. But 25 states have testing or evaluation requirements for home-schooled children. Home school families often participate in networks of similar families for socialization and group activities such as choir.[54]

Some parents decide to unschool their children. **Unschooling** is unstructured schooling, usually home based, in which children are allowed to explore their worlds unencumbered by the usual teacher-student relationship. Unschooling tends not to be religiously motivated. It encourages students to follow their own interests and abilities and choose how and what to learn.[55]

HIGHER EDUCATION

Although the federal government has dramatically increased its financial support for student financial aid and faculty research since the end of World War II, higher education still remains primarily a function of state government. Almost 13 million of the nearly 17 million students enrolled in all of America's 4,276 two-year and four-year colleges and universities attend public institutions. The $182 billion expended on higher education by state and local governments in 2005 represented 9.1 percent of all expenditures by these governments.[56]

Funding Issues

The tension between the academy and government is part of the struggle for power inherent in higher education. The campuses struggle to maintain the independence considered critical to the unfettered search for truth and discovery. The governments that fund the campuses seek accountability for their tax dollars. While college campuses want more government dollars with fewer strings attached, elected officials want higher "productivity," low tuitions, and an absence of public controversy from sometimes provocative faculty and unruly students.

The stakes in higher education are high for students and their families as well as for faculty and staff. In 2006, for example, the average cost of tuition, room and board, and fees at public higher education institutions averaged $12,100. Although these expenditures are lower than the costs of private higher education, the $182 billion in public funds for higher education in 2005 divided by about 13 million public higher education students equals about $14,000 per student in additional spending, primarily because of the appropriations of tax-payer dollars to support the public institutions. Although not directly paid by the student, these average dollar amounts bring actual public education costs close to the $27,317 paid in average tuition, room and board, and fees at private colleges and universities in 2006. With both rising college costs and an economic recession to deal with, the president's 2010 budget request included making more low-interest loan funds available to students with greater financial need and increasing the maximum Pell grant awards.[57] (See Table 11.3.)

A report from the Carnegie Foundation for the Advancement of Teaching set the parameters for the roles of government this way: "States have an obligation to develop long-range plans, authorize new campuses, develop missions for each institution, set enrollment goals and provide adequate support." The report also states that the academy "has had a long, successful tradition of self-regulation," and that colleges and universities have been "good stewards of public funds."[58] Nevertheless, increasing state taxes to fund colleges and universities is often controversial.

Author Jim Nowlan recalls an incident from his time as a member of the Illinois house of representatives. Back home in a coffee shop in his rural hometown shortly after having voted to enact a new income tax, Nowlan heard loud, angry complaints from some of his constituents about a multimillion-dollar home just constructed for the president of Southern Illinois University. The mansion had steel-beam construction and electrically heated driveways, far beyond what the voters considered reasonable during tough times. "So that's why we need the new income tax so badly!" the exercised taxpayers exclaimed. The lawmakers had been unaware of the extravagance. The mansion had been built with funds from federal grants, so the expense had not gone through the state budget process. Nevertheless, Nowlan and his colleagues paid a political price for the university's project, which unfortunately coincided with a new tax that would provide more money for education.

Campus Independence

Public colleges and universities have been able to maintain substantial independence from state and local governments. While state statutes for elementary and secondary education usually cover what will be taught, and who is certified to teach in public schools, colleges and universities largely determine who will teach in their institutions, what qualifications are acceptable, and what will be taught.

This level of independence stems in large part from assiduous efforts first by private colleges and later by public universities to protect their campuses from government control. In the historic *Dartmouth College* case of 1819, Daniel Webster argued successfully before the U.S. Supreme Court that if the state of New Hampshire could alter the charter of the private college, "College halls will be deserted by all better spirits, and become a theatre for the contentions of politics. Party and faction will be cherished in the places consecrated to piety and learning."[59]

TABLE 11.3 Per Capita State and Local Government Expenditures for Higher Education in 2005, by State Rank

National per Capita = $616

Rank	State	Per Capita
1	North Dakota	$997
2	Vermont	945
3	Wyoming	937
4	Delaware	929
5	New Mexico	890
6	Utah	872
7	Iowa	857
8	North Carolina	853
9	Kansas	833
10	Alaska	816
11	Wisconsin	790
12	Nebraska	786
13	Michigan	781
14	Alabama	769
15	Oregon	752
16	Washington	722
17	Oklahoma	709
18	Mississippi	698
19	Hawaii	695
20	Maryland	694
21	Indiana	678
22	Montana	656
23	Texas	655
24	California	647
25	Virginia	645
26	Colorado	624
27	Arkansas	615
27	Kentucky	615
29	South Carolina	612
30	Minnesota	610
31	West Virginia	608
32	Idaho	602
33	Illinois	590
34	Ohio	589

(continued on next page)

TABLE 11.3 Per Capita State and Local Government Expenditures for Higher Education in 2005, by State Rank *(continued)*

National per Capita = $616

Rank	State	Per Capita
35	Arizona	579
36	Connecticut	561
37	New Jersey	552
38	South Dakota	543
39	Louisiana	529
39	Pennsylvania	529
41	New Hampshire	517
42	Maine	510
43	Missouri	501
44	Massachusetts	494
45	New York	490
46	Georgia	487
47	Rhode Island	484
48	Tennessee	457
49	Nevada	440
50	Florida	404
District of Columbia		174

Source: *State Rankings 2008.*

In the 19th century, the Michigan state legislature and the University of Michigan battled over governance. In 1840, a legislative committee that looked into the slow progress of the new university harshly criticized its fellow lawmakers: "The argument by which legislatures have hitherto convinced themselves that it was their duty to legislate universities to death is this: 'It is a state institution, and we are the direct representatives of the people, and therefore it is expected of us; it is our right. . . .' As if, because a university belongs to the people, that were reason why it should be dosed to death for fear it would be sick if left to be nursed . . . by its immediate guardians."[60] In 1850, the lawmakers decided to establish the University of Michigan as an independent corporation, protected within the state constitution of that state.[61] A number of other states followed the example of Michigan.

In 1915, the newly founded American Association of University Professors (AAUP) adopted a General Declaration of Principles. One of these principles was tenure for college professors, similar to the life-time tenure for federal judges. For educators, **tenure** is a status granted after a trial period that gives protection from arbitrary dismissal. The AAUP declared that professors' tenure should be unconditional, and departures from the "scientific spirit and method" should be judged only by faculty peers.[62] Tenure helps to protect teachers from being dismissed because of political motivations.

Independence in a context of rapid growth is sometimes a recipe for chaos. In the 19th and early 20th centuries, hundreds of new colleges came into being, each with its own standards and degree requirements. To bring order from chaos, college leaders generally supported creation of regional accrediting agencies such as the New England Association of Colleges and the Western Association of Colleges and Schools. Specialty accrediting organizations proliferated as well, ranging from the American Medical Association to the American Association of Colleges and Schools of Business and the National Association of Schools of Music. Accrediting agencies establish minimum standards and other criteria that colleges must meet in return for the valuable stamp of professional **accreditation**. As the AAUP warned in 1915 in its declaration, if the academy failed to oversee its own affairs effectively, "it is certain that the task will be performed by others."

Taxpayer dollars invite calls from elected officials for oversight and accountability. Equal opportunity requirements of federal programs of the 1960s, for example, forbade federal dollars to higher education institutions that discriminated on the basis of race, sex, color, and national origin. The well-known Title IX of the Civil Rights Act of 1964 requires, among other things, parity in spending for men's and women's sports programs, initially over the opposition of many universities.

In a joint response between campuses and capitols, university leaders and elected officials in most states have come together since the 1960s to create statewide governing or coordinating boards, some with great powers over higher education. In concept, the statewide boards could make grand plans on a rational basis, rather than allow the competing higher education interests to fight it out in the legislature on the basis of raw political bargaining among governors and lawmakers who have limited knowledge of higher education.

Today, each state has a unique structure and policy-making environment for higher education. The state of Florida, for example, has strong input from lawmakers on higher education, a weak planning unit, and a strong chancellor who oversees the state's university system. In 2009, Florida had 11 universities overseen by the State University System (SUS), 28 community colleges, and a large aggregation of private colleges represented in the state capital by a strong association.[63] The community college system provides the primary access to higher education in Florida; 83 percent of Floridians start their higher education in community colleges.

"The legislative process drives higher education policy in Florida," according to one study of higher education systems. "Most important decisions are made by legislative committees." The governor's cabinet is all elected, including the commissioner of education, and they comprise the state board of education. The governor appoints members of the State University System and community college boards. The SUS board operates like a corporate board, with the chancellor as chief executive officer. The legislature may limit enrollment or tell the SUS to take more students, but lawmakers may not tell the SUS what new programs will be approved or where new programs should be located.[64]

In contrast, California's higher education has been able to maintain some distance from state lawmakers. Overseen by a weak coordinating California Postsecondary Education Commission, the sprawling "system of systems" includes the 10 campuses of the University of California (UC), 23 California state universities (CSU), 110 community colleges, and almost 200 private institutions. The governor appoints most members of the UC board of regents and members of the CSU system board, while community college boards are locally elected.

Chartered by the California constitution of 1879, the University of California prizes and jealously protects its independence. "While the state legislature passes bills affecting the university, lawmakers usually include a clause explaining that the statute will go into effect only if the board of regents passes a comparable resolution."[65] The independence of the UC System has induced some observers to call it a fourth branch of government in California. Faculty unions play a large role in decision-making in the CSU and community college systems. In the Los Angeles district, for example, faculty unions contribute heavily to favored candidates and often elect them to the board. These local boards negotiate collective bargaining agreements for the faculty and staff as well as setting education policies and guidelines.

Citizens and the courts also affect higher education policies, sometimes against the wishes of universities. Affirmative action provides a good illustration. In the 1970s, a white applicant who was rejected for admission to the University of California medical school at Davis sued the board of regents for discrimination on the basis of race. He had scored higher than several minority applicants who had been admitted because of the school's admission quotas. In its decision in 1978, the U.S. Supreme Court struck down the use of racial quotas but upheld race as an acceptable criterion in making admissions decisions.[66]

Not satisfied with the Court's decision, one UC regent, Ward Connerly, led a citizen's initiative drive in California that resulted in enactment in 1996 of Proposition 209, which prohibits preferential treatment based on sex, color, ethnicity, or national origin.[67] Two years later, Washington State enacted a similar proposition. In 2002, the U.S. Supreme Court decided to hear simultaneously two affirmative action cases from Michigan. In a 5–4 decision, the Court upheld the law school's practice of considering race, but in a 6–3 decision the Court struck down the formula-driven appraisal system for admitting undergraduate freshman.[68] One of the plaintiffs then led a successful fight in 2006 (with assistance from Ward Connerly) for enactment in Michigan of a citizen initiative that bars use of race in higher education admissions. Clearly, public participation has an impact on higher education policies.

Colleges and universities are active combatants in political processes. Major universities employ lobbyists. A large graduate research university will often have a lobbyist in Washington, D.C., who is focused on maximizing research grant awards from federal agencies, which can amount to hundreds of millions of dollars a year for a major university. Established research institutions favor grant-making by **peer review,** that is, review by a panel of professionals in the same field. Newer colleges and universities that feel slighted in the peer review process tend to focus on securing federal earmarks, the targeted appropriations sponsored by members of Congress from their state and congressional district.

At the state capitols, university and college lobbyists often play defense, working to protect higher education from what they consider onerous and intrusive regulations. Illinois state legislators once tried to require faculty members to teach a minimum of 12 "contact hours" per week in the classroom when 6 hours was the standard load at graduate research universities such as the University of Illinois. After much wrangling, the legislation was amended from a strict requirement to simply reporting university teaching activity.

In many capitols, the associations of independent (private) colleges and universities spar with public institutions over state aid to their respective institutions. Some states provide direct aid to private colleges on a per student basis. Independent colleges in states with well-funded financial aid grants seek to increase the maximum size of the per student grant, because their tuition charges are higher than the subsidized tuition levels at public universities. In contrast, the public colleges and universities prefer to keep the maximum grant low and make more students eligible for the grants.

Much of the politics of higher education comes down to money. In many states, governors and legislators are becoming less generous with dollars for higher education than in decades past. Increasing shares of state budgets go for Medicaid and to fund state pensions. This squeezes higher education, which is highly regarded yet generally lacks the political clout in the capitols that the teachers unions and strong public support provide for elementary and secondary education.

According to longtime education observer Aimes McGuinness, "The period of the early years of the 21st century is likely to be one of the most troubling in the history of the nation's higher education enterprise." He cites exacting demands being made on higher education to do more for less, the economic constraints of state budgets, resistance to change in higher education, and the instability in leadership that results from term limits in the legislature.[69]

CHAPTER SUMMARY

Schools have been part of the struggle for power and influence since colonial times. Today public schools exist in every part of the country, financed by tax dollars, and there is constant debate over who decides what goes on in our schools and how schools should be funded.

Public schools in most states are mainly funded through local property taxes. States also provide some of the funding for public schools; this is especially important for poorer school districts. In addition, the federal government provides funding for education through mandates set up by laws such as the Individuals with Disabilities Education Act and the No Child Left Behind Act. Unfortunately, such laws do not provide all the funding necessary to implement them. States also sometimes provide mandates for local school boards to implement and, like the federal government, do not always provide the necessary funds to go with them.

Some of the most intense debates surround local control over what is taught. Other debates concern teachers' rights and benefits, especially as put forth by the powerful teachers' unions. One ongoing debate concerns whether increased expenditures for education results in better student performance.

Alternatives to the established public school system include school choice, magnet schools, charter schools, home schooling, and unschooling. Another issue is whether to provide school vouchers to families to help pay for alternative schooling.

Colleges and universities have maintained a degree of independence from government control through their systems of tenure, accreditation, and peer review. The cost of higher education has been increasing, and state universities have sought more government funding to provide quality education comparable to private universities that young citizens can afford.

Elementary, secondary, and higher education may appear to be apolitical endeavors. The stakes are, however, too big to be free of the legitimate struggle among teachers, school boards, mayors, legislators, governors, Congress, and business and other advocacy groups for control, or at least influence, over the size and allocation of the dollars and the directions of the enterprises.

KEY TERMS

accreditation The approval from a national or regional agency that an institution is providing a professionally acceptable level of education, p. 297

American Federation of Teachers (AFT) The second largest teachers union in the United States, with over 1.3 million members, p. 282

charter school A public school that has legal authority to provide a different type of education within a public school district, often without the restrictions under which the regular schools operate, p. 292

foundation level The minimum amount of combined local and state financial support that will be provided each pupil in all school districts in a state, p. 288

home schooling Schooling that is carried out in the home, usually by a parent, rather than in an established school, p. 293

local control The concept that school affairs should be governed primarily by the local school district rather than by state and federal government laws and regulations, p. 280

local property tax A tax assessed on residential, commercial, industrial, and farm property according to the value of the property, p. 288

magnet school A specialized public school that focuses on a particular field in education, such as the arts, science, or mathematics, p. 292

National Education Association (NEA) The largest teachers union in the United States, with 3.2 million members, p. 282

peer review Review by a panel of professionals in the same field, p. 298

school choice An alternative to conventional public schooling that lets parents choose among the schools in a district, including magnet and charter schools, p. 292

school code That part of a state's statutes devoted to education, p. 280

school mandate A law or regulation that local school districts must follow, p. 280

school voucher A grant of money to a family to assist in paying tuition for a child to attend a private school, p. 292

tenure A status granted after a trial period that gives protection from arbitrary dismissal, p. 296

unschooling Unstructured schooling, usually home based, in which children are allowed to explore their worlds unencumbered by the usual teacher-student relationship, p. 293

QUESTIONS FOR REVIEW

1. How did passage of the Land Ordinance of 1785 and the Morrill Act of 1862 encourage the growth of public education in the United States?

2. How do school mandates, school codes, and local control affect public education? At which level of government can the responsibility for each be found?

3. What is a recent example of a state implementing a school reform program?

4. What are the two largest national teachers unions in the United States? How do they affect local education policy?

5. How have the National Defense Education Act, the Elementary and Secondary Education Act, the Individuals with Disabilities Act, and the No Child Left Behind Act extended the influence of the federal government on education?

6. How do state and local school boards find the money to pay for education? What are the main sources of school funding?

7. What are the various alternatives to the public school system? How do they differ from each other and from the public school system?

8. What are some of the conflicts between state universities and state governments?

DISCUSSION QUESTIONS

1. What percentage of pupils in your state are enrolled in private schools? What are the differences between public and private school educations? Should private schools be provided with taxpayer support? Why or why not?

2. Should elementary and secondary school students have choices in the schools they attend? Do specialized charter schools within the public school districts offer meaningful benefits and choice?

3. Do you favor providing vouchers to students so they can attend private schools? Why or why not?

4. Think back to your own experiences in elementary school and high school. What reforms would you make in your local schools? Do you think such changes would generate improved test scores?

5. Is there currently too much or too little testing of students in elementary and secondary schools? How valuable are the results of the tests? Is "teaching to the test" a good educational strategy?

6. Why do students from low-income backgrounds tend to score lower on standardized tests than those from middle- and higher-income families? What do you think could be done to close this achievement gap?

7. Per pupil funding varies dramatically across the different states and across different school districts. Should

school funding be equalized? Or can different levels of funding be equitable?

8. How much regulation should be provided by state government or local school districts for home schooling? Should testing be required as it is for public school students? Is testing required for home schooling in your state?

9. Should state government provide direct financial assistance to private colleges, say on a per student basis? Should financial aid in the form of grants and loans be available to students at private colleges and universities on the same basis as for students at public colleges and universities? Why or why not?

10. How independent should public universities be from direct control by the state legislature and the governor? Should state elected officials have a say in what courses are taught or the workload of college professors?

PRACTICAL EXERCISES

1. Find out what special considerations your state gives to residents applying to attend state-funded colleges and universities. For example, are there tuition tax breaks, lower admission standards, free tuition or tax breaks for good grades? Do you think giving such breaks enhances or harms the reputation of such universities? Why or why not?

2. Make a list of what you consider the necessary qualities of an ideal elementary or secondary school system. Your list can include required courses, extracurricular activities, buildings, programs, special needs, and anything else you want. After you have made up your list, write down who should be responsible to run these programs and who should finance them. For example, should they be overseen by private individuals, or federal, state, or local government? Share your list and recommendations with other students in the class.

3. Find out what teachers in schools and colleges earn in your state. Compare their salaries with those of professions requiring equal education and training.

WEB SITES FOR FURTHER RESEARCH

The Web site for *Education Week*, www.edweek.org, contains articles about how school districts are affected by state, federal, and local legislation and court rulings.

For information about the National Education Association (NEA), visit their Web site at www.nea.org.

For the American Association of School Administrators, go to www.aasa.org.

The federal Department of Education site is www.doe.gov.

You may also wish to check out the Department of Education of your state and local governments, which can generally be found through a link on your state's or city's home page.

SUGGESTED READING

Alexander, Kem, and Richard G. Salmon. *Public School Finance*. Boston: Allyn and Bacon, 1995.

Conley, David T. *Who Governs Our Schools: Changing Roles and Responsibilities*. New York: Teachers College Press, 2003.

Leyden, Dennis Patrick. *Adequacy, Accountability, and the Future of Public Education Funding*. New York: Springer, 2005.

Kimmelman, Paul L. *Implementing NCLB: Creating a Knowledge Framework to Support School Improvement*. Thousand Oaks, CA: Corwin, 2006.

Mitchell, Samuel. *Tidal Waves of School Reform*. Westport, CT: Praeger, 1996.

Nasaw, David. *Schooled to Order: A Social History of Public Schooling in the United States*. New York: Oxford University Press, 1979.

Peterson, Paul E., and Martin West, eds. *No Child Left Behind? The Politics and Practice of School Accountability*. Washington, DC: Brookings Institution, 2003.

Ravitch, Diane, ed. *Debating the Future of American Education*. Washington, DC: Brookings Institution, 1995.

Wirt, Frederick, and Michael Kirst. *Schools in Conflict: The Politics of Education*, 2nd ed. Berkeley, CA: McCutchan, 1989.

Illinois Governor
Blagojevich after
being indicted for
public corruption.

DEMOCRACY AND ETHICS
AT THE GRASSROOTS

In this concluding chapter, we have the opportunity to review and reflect. Throughout this book, we have sought to show you different aspects of the struggle for power and influence in cities and states. We have tried to convey, among other lessons, the central point, that power in itself is neither good nor evil. Power can be used for either good or evil purposes. It can ennoble or corrupt. To refuse to seek power and influence can be an even greater evil. Nothing is required for evil to prevail but for good women and men to stand aside and do nothing.

Our hope is that you will become engaged not only as knowledgeable citizens but also as democratic leaders forging more successful public policies. To that end, we have sought to provide the tools, knowledge, and perspective necessary. The rest that you need to learn we are confident you will learn for yourselves. We share, with the ancient Greeks and the early Americans who founded our nation more than 200 years ago, the belief that we learn about democracy by participating in the democratic process. It cannot be learned from a textbook alone. We hope we have inspired you to want to be full citizens and political leaders in your community, and we have attempted to give you the best perspective from which to begin the process.

In *The Struggle for Power and Influence in Cities and States*, we have presented information about institutional structures such as constitutions, charters, separate branches of government, and laws governing elections and lobbying. But more than most textbooks, we have focused on the political processes. Institutional structures set up the ballpark for playing the game of politics, but it is by mastering the processes that you win power and enact policies. Of course, luck or fate plays a hand as well. We hope by our stories, special features, and illustrations that we have given you some sense of the political process and allowed you to identify closely enough with some politicians and political leaders to sense the exhilaration and thrill of politics.

MAINTAINING A REPRESENTATIVE DEMOCRACY

Throughout this book, we have stressed how democracy is challenged in the 21st century. Democracy is never guaranteed. It has to be maintained, not only by eternal vigilance, but also by citizens unwilling to give up their rights and leaders willing to provide democratic leadership. The courts, legislatures, and chief executives all have key roles to play if we are to maintain and enlarge our representative democracy. These institutions and governmental

Electronic scoreboard used in many state legislatures and city councils.

officials have to adapt to the challenges and new technology of this century while preserving the democratic principles on which these branches of government were founded.

The fundamental process of making laws has remained unchanged. A state legislator, alderman, or county board member from a century ago would recognize many of the rules of procedure, the process of handing legislative proposals to the clerk, having the legislation considered in committee, and then the vote by the entire legislative body after floor debate. The types of legislators—experts, brokers, lawmakers, and ritualists—are also the same. And the tension between representing constituents as a trustee or delegate is unchanged.

In the 21st century, we have electronic scoreboards to tally votes in the legislature, e-mails from constituents, webcasting of legislative sessions, and professional staffs for legislators. It is easy to adjust to these innovations, but handling 2,000–10,000 pieces of legislation a year in year-round legislative sessions is more intense than the good old days of 30–60 days in an annual session. And we all struggle with information overload.

A major challenge in the 21st century is to make the state and local legislative branches full partners—co-equal branches—in government. Representing thousands of constituents according to complex democratic impulses is an equally difficult challenge. Legislators have to provide citizens a voice and make their voices heard in the halls of government. Legislators must also provide leadership to promote what is best for the public good rather than to merely reflect constituents' prejudices, greed, or personal self-interest. Legislators must hold corruption at bay by refusing bribes and by creating honest, efficient, and effective government. Legislators must also prevent the chief executive of the city, county, or state from becoming autocratic and tyrannical. And together, legislators must create a strong government that can overcome the disorder of our sprawling metropolitan regions.

Lawmaking, budget-making, and approval of executive appointments under modern conditions would seem to require Solomon-like wisdom. It is easier to succumb to the wishes of lobbyists, the lure of massive campaign contributions, and subtle bribes such as the promise of future jobs or future clients from big businesses. Legislators are often tempted to follow group thinking in the legislature, or may be driven too easily by the latest public opinion poll, or by the latest talking point in the media instead of taking an unpopular but well-thought-out stand on key issues. One of the questions this book poses is: "Could *you* be a courageous legislator facing all these challenges and temptations?"

Yet a good legislative branch by itself is not enough. We need chief executives able to play multiple roles—visionary leader, legislator in chief, political party leader, spokesperson, and wise head of a complicated, modern bureaucracy. At times, we need a strong leader who can take actions that inevitably will generate conflict. In other periods, we need a calm leader who can dampen conflicts in order to provide moderation, stability, and tranquility. At all times, we need chief executives who can be forceful without becoming dictatorial. We need leaders who can sense the desires and needs of constituents and shape their often unarticulated aspirations into principles, visions, and solutions that are both creative and practical. Particularly inspiring chief executives can articulate the needs of the weak, sick, elderly, poor, and powerless members of our society. They can provide these citizens a champion within government where frequently they have no other voice or power.

State and local chief executives face difficult problems. How do they cope with global warming, international economic downturns, high levels of unemployment and housing foreclosures, thousands of newly arrived immigrants, outmoded businesses losing out to transnational corporations, racial and ethnic conflicts, and the threat of terrorism? How can they control the epidemic of drugs, turn around schools that are failing to educate our children, strengthen family structures that are under stress, and respond to polarized conflicts over social issues like abortion or same-sex marriage?

Imagine being a chief executive who struggles politically for decades to reach a position of power, then faces not only these seemingly intractable problems but also hundreds of millions of dollars in revenue shortfalls and new federal cutbacks that bring the loss of jobs for hundreds of people. You must decide who won't be allowed health services or job training, or how overcrowded classrooms can still make learning possible. You must decide how many college students won't receive the loans they need to graduate, which roads or streets won't be paved, which parks won't be built, and whether the number of police officers and fire fighters can be cut without jeopardizing public safety. As Shakespeare wrote in *Henry IV*, "Uneasy lies the head that wears the crown." Being a chief executive in our time, as in Shakespeare's, is a daunting task with heavy and difficult responsibilities.

The judicial system is also essential. Lawyers and judges determine whether people are incarcerated or given their liberty, whether fines are levied or defendants are given a warning, and whether justice or injustice is done. Juries on which you will sit as a citizen have many of the same powers. Executives in the bureaucracy and legislators in both state and local government may make the law and regulations, but their constitutionality is in the hands of the judicial system. Courts also adjudicate conflicts between citizens, corporations, and government. If the judicial branch of government cannot be relied on to guarantee civil rights, liberties, and impartial justice, our representative democracy is doomed no matter how well our other institutions play their roles. The twin dystopias of a lawless society and an unjust police state are equally threats to representative democracy.

A Balance of Power

State and local elected and appointed officials do not act in a vacuum. First of all, there is a tug of war between national, state, and local government, each of which has a separation of powers between their executive, legislative, and judicial branches.

Often state and local government officials are struggling to get more federal funds to carry out their programs. They further seek to prevent unfunded mandates and undue federal regulations from sapping resources and preventing them from acting more effectively. Most day-to-day laws and programs that affect you most directly are made by your state and local governments, such as school funding, public transportation, road repair, crime prevention, emergency response, taxes, and fees. But much of the legal framework, funding, and regulations come from the national government. Currently, there is a great effort to ensure that the hundreds of billions of dollars being allocated by the federal government to help lift the country out of an international economic recession are being well spent by state and local governments on infrastructure, education, and health care.

There is a legitimate struggle for power between cities, states, and the federal government. Likewise, there is a level of cooperation and competition between cities and between states to attract residents, businesses, and federal grants. When Kansas City gets more federal funds for homeland security, New York City probably gets less. When Boeing shifts its corporate headquarters to Chicago, Seattle loses hundreds of jobs.

State and local governments are in a constant tug of war with the federal government. Local governments lack the funding they need to meet all legitimate demands and needs of their citizens. They face major new challenges of metropolitan sprawl, globalization, education, health care, public safety, environmental protection, and economic development. Through unfunded mandates and superior resources, the federal government tends to dominate American policy-making. At the same time, when the extreme partisanship in Washington during the first decade of the 21st century led to a stalemate, states and cities had to become more creative in problem solving. With a new president and a new partisan change in Congress, states and cities must establish a new balance of power with the federal government. In a time of recession, they look for more resources and hope for fewer restrictions so they can have more flexibility in spending them.

For democracy to work, there must be a balance of power. No one branch of government, no one level of government, and no one political faction can be allowed absolute power. The divisions of power and authority were created so the branches and levels of government could check one another. The tug of war that ensues must, in the end, benefit the citizens.

Political Processes

The branches of government and the institutions of government are embedded in the political processes that animate them. In this book, we focus particularly on elections and pressure politics. In these arenas, we face perhaps the most daunting challenges in our time.

Elections have become too dependent on the personal wealth of candidates or their ability to raise massive sums of money from powerful special interests—business corporations, labor unions, and organizations such as the National Rifle Association. Skilled campaign professionals can use the results of public opinion tracking polls and focus groups to shape false candidate personas and fake issue stands. Worst of all, they launch unfair and unanswerable negative campaigns on political opponents. Much of this is made possible by manipulating the mass media and through Web sites, e-mail videos, paid network and cable TV ads, and direct mail and phone campaigns. But one campaign's manipulative techniques can be overcome by a strong campaign on the other side, by good political reporting in the mass media, and by vigilant work by government agencies to prevent vote fraud or illegal campaign financing. But the possibility of Orwellian manipulation remains, and there is also the possibility of electronic vote fraud at the ballot box.

Negative campaigning undermines democracy in two ways, especially if not responded to immediately by the assailed candidate. First, the less worthy and ethically questionable candidate can be elected and put into a position of power. Negative advertising is often effective, and the good candidates do sometimes finish last—particularly if voters are apathetic and uninformed. Second, negative campaigns create disgust among voters, who are turned off by negative campaigns and by the sense that they are being manipulated. Negative campaigning convinces too many voters that no one cares about them, their needs, and their problems. As a result, voter turnout is low, often only 20 percent in local elections. That translates to as few as 10 percent of the potentially eligible voters. We can't maintain democracy with the participation of such a small fraction of the electorate.

While political parties, public interest groups, and political candidates seek to remain viable and to control the election process so that democracy is maintained, challenges always remain. Although 21st-century technology has the potential to allow average citizens to participate more effectively in elections and to be a better informed electorate, it also has a dark side—the possibility of violating a voter's privacy, manipulation, disinformation, and unfair, negative tactics. Misuse of our advanced technology can defeat good candidates and encourage greater voter antipathy.

Democracy depends on candidates being able to communicate effectively and directly with voters. The ideal model of election communication is a debate in which the voters can hear all the candidates and carefully weigh the strengths and weaknesses of each and their stands on issues. Further, new laws can provide greater transparency by requiring candidates to disclose their financial contributors. However, it is uncertain that laws which require more public debate and provide more information will be adopted. Those in elected office now have a strong incentive to keep the system that elected them unchanged. And the current system provides many advantages to incumbents.

In the end, political processes will not be safeguarded by new laws alone. The public must use the information made so easily available on the Internet to elect the best possible candidates. Most of all, we need aware and informed citizens who will elect candidates who support our representative democracy and defeat those who would subvert it.

The most hopeful sign is the level of participation in the 2008 elections, when 131 million people voted, 63.3 percent of eligible voters. Most importantly, it was the most diverse electorate in American history. African American voters increased their turnout by 4.9 percent, Hispanics by 2.7 percent, and Asian Americans by 2.4 percent. The turnout among younger voters age 18 to 29 was especially strong. If voters continue to be involved in less high-profile state and local elections, this is a particularly hopeful omen that election excesses can be curbed.[1]

Lobbying

Pressure politics—the efforts to pass laws and shape regulations no matter who is in office—has its own challenges. Contrary to popular images, interest groups and lobbying are good in principle. When we elect legislators to represent us, we don't give them a four-year mandate and cut them loose to do as they please. As citizens, we have a right and a duty to be heard in the lawmaking process. We have a constitutionally protected right to petition our government for redress—to pass laws, rules, and regulations that favor us and our point of view on public policies. Our democratic faith is that better governmental decisions, more acceptable to the majority of the people, are produced through the clash of differing views than by the dictates of an all-powerful ruler.

We lobby our public officials when we speak, write, or e-mail them personally. Or the organizations to which we belong may hire a professional lobbyist. Professional lobbyists can provide officials with factual analyses of problems and government solutions that we favor. We may join grassroots efforts, social movements, or political protests at the state capitol or city hall in order to get public officials to do what we believe should be done. All of these activities are worthwhile parts of the democratic process in addition to voting.

The fear, of course, is that the lobbying process, just like the election process, will be distorted and manipulated in ways that undermine democracy and create bad public policies. Generally speaking, the older image of a fat lobbyist wining and dining elected officials is no longer valid. Lobbying today is a much more sophisticated operation that stays carefully away from outright bribery. Today's rules and restrictions make most forms of corruption and payoffs illegal, but the convictions of government officials, lobbyists, and business people caught in trying to bribe public officials remind us that corruption still occurs.

However, it is more likely that a trade group or single issue organization will orchestrate a flood of phone calls and e-mails to public officials to suggest overwhelming support when it does not exist. This can result in one-sided laws and regulations that favor the rich and the powerful, not the public interest. Lobbyists and others may "bundle" legal election campaign contributions, creating sizable amounts of money in order to have more influence with candidates running for office. Large contributions bundled from wealthy professionals, like lawyers or doctors, for example, could determine the outcome of tort reform legislation. They can even help to elect individual judges who are likely to decide tort court cases in ways that favor one side or the other.

Individuals, firms, or organizations wealthy enough to hire well-connected lobbyists are more likely to influence lawmaking than the poor, unorganized, apathetic, or uninformed. Thus, some people have more influence over public policy than others, although citizens in a democracy should have an equal voice in government decision-making. If you are to have an effective voice in government decision-making, you have to join and support those interest groups that speak on your behalf.

Media Influence

To redress political imbalances, journalists and others in the media—from mainstream media to lone bloggers—play an important role as watchdogs for all citizens over their governments. However, the need for entertainment to enlarge audiences, the weight of the advertising dollar in media decisions, and the need to cut costs by cutting investigative reporters have undermined some news reporting. There are fewer investigative reporters today who can bring to light the inner workings of government and businesses, deal-making, and corruption. They are also likely to be less well paid and to have less technical training and time on the job. Television news is especially weak in providing a check on government, because TV news rarely has more than 30 seconds of state and local government news stories other than reports of personality clashes or canned press announcements. Even all-news cable channels don't cover government and politics at the state and local levels very well.

Newspapers are cutting back their news departments, so many Americans get more news from late-night comedy routines on television than from more traditional sources. Newspaper readership has been declining at alarming rates, and many newspapers have gone out of business. Fewer communities have more than one newspaper anymore. While Internet usage is up among the American public, it is unclear how much political and governmental news the average Internet user gets from Web sites and blogs. Even as the number of blogs and the number of people reading them increase, the level of political awareness in the general public is decreasing. For our democracy to be maintained, we must find new ways to strengthen journalism and the media. Without strong media, whatever technological form they take, we cannot obtain the information needed to hold government officials accountable for their actions.

Community Groups and Activists

Community and civic organizations and community activists also serve as a check on governments. They help citizens join together in promoting a better life for themselves and their fellow citizens at the most local level. They can stop expressways or other government projects from destroying their neighborhoods. They promote positive programs to eradicate hunger, build affordable housing, and promote harmonious local economic development. But in the battle to affect policies made in city hall and the state capitol, local organizations and activists are frequently outgunned by better-financed groups that can hire paid lobbyists and make large campaign contributions.

Viable community, civic, and interest group organizations that represent the average citizen require paid, well-trained, and dedicated staff. Bake sales or cheap membership dues won't, unfortunately, pay those bills. So these groups are inevitably tempted to accept government grants to provide community services and then to use some of these government funds to pay staff. Soon community groups get co-opted. They are no longer free to criticize the government that now funds them.

THE FISCAL CRISIS

In addition to the political and institutional restraints on state and local government, we face a series of major problems today. We face the perennial problem of funding state and local government. In 2008–2009, we struggled with an economic recession in which the tax revenues available to state and local governments were declining but the problems with which our governments had to cope were multiplying. A crisis in mortgage foreclosures combined with a shortage in affordable housing. The climbing unemployment rate added to the crisis, as many needed unemployment benefits, job retraining, and public works jobs.

At the same time, the costs of wars abroad in Iraq and Afghanistan, federal tax rebates, and tax cuts have increased the national debt. In future years, even with economic recovery, the need to pay off the debt we have incurred will mean that the national government will have less money to give cities and states as grants-in-aid.

State and local governments have been facing deficits, exhausting their rainy day funds, and tightening their budgets. State and local politics during the next few years will inevitably revolve around city, county, and state budgets that will decide which government services will continue and which will suffer. This is not all bad. It is a good opportunity to rethink priorities and cut waste, for example. But it is also a critical focus of the struggle for power in cities and states.

Traditional state and local government functions like education, health care, and public welfare will come under intense scrutiny in the immediate future. Is it better to fund Head Start and kindergarten programs, or community colleges and universities? Which can better deliver education programs—government agencies or private companies? How are our cities to find their niche in the global economy, and how are we to do a better job in governing our metropolitan regions? The debate goes on, and the outcomes will directly affect us all. We have a real stake in these policy decisions.

For our democracy to evolve and adapt, we must solve fiscal crises, tackle traditional local and state government problems more successfully, and invent a new politics adequate for the metropolitan cities and global society in which we live. That is the challenge which your generation faces.

JUST OUTCOMES

The cost of providing government services at the state and local levels has increased exponentially. State and local governments have had to make hard choices the last few years, and those tough choices will continue for the foreseeable future because of the effects of recession.

A small suburb has a budget of a few million dollars, a major city has a budget of a few billion dollars, and large states have multibillion-dollar budgets, but those amounts are still not enough to meet all the government services that citizens demand, especially at a time when unemployment and poverty are increasing. One of the tasks of every government is to figure out how to cut waste and lower taxes and government fees. But what follows is the painful task of deciding which government programs will have to be cut back and which citizens won't get services.

Since every conceivable legitimate government service cannot be fully funded every year, public officials must decide which programs get priority. This can be achieved in a balanced, cost-effective way by using technology to the fullest, employing special budget techniques like zero-based budgeting and instituting public quality management. But, in the end, hard choices will have to be made. Which programs will get fully funded and which programs will get less? Should we spend more money on primary and secondary education? Is higher university education, health care, or job creation more important? Should college students pay higher or lower tuition and fees? In general, the ideal approach is to involve the public in making those decisions and to make sure that priority is given to those who need the services most and who lack clout in state capitols and city halls.

Public involvement is important because then citizens are more likely to see the decisions as fair and just, even if they or their group did not get everything they wanted from the budget. Imagine, for instance, that streets, curbs, and sidewalks in your neighborhood are in need of repair. Suppose there is less than a million dollars available for street, curb, and sidewalk replacement. Suppose that there are 40 blocks of streets, and a million dollars will allow 4–5 blocks to be completely resurfaced plus temporary patching on a limited number of additional streets. If the entire community is involved in choosing which blocks should be resurfaced—because they are in most need of repair, because they are main streets that everyone uses, or because they have gone the longest without repair—then citizens will be willing to say, "Yes, we will repair Main Street this year, and my street will get repaired in two or three years. I agree this is a good decision, and best for the overall health of our community." If instead the mayor and aldermen decide to repair only the blocks where they live

Volunteers rallying for Barack Obama.

and not to consult the community, citizens will be unhappy and will complain loudly about the misuse of power.

All budgetary decisions, however, create conflict and differing opinions. Some citizens will strongly believe that it is important to spend $1 million of government funds to provide food pantries for the poor and shelters for the homeless. Other citizens may believe just as strongly that it is more important to give that amount to the Chamber of Commerce to promote economic development which will provide jobs to the unemployed. Fights over the government's budget and public policies are often perceived as fights over justice.

You need to become informed enough about the operations of state and local government to participate in the debate about how government money should be spent, how taxes should be raised, what government taxes and fees are legitimate, and which government services should be delivered, to whom, and how. We have tried to provide you with the perspective, tools, and understanding to be effective in influencing state and local government if you choose to do so.

And you can become involved right now! The democratic impulse offers many opportunities to be heard and to begin to develop your influence. For example, if you have a preferred political party, you can seek to become a precinct committee worker in your county. Although state laws vary, in many counties across America the county party chair has precinct vacancies to which you can be appointed. In return for helping your party turn out votes in your designated precinct, you can become an official part of the political party organization.

Volunteering in a political campaign is highly appreciated by candidates and their campaign staff. Making phone calls, going door to door, and helping make a campaign office hum are tasks that money cannot easily buy. You can play a key role in electing a candidate in whom you believe.

You can start your own grassroots group to lobby for change in the city council, county board, or state legislature. Or you can join an existing organization. Recruiting like-minded people to your cause can be aided by facebook.com and other Internet resources. Testifying at a committee hearing or meeting directly with a city council member or state legislator can be a very empowering experience.

Letters to the editor represent an important tool for putting your views before the public. The letters page is one of the best read sections of a newspaper. Blogs provide another outlet for your views. Politics is about communicating with people and trying to influence them. And in just a few years, as with 25-year-old Illinois state legislator Aaron Schock, now a member of Congress, you may have enough political influence to be the lobbied rather than the lobbyist.

ETHICS IN POLITICS

Whatever your role in years to come in politics, business, or other pursuits, always have your moral compass turned on. We know from firsthand observation that too many people enter the political arena without thinking about issues of right and wrong, of the ethics of legitimate versus questionable, even illegal exchanges of favors that are the stock in trade of politics.

Rare is the person who enters politics and elective office planning to be unethical, yet across our state and local governments every year, scores of elected officials are charged with just that. Money, greed, and the human desire to do favors, especially when the favors may aid an official's political program, can entangle officials who never thought of themselves as unethical.

We define political corruption as "personal gain at public expense." The issue of ethical behavior and corrupt practices can become complex as the powerful cocktail of money and favors becomes part of decision-making. Yet avoiding private gain at public expense suffices in most cases to guide ethical behavior.

In recent years, three of author Jim Nowlan's former college roommates have been measured for striped suits as a result of convictions for unethical behavior. All three were successful men who had been active in church work. Why, then, years later, would these otherwise good people abuse the trust of their colleagues and the rules of the game? Possibly because they never gave their actions a thought, and possibly through hubris, these seemingly very smart men abused the public trust.

Recently a young friend of Nowlan's ran for mayor of his hometown as a reformer against the ossified "old guard." A few weeks after he won the post, this new mayor was at a national mayors' conference in Washington, D.C., hobnobbing with famous people like Rudy Giuliani of New York. Realizing that he looked shabby in the company of such sophisticated big-city politicians, the new mayor bought a suit in D.C., using his city's credit card, the only one he had. He planned to pay the city back, but he had not done so by the time the purchase became public. It tarnished his otherwise fine reputation.

Many public officials act unethically at some time, whether intentionally or not. Every year, each Illinois legislator gets to award one four-year scholarship to the University of Illinois and another to any one of the regional state universities. Some members of the Illinois general assembly awarded scholarships to the children of relatives and political friends. But that is unfair and unethical.

We have seen many elected officials become snarled in the net of public criticism for unethical behavior. Here are some of the reasons:

1. "This is the way it's always been done." For instance, the Illinois state legislative scholarship program is a century old, and for decades many legislators viewed the scholarships as one of their few perquisites of office.

2. Changing ethical standards cause many longtime public officials to commit ethical errors. What might have been acceptable two generations ago may not be today.

3. In some local political cultures, an elected official would be criticized for failing to help a friend or supporter with a scholarship, patronage job, or governmental service delivered as a favor. Chicago Mayor Richard J. Daley once argued that if you couldn't help your relative, whom should you help? A lot of voters in the mid-20th century agreed with him.

4. The psychological and career pressures to win election and reelection push officials to do favors to win support. People who run the gauntlet of a campaign for office want to win, badly. And for a growing number of officials developing careers in politics, reelection is so important that the ethics of decisions can be blocked out by the demands of the campaign.

5. "I seen my opportunities and I took 'em," as Boss George Washington Plunkitt of the old Tammany Hall machine of New York City put it. We cannot ignore greed as a major reason for unethical behavior.

6. "I never gave it a thought." One Illinois legislator who was convicted for taking small bribes died in prison, a broken man. A decade later, his son said to Nowlan, "I was just a child when Dad had his problems. What kind of man was he? Was he a bad man?" Nowlan replied that his father was a man of good will, much liked by his colleagues. He just wasn't thinking. "He went along, he did what some around him had been doing for a long time, and they rationalized the bribes as campaign contributions."

In a society focused on the individual and on how things have always been done, personal gains from gaming the system may outweigh for many the amorphous societal benefits of supporting a political system that is open, fair, and follows "the rule of law." If history serves as a guide, personal gain at public expense in the declining Roman Empire trumped the selfless virtue of the Roman Republic's founders. This led to its demise. So what are we to do to prevent such a decline in our country? Studies of corruption in Chicago have led author Dick Simpson and his team of researchers to conclude that the cost of corruption, or the "corruption tax," in Chicago and Cook County may be as high as $500 million a year. Further, Simpson's research team discovered that 1,500 Illinois public officials and business people have gone to jail for public corruption since 1970.[2]

In recent years, several states and cities have adopted ethics legislation that should help by keeping government officials away from temptation. But that won't be enough. Here are some practical suggestions to guide your own ethical decision-making in public life:

1. Think! Is there any dimension of a decision that could be seen by others as an abuse of the public trust? As former Illinois Auditor General Robert Cronson put it: "How would this decision be viewed by a grand jury?"

2. Never justify a decision on the basis that "this is the way it has always been done." Times change. Ask former U.S. Congressman Dan Rostenkowski of Chicago, convicted a few years ago for actions that were standard operating procedure decades before, when he started in politics.

3. Identify a friend or acquaintance that you respect for integrity. This person, or at least your image of her or him, can serve as your second opinion. How would that person view your pending decision?

4. How do you want your children to remember you? Could this decision conceivably hurt them or diminish you in their eyes?

Biographies, movies, history books, and personal examples of ethical and moral behavior can serve as a guide for our own behavior and as a standard that we seek to meet. When author Nowlan was first elected to the Illinois house of representatives at age 26, a fellow freshman gave his new colleagues copies of Robert Bolt's play *A Man for All Seasons*, about Thomas More's unwillingness to capitulate, at the cost of the supreme sacrifice, when Henry VIII dissolved the Catholic Church in England. Nowlan was thunderstruck. You mean there are matters of principle and ethics, right and wrong, moral and immoral in the venture we are embarking upon! He had never thought about it. Nowlan is still indebted to the late U.S. Representative George O'Brien for that gift of awakening.

When author Simpson was in college, he read John F. Kennedy's book *Profiles in Courage* about the brave decisions of U.S. senators that sometimes cost them reelection. The book provided him with useful images to keep in mind in the Chicago city council when he was often in the minority facing heavy pressure and opposition from the dominant ruling faction.

Author O'Shaughnessy remembers that during the 1968 election, her mother told her she supported Robert Kennedy because "he is always trying to help people that need it." O'Shaughnessy recalled Kennedy's actions during the civil rights movement and how he marched with Cesar Chavez and the United Farm Workers union of migrant workers. O'Shaughnessy's grandfather had been an alderman during the Great Depression. When he died, he had not been in politics for over 30 years, but at the funeral, there was a line around the block of people she had never seen before. When they came up to her mother to pay their respects, dozens simply said, "Your dad got me a job in the '30s." O'Shaughnessy's family saw politics as public service, a value that she tries to instill in her students and in her role as a township trustee.

Most ethical decisions don't take high courage or always have negative consequences for a public official. Most of the time, they just require common sense, humility, and remembering that you are serving in office as a representative of other citizens. You are not there to enrich yourself, your family, or your friends. A former Chicago alderman and U.S. senator, Paul Douglas, had a simple rule: never accept any gift worth more than $5. Having simple standards before you are tempted by bribes of money or power is very helpful.

As we begin the second decade of the 21st century, the speed of change, the complexity of society, and the challenges we face are great. If democracy is to prevail, we must have ethical, democratic leaders and an informed, critical electorate.

In politics and in the struggle for power, ethical behavior is an assumed prerequisite of the democratic impulse. Violate ethical behavior and the democratic process is compromised, as will be your lofty goals and purposes. If you remember your ethics, we know you will become a valued citizen in our state and local governments. For your ideals to win, you must join the struggle for power, but remember to take your ethics with you.

PRACTICAL EXERCISES

Now that you have read this book, repeat the exercise from Chapter 1. The questions are repeated here:

1. Suppose in a few years you wanted to run for a seat on the city council or in the state legislature. How would you begin your campaign for the office? What platform would you run on? How would you win enough votes to get elected? Why might you decide to run for this office?

2. Suppose you were elected to the city council or the state legislature. What would be your goals, in the same way that Simpson, Nowlan, and O'Shaughnessy had goals when they served in public office? How would you go about accomplishing those goals?

Now answer this question:

3. How have your answers changed from several months ago when you began your study of the struggle for power and influence in state and local government?

NOTES

Chapter 1

1. Michael J. Dear, *From Chicago to L.A.: Making Sense of Urban Theory* (Thousand Oaks, CA: Sage, 2002).
2. James McFarrand, in *The Records of the Federal Convention of 1787*, vol. 3, ed. Max Farrand (1911, reprinted 1934), Appendix A, 85.

Chapter 2

1. Donald Lutz, "The Purposes of American State Constitutions," *Publius* 12 (Winter 1982): 32.
2. This sketch of local government history is drawn largely from Clyde F. Snider, *Local Government in Rural America* (New York: Appleton-Century-Crofts, 1957), chapter 1.
3. U.S. Census Bureau, *Census of Governments, Vol. 1, No. 1, Government Organization*, www.census.gov/govs/www/cog2007.html.
4. G. Alan Tarr, "Introduction," *Constitutional Politics in the States*, ed. G. Alan Tarr (Westport, CT: Greenwood, 1996).
5. *The Book of the States, 2005*, (Lexington, KY: Council of State Governments), 2005.
6. Albert Sturm, "The Development of American State Constitutions," *Publius* 12 (Winter 1982): 61.
7. Sturm, "The Development of American State Constitutions," 63–73.
8. Sturm, "The Development of American State Constitutions," 73.
9. Paul C. Reardon, "The Massachusetts Constitution Marks a Milestone," *Publius* 12 (Winter 1982): 45–55.
10. *The Book of the States, 2006* (Lexington, KY: Council of State Governments, 2006), 9.
11. *The Book of the States*, Table 1.2, Constitutional Amendment Procedure: By the Legislature, 11–12.
12. Clyde F. Snider, *American State and Local Government* (New York: Appleton-Century-Crofts, 1965), 27–28.
13. Sean Wilentz, *The Rise of American Democracy: Jefferson to Lincoln* (New York: W. W. Norton, 2005), 716–18.
14. *Model State Constitution* (New York: National Municipal League, 1963).
15. *Model State Constitution*, 6th ed. (New York: National Municipal League, 1968).
16. *The Book of the States*, Table 5.6, Selection and Retention of Judges, 251–54.
17. *The Book of the States*, Table 4.10, Selected State Administrative Officials: Methods of Selection, 169–72.
18. *The Book of the States*, Table 1.2, Constitutional Amendment Procedure: By the Legislature, 11–12.
19. *State Constitutions in the Twenty-first Century*, vol. 1, ed. G. Alan Tarr and Robert F. Williams (Albany: State University of New York Press, 2006), 1–2.
20. H. Bailey Thomson, "Constitutional Reform in Alabama: A Long Time in Coming," in Tarr, 113–43.
21. Thomson, "Constitutional Reform in Alabama," 121.
22. Thomson, "Constitutional Reform in Alabama," 122.
23. Thomson, "Constitutional Reform in Alabama," 138.
24. Rebecca Mae Salokar, "Constitutional Revision in Florida: Planning, Politics, Policy, and Publicity," in Tarr, 19–58; and A. E. Dick Howard, "Adopting a New Constitution: Lessons from Virginia," in Tarr, 73–112.

25. *The Book of the States*, Table 1.3, Constitutional Amendment Procedure: By Initiative, 13.

26. Anne G. Campbell, "Direct Democracy and Constitutional Reform: Campaign Finance Initiatives in Colorado," in Tarr, 175–95.

27. Campbell, "Direct Democracy and Constitutional Reform," 176.

28. Campbell, "Direct Democracy and Constitutional Reform," 192.

29. Lawrence F. Keller, "Municipal Charters," *National Civic Review* 91, no. 1 (2002): 60.

30. Samuel K. Gove and James D. Nowlan, "Loosening the Constitutional Framework," in *Illinois Politics and Government: The Expanding Metropolitan Frontier* (Lincoln: University of Nebraska Press, 1996).

31. City of New York, *New York City Charter*, as amended through July 2004. Also see Fred Siegel, *The Prince of the City: Giuliani, New York, and the Genius of American Life* (San Francisco: Encounter Books, 2005).

32. Keller, "Municipal Charters," 55.

33. Keller, "Municipal Charters," 56.

34. The National Civic League presents seven different options for the composition of a mayor-council government, based on how the council members and mayor are elected or selected for office. See Model Cities Charter, sec. 2.02c.

35. For a comprehensive analysis of the history of charters for New York City, see Joseph P. Viteritti, "The Tradition of Municipal Reform: Charter Revision in Historical Context," in *Proceedings of the Academy of Political Science* 37, no. 3, *Restructuring the New York City Government: The Reemergence of Municipal Reform* (1989): 16–30.

36. City of New York, *New York City Charter*, as amended through July, 2004, http://www.nyc.gov/html/charter/downloads/pdf/citycharter2004.pdf.

37. H. George Frederickson, Curtis Wood, and Brett Logan, "How American City Governments Have Changed: The Evolution of the Model City Charter," *National Civic Review* 90 (Spring 2001): 3–18.

Chapter 3

1. Thad Beyle, *State and Local Government, 2003–04* (Washington, DC: CQ Press, 2003).

2. Alan Ehrenhart, "Honk If You Want to Be in the Arizona Legislature," *Governing*, September, 1998.

3. Thad Beyle, *State and Local Government, 2004–05.* (Washington, DC: CQ Press, 2004), 87.

4. This form of representation is called descriptive representation and is fully discussed in Hanna Fenichel Pitkin, *The Concept of Representation* (Berkeley: University of California Press, 1967), 11–12, 44, chap. 4. See also Edmund Burke, "Letters to Langriche (1792)" *Burke's Politics*, ed. Ross J. S. Hoffman and Paul Levack (New York: Alfred A. Knopf, 1949), 495.

5. Beyle, *State and Local Government, 2004–05*, 87.

6. Robert Tanner, "Female State Leaders Double Since 2000," *Chicago Sun-Times*, April 3, 2007, 22.

7. *Baker v. Carr*, 369 U.S. 186 (1962).

8. Dick Simpson, Socrates Harisiadis, Sharmeen Huyssain, Sumaira Hussain, and Tom Sdralis, "Chicago City Council Report, April 9, 2003–November 15, 2004," www.uic.edu/depts/pols/AldermanicVotingRecords/CityCouncilVotes2003-2006.htm.

9. Pitkin 119–23, chap. 8; see also Edmund Burke, "Speech to the Electors at Bristol (1774)," *Burke's Politics*, 115.

10. James David Barber, *Lawmakers: Recruitment and Adaptation to Legislative Life* (Westport, CT: Greenwood, 1980).

11. New York City before it adopted a new city charter in the 1990s had a very complicated two-chamber system of local government.

12. Thomas R. Dye and Susan A. MacManus, *Politics in States and Communities*, 11th ed. (Upper Saddle River, NJ: Prentice Hall, 2003), 207.

13. Richard Locker, "A True Tennessee Titan," *State Legislatures*, July/August 2005, www.ncsl.org.

14. Daniel Weintraub, "The Rise and Fall of the California Governor, *State Legislatures* 31 (October/November 2005): 28–30.

15. Pamela Prah, "Legislators Balance the Books, Head Home," Stateline.org., June 8, 2007, http:// pewresearch.org.

16. Brenda M. Erickson, "Circumventing Stalemate," *State Legislatures* 24 (July/August 1998): 46–49.

17. Dye and MacManus, *Politics in States and Communities*, 203.

18. Dye and MacManus, *Politics in States and Communities*, 194.

19. Samuel K. Gove and James D. Nowlan, "Loosening the Constitutional Framework," in *Illinois Politics and Government: The Expanding Metropolitan Frontier*, Samuel K. Gove and James D. Nowlan (Lincoln: University of Nebraska Press, 1996), 87.

20. Tribune Staff Report, "Speaker Kills Referendum Hopes," *South Bend Tribune*, January 19, 2006.

21. Bill Mears, "High Court Upholds Most of Texas Redistricting Map, "*CNN*, June 28, 2006, www.CNN.com.

22. See Larry Gerston and Terry Christensen, *Recall: California's Political Earthquake* (Armonk, NY: M. E. Sharpe, 2004), for an account of the recall election in California.

23. For a chart of states that use initiative and recall, see Dye and MacManus, *Politics in States and Communities*, 41.

24. John M. Carey, Richard G. Niemi, and Lynda W. Powell, *Term Limits in State Legislatures* (Ann Arbor: University of Michigan Press, 2000), 24–27; and Karl T. Kurtz, Bruce Coin, and Richard G. Niemi, *Institutional Change in American Politics: The Case of Term Limits* Ann Arbor: University of Michigan Press, 2007), 187.

25. Tom Loftus, *The Art of Legislative Politics* (Washington, DC: CQ Press, 1994), 167.

26. Alan Greenblatt, "Term Limits Aren't Working," *Governing*, April 2005, 13–14.

27. Greenblatt, "Term Limits Aren't Working," 14.

28. Edmund Burke, "Speech to the Electors at Bristol (1774)," *Burke's Politics*, 115ff.

29. James D. Nowlan, *Glory, Darkness, Light: The History of the Union League Club of Chicago* (Evanston, IL: Northwestern University Press, 2004), chap. 7; and Joel A. Tarr, *A Study in Boss Politics: William Lorimer of Chicago* (Urbana: University of Illinois Press, 1971).

30. Thomas J. Gradel, Dick Simpson, Andres Zimekis, Kirsten Byers, and Chris Olson, "Curing the Corruption in Illinois," Anti-Corruption Report Number 1, February 3, 2009, University of Illinois at Chicago, www.uic.edu.

31. Scott Andron, "Gift Law Confusing, Experts Say," *Miami Herald*, February 8, 2004, 1 and 8BR.

32. Andron, "Gift Law Confusing," quoting Philip Claypool, deputy executive director and general counsel for the Florida State Ethics Commission.

33. Deron Schreck, "Suburban Politics: The Challenge of Morality Politics," doctoral dissertation, University of Illinois at Chicago, 2007.

34. Miami-Dade County Commission on Ethics, "A Community's Resolve to Restore Integrity, Accountability, and Public Trust: The Miami-Dade Experience," excerpted in the *Miami Herald*, February 21, 2004. See www.miamidade.gov/ethics.

35. Barry Rundquist and Gerald Strom, "Citizen Evaluations of Chicago City Government," in *Chicago's Future in a Time of Change*, ed. Dick Simpson (Champaign, IL: Stipes, 1993), 263–68.

36. Marion Orr and Darrell M. West, "Citizen's Views on Urban Revitalization: The Case of Providence, Rhode Island," *Urban Affairs Review* 37 (January 2002): 397–419.

37. See, for example, the conclusions by Dye and MacManus, *Politics in States and Communities*, 211–12; Ann Bowman and Richard Kearney, *State and Local Government* (Boston: Houghton Mifflin, 2002), 167; and John Harrigan and David Nice, *Politics and Policy in States and Communities* (New York: Longman, 2004), 208.

38. Alan Ehrenhalt, "In Search of the Ideal Legislature," *Governing*, September 2004, 4.

39. Ehrenhalt, "In Search of the Ideal Legislature," 4.

40. Quoted in Ehrenhalt, "In Search of the Ideal Legislature." See also Alan Rosenthal, *Heavy Lifting: The Job of the American Legislature* (Washington, DC: CQ Press, 2004).

41. Rob Gurwitt, "Are City Councils a Relic of the Past?" *Governing*, April 2003, 20–24.

42. Vera Vogelsang-Coombs, "Inside an American City Council: Democracy, Leadership, and Governance," *Journal of Public Management and Social Policy* 7 (Summer 2001): 43–64.

Chapter 4

1. Amanda Ripley et al., "4 Places Where the System Broke Down, *Time*, September 11, 2005, www.time.com.

2. Michelle Roberts, "New Orleans Re-elects Nagin," *Chicago Sun-Times*, May 21, 2006, 29A.

3. John Zogby, "Will Katrina Be Our Defining Moment?" *Campaigns and Elections*, June 2007, 64.

4. For an account of the key court decisions, see "Patronage: *Shakman v. Democratic Organization of Cook County* and *Cynthia Rutan et al. v. The Republican Party of Illinois, et al.*," in *Chicago's Future in a Time of Change*, ed. Dick Simpson (Champaign, IL: Stipes, 1993), 147–54.

5. Michael Grunwald, "The New Action Heroes," *Time*, June 14, 2007.

6. Neal Peirce, "California Versus New York: Grappling with the Prison Dilemma," *Washington Writers Group*, February 18, 2007, citistates.com.

7. See especially Thad Beyle, "Gubernatorial Power: The Institutional Power Ratings for the 50 Governors of the United States," www.unc.edu/~beyle/gubnewpwr.html.

8. Pamela M. Prah, "Massachusetts Gov Rated Most Powerful," www.stateline.org.

9. Alan Rosenthal, *Heavy Lifting: The Job of the American Legislature* (Washington, DC: CQ Press, 2004), 169.

10. Daniel C. Vock, "Dems Take Full Control in 3 More States," November 7, 2008, www.stateline.org.

11. Dick Simpson and Tom Kelly, "Chicago City Council's Increasing Independence, *Chicago City Council Report, May 7, 2003–November 15, 2006*, University of Illinois at Chicago, Department of Political Science, www.uic.edu.

12. Tim Novak, "Some Critical of Governor's Call for Another Death Penalty Study," *Chicago Sun-Times*, February 1, 2000, 8.

13. "Illinois Suspends Death Penalty," January 31, 2000, www.CNN.com.

14. Ken Armstrong and Steve Mills, "Death Penalty Support Erodes; Many Back Life Term as an Alternative," *Chicago Tribune*, March 7, 2000, 1.

15. T. Harry Williams, *Huey Long* (New York: Knopf, 1969).

16. Lyle W. Dorsett, *The Prendergast Machine* (New York: Oxford University Press, 1968).

17. Williams, *Huey Long*, 751.

18. Jessica Winski, "Don't Get No Respect, *Illinois Issues*, March 1998, 18–22.

19. Jonathan Walters, "The Taming of Texas," *Governing*, July 1988, 18–20.

20. Walter Bagehot, *The English Constitution* (London: Kegan Paul, Trench, Trubner, 1891), 200.

21. James D. Nowlan, *Inside State Government in Illinois* (Champaign: University of Illinois Institute on Government and Public Affairs, 1982), 23.

22. Terry Sanford, *Storm over the States* (New York: McGraw-Hill, 1967).

23. Ronald D. Michaelson, "An Analysis of the Chief Executive: How the Governor Uses His Time," *Public Affairs Bulletin*, Southern Illinois University, September 1971.

24. Author's interview with Mayor Shawn Gillen.

25. International City/County Management Association, www.icma.org.

26. International City/County Management Association.

27. Rob Gurwitt, "Mayor Brown & Mr. Bobb," *Governing*, January 2000, www.governing.com.

28. International City/County Management Association.

29. Author's interview with City Manager Cameron Benson at Hollywood, Florida, City Hall, March 25, 2004.

30. Author's interview with Mayor Mara Giuliante at Hollywood, Florida, City Hall, March 26, 2004.

31. Author's interview with Mayor Mara Giuliante.

32. Author's interview with Mayor Mara Giuliante.

33. "Total Local Government Employees, 2008" and "Total State Government Employees, 2008," *Governing State and Local Sourcebook 2008*, http://sourcebook.governing.com.

34. California Performance Review, "SO43 Creating a Workforce Plan for California State Employees," 2004, http://cpr.ca.gov.

35. *Rutan et al. v. Republican Party of Illinois* (110 S. Ct. 729), 1990.

36. *The Book of the States*, Council of State Governments, www.stateline.org.

37. Gary Washburn and Laura Cohen, "Hiring without Politics? No Way," *Chicago Tribune*, Oct. 3, 2004. Section: Metro, 1*ff*.

38. Nowlan, *Inside State Government in Illinois*, chap. 10.

Chapter 5

1. "Call to Action," National Center for State Courts, 2002, www.judicialcampaignconduct.org.

2. "Examining the Work of State Courts: Overview," Court Statistics Project, National Center for State Courts, 2004, www.ncsconline.org.

3. "Courts and Judges," Bureau of Justice Statistics, 2008, www.ojp.usdoj.gov.

4. "State Court Organization, 2004," Bureau of Justice Statistics, www.ojp.usdoj.gov.

5. "Examining the Work of State Courts: Overview."

6. Author's interview with panel of Indiana judges and lawyers, July 10, 2007.

7. "State Court Organization, 1998"; "Appellate Courts: Jurisdiction, Staffing, and Procedures," Bureau of Justice Statistics; "How Courts Work: Steps in a Trial. Officers of the Court," American Bar Association, www.abanet.org; "Examining the Work of State Courts: Overview."

8. "State Court Organization, 2004."

9. "Examining the Work of State Courts: Appellate," Court Statistics Project, National Center for State Courts, 2004; National Center for State Courts, "Appellate Court Innovative Practices, FAQ," CourTopics, www.ncsconline.org.

10. "State Court Organization, 1998"; "Appellate Courts: Jurisdiction, Staffing, and Procedures"; "Examining the Work of State Courts: Appellate."

11. National Center for State Courts, "Appellate Court Staff and Administration," CourTopics, www.ncsconline.org.

12. Caroline Wolf Harlow, "Defense Counsel in Criminal Cases," Bureau of Justice Statistics, November 2000.

13. G. Alan Tarr, "The State Judicial Article," www.camlaw.rutgers.edu.

14. "Rochester Police Department's Youth Services Section," and "Rochester's Community Response at Curbing Youth Violence," U.S. Conference of Mayors, www.usmayors.org;.

15. John Buntin, "Gangbuster." *Governing*, December 2003, www.governing.com; "City Attorney Belgadillo, Police Chief Bratton Announce Gang Injunction Against Black P Stones," press release, Office of City Attorney Rockard J. Delgadillo, June 1, 2006.

16. Kevin Giles, "A Back-Alley Approach to Fight Crime in Minneapolis," *Star Tribune*, April 13, 2005.

17. C. David Kotok, "The Mayor Rejects a Measure Calling for Tougher Penalties for Prostitutes," *Omaha World-Herald*, December 9, 2005, 1B.

18. "Facts About the American Judicial System," American Bar Association; "Judicial Careers: Recruitment, Tenure, Longevity, FAQ," National Center for State Courts www.ncsconline.org.

19. "Judicial Careers: Recruitment, Tenure, Longevity, FAQ."

20. Madelynn Herman, "Race and Ethnic Bias Trends in 2002: Diversity in the Courts," National Center for State Courts.

21. "Trial Court Administration, Procedures, and Specialized Jurisdiction," and "Judicial Assignment, FAQ," National Center for State Courts.

22. "Judicial Ethics, Conduct, and Discipline, FAQ," National Center for State Courts.

23. "A Call to Action: Statement of the National Summit on Improving Judicial Selection," National Center for State Courts, 2002.

24. Author's interview with panel of Indiana judges and lawyers, July 10, 2007.

25. "A Call to Action."

26. David Herszenhorn, "New York Cuts Aid Sought for City Schools," *New York Times*, November 21, 2006, A1.

27. "A Call to Action."

28. "A Call to Action."

29. Zach Patton, "Robe Warriors," *Governing*, March 2006, 106–10.

30. Wolf Harlow, "Defense Counsel in Criminal Cases."

31. "Contracting for Indigent Defense Services: A Special Report," U.S. Department of Justice, Bureau of Justice Assistance, April 2000, www.ncjrs.gov.

32. "Keeping Defender Workloads Manageable," U.S. Department of Justice, Bureau of Justice Assistance, January 2001, www.ncjrs.gov.

33. "Gideon House Report," Recommendation Adopted by the House of Delegates, American Bar Association, August 2005, www.abanet.org.

34. Eyal Press, "Keeping Gideon's Promise," *Nation*, March 16, 2006, www.thenation.com.

35. Wolf Harlow, "Defense Counsel in Criminal Cases."

36. Wolf Harlow, "Defense Counsel in Criminal Cases."

37. "In Profile: Gary L Walker," National District Attorney's Association, www.ndaa.org.

38. "Prosecution in the 21st Century: Goals, Objectives, and Performance Measures," American Prosecutors Research Institute, February 2004, www.ndaa.org.

39. Steven Perry, "Prosecutors in State Courts, 2005," Bureau of Justice Statistics, July 2006, www.ojp.usdoj.gov; Matthew R. Durose and Patrick A. Langan, "Felony Sentences in State Courts, 2002," Bureau of Justice Statistics, December 2004, www.ojp.usdoj.gov.

40. Alan Ehrenhalt, "Jurors' Prudence," *Governing*, May 8, 2004.

41. Radley Balko, "Justice Often Serviced by Jury Nullification," FOX News, August 1, 2005, www.foxnews.com.

42. Ehrenhalt, "Jurors' Prudence."

43. Nancy Jean King, "The American Criminal Jury Source," *Law and Contemporary Problems*, Spring 1999.

44. Ted. M. Eades, "Revisiting the Jury System in Texas: A Study of the Jury Pool in Dallas County Source," *SMU Law Review* 54, Special Issue 1813–26 (2001).

45. Penelope Lemov, "The Juror's Best Friend," *Governing*, 1996, www.governing.com.

46. "Corrections Statistics," Bureau of Justice Statistics, www.ojp.usdoj.gov; Paige M. Harrison and Allen J. Beck, "Prison and Jail Inmates at Midyear 2004," Bureau of Justice Statistics, April 2005, www.ojp.usdoj.gov.

47. Roy Walmsley, "World's Prison Population List," International Centre for Prison Studies, www.prisonstudies.org; "Criminal Offender Statistics," Bureau of Justice Statistics, www.ojp.usdoj.gov.

48. Harrison and Beck. "Prison and Jail Inmates at Midyear 2004"; "Additional Corrections Facts at a Glance," Bureau of Justice Statistics.

49. "Crime in the United States, 2005: Table 32, Ten-Year Arrest Trends," Federal Bureau of Investigation; Caroline Wolf Harlow, "Educational and Correctional Populations," Bureau of Justice Statistics, January 2003; Christopher J. Mumola, "Incarcerated Parents and Their Children," Bureau of Justice Statistics, August 2000.

50. Durose and Langan, "Felony Sentences in State Courts, 2002."

51. "Confronting Confinement: A Report of the Commission on Safety and Abuse in America's Prisons," Vera Institute of Justice, June 2006, www.prisoncommission.org.

52. Rodney F. Kingsnorth, Randall C. MacIntosh, and Sandra Sutherland, "Criminal Charge or Probation Violation? Prosecutorial Discretion and Implication for Research in Criminal Court Processing," *Criminology*, August 2002.

53. "State Court Organization, 1998"; "The Sentencing Context," Bureau of Justice Statistics.

54. "The Justice System," Bureau of Justice Statistics.

55. "How Courts Work: Steps in a Trial. Plea Bargaining," American Bar Association, www.abanet.org.

56. Author's interview with panel of Indiana judges and lawyers, July 10, 2007.

57. Dirk Olin, "Plea Bargain," *New York Times Magazine*, September 29, 2002.

58. Timothy Lynch, "The Case Against Plea Bargaining," Cato Institute, Fall 2003, www.cato.org.

59. William J. Stuntz, "*Bordenkircher v. Hayes*: The Rise of Plea Bargaining and the Decline of the Rule of Law," *Harvard Law Review* Working Paper No. 120, papers.ssrn.com.

60. Durose and Langan, "Felony Sentences in State Courts, 2002."

61. Durose and Langan, "Felony Sentences in State Courts, 2002."

62. Don Stemen and Jon Wool, "Changing Fortunes or Changing Attitudes? Sentencing and Corrections Reform in 2003," Vera Institute of Justice, March 2004, www.vera.org.

63. Paula M. Ditton and Doris James Wilson, "Truth in Sentencing in State Prisons," Bureau of Justice Statistics, January 1999, www.ojp.usdoj.gov.

64. Ditton and Wilson, "Truth in Sentencing in State Prisons."

65. John Clark, James Austin, and D. Alan Henry, "Three Strikes and You're Out: A Review of State Legislation," National Institute of Justice, September 1997, www.ncjrs.gov.

66. Franklin E. Zimring, "Imprisonment Rates and the New Politics of Capital Punishment," Sage Publications, http://pun.sagepub.com.

67. "The Nation's Prison Population Continues It's Slow Growth: Up 1.9 Percent Last Year," press release, Department of Justice, October 23, 2005, www.ojp.usdoj.gov.

68. "Jail Statistics," Bureau of Justice Statistics, www.ojp.usdoj.gov.

69. "Confronting Confinement," Vera Institute of Justice.

70. "Confronting Confinement," Vera Institute of Justice.

71. Wolf Harlow, "Educational and Correctional Populations."

72. "In the Name of Justice: Ex Offenders and Employment Rights Petition," www.petitiononline.com.

73. "Invisible Punishment," Public Broadcasting Services, June 27, 2003, www.pbs.org.

74. John Zarrella and Patrick Oppmann, "Florida Housing Sex Offenders Under Bridge," CNN.com, April 4, 2007, www.cnn.com.

75. Matthew Hickman and Brian Reaves, "Community Policing in Local Police Departments, 1997 and 1999," Bureau of Justice Statistics, February 2001, www.ojp.usdoj.gov.

76. "Community Oriented Policing Services," FY2004 Budget Summary, Department of Justice, www.usdoj.gov.

77. Paul Rosenzweig, "The Death Penalty, America, and the World," Heritage Foundation, October 23, 2003, www.heritage.org.

78. Eric M. Freedman, "The Case Against the Death Penalty," *USA Today*, March 1997.

79. "Deterrence News and Developments—Previous Years," Death Penalty Information Center, www.deathpenaltyinfo.org.

80. Thomas Bonczar and Tracy Snell, "Capital Punishment, 2004," Bureau of Justice Statistics, November 2005, www.ojp.usdoj.gov.

81. *Roper v. Simmons* (03-633) 543 U.S. 551 (2005) 112 S. W. 3d 397, affirmed. Also see "Crime in the United States 2005: Table 32, Ten-Year Arrest Trends," Federal Bureau of Investigation.

82. Callie Rennison, "Criminal Victimization 2001: Changes 2000–2001 with Trends 1993–2001," Bureau of Justice Statistics, September 2002, www.ojp.usdoj.gov.

83. Paige Harrison and Allen Beck, "Prisoners in 2001," Bureau of Justice Statistics, July 2002, www.csdp.org.

84. "Offenses Cleared: Crime in the United States in 2004," Federal Bureau of Investigation, www.fbi.gov.

85. Rennison, "Criminal Victimization 2001."

86. "Table 102, Personal and Property Crimes, 2004, Percent of Reasons for Not Reporting Victimization to the Police, by Type of Crime," Bureau of Justice Statistics, www.ojp.usdoj.gov.

87. "Confronting Confinement," Vera Institute of Justice.

88. Stemen and Wool, "Changing Fortunes or Changing Attitudes?"

89. Christopher Mumola and Jennifer Karberg, "Drug Use and Dependence, State and Federal Prisons, 2004," Bureau of Justice Statistics, October 2006, www.ojp.usdoj.gov.

90. Stemen and Wool, "Changing Fortunes or Changing Attitudes?"

91. Donna Lyons, ed., "Tough Times to Be Tough on Crime," National Conference of State Legislators, June 2003, www.ncsl.org.

92. "Examining the Work of State Courts: Civil," Court Statistics Project, National Center for State Courts, 2004, www.ncsconline.org.

93. Thomas H. Cohen and Steven K. Smith, "Civil Trial Cases and Verdicts in Large Counties, 2001," Bureau of Justice Statistics, April 2004, www.ojp.usdoj.gov.

94. Thomas H. Cohen, "Tort Trials and Verdicts in Large Counties, 2001," Civil Justice Survey of State Courts 2001, Bureau of Justice Statistics, November 2004.

95. Cohen, "Tort Trials and Verdicts in Large Counties, 2001."

96. "Examining the Work of State Courts: Civil."

97. Cohen, "Tort Trials and Verdicts in Large Counties, 2001"; "Examining the Work of State Courts: Civil."

98. Thomas H. Cohen, "Contract Trials and Verdicts in Large Counties, 2001," Civil Justice Survey of State Courts 2001, Bureau of Justice Statistics, January 2005.

99. Author's interview with panel of Indiana judges and lawyers, July 10, 2007.

Chapter 6

1. Eric Kelderman, "Real ID Showdown Averted," April 24, 2008, www.stateline.org; Pam Belluck, "Mandate for ID Meets Resistance from States," *New York Times*, May 6, 2006; and William T. Pound, "Real ID—Real Questions," February 24, 2006; John Gramlich, "Obama, Congress to Revisit Real ID," April 23, 2009, www.stateline.org.

2. Greg Lucas, "Beale Community Stands Down; Citizens Prevail in Fight to Keep Air Force Base," *San Francisco Chronicle*, May 14, 2005, B; Bob von Sternberg, "Cheers All Around as SD Keeps Air Base," *Minneapolis Star Tribune*, August 27, 2005, 9A; and Philip Dine, "Too Little Too Late," *St. Louis Post Dispatch*, August 28, 2005, B1.

3. Ann O'M. Bowman and Richard C. Kearney, *The Resurgence of the States* (Englewood Cliffs, NJ: Prentice-Hall, 1986).

4. Jeff Archer, "Connecticut Governor Backs NCLB Suit," *Education Week*, August 10, 2005, www.edweek.org.

5. James Madison, Federalist Paper No. 39, http://thomas.loc.gov; Samuel H. Beer, *To Make a Nation: The Rediscovery of American Federalism* (Cambridge, MA: Belknap Press, 1993); Aaron Wildavsky, *Federalism and Political Culture*, ed. David Schleicher and Brendon Swedlow (New Brunswick,

NJ: Transaction, 1998); Martha Derthick, *Keeping the Compound Republic: Essays on Federalism* IWashington, DC: Brookings Institution, 2001); and Jack N. Rakove, *Original Meanings: Politics and Ideas in the Making of the Constitution* (New York: Vintage Books, 1997).

6. "John deWitt #1," Pennsylvania Minority, 1787. May 31, 2008, www.constitution.org/afp/dewitt01 .htm; "John deWitt #2," Pennsylvania Minority, 1787. May 31, 2008, www.constitution.org/afp/ dewitt02.htm. "Letters from the Federal Farmer to the Republican, #1," 1787. *Antifederalist Papers.* May 31, 2008, www.constitution.org/afp/fedfar01.htm.

7. See, for example, Federalist Papers Nos. 10, 39, 45, and 46, http://thomas.loc.gov.

8. Richard Hofstadter, *The Age of Reform: From Bryan to FDR* (New York: Vintage Books, 1955); Samuel H. Beer, "The Idea of the Nation," in *American Intergovernmental Relations*, ed. Laurence O'Toole (Washington, DC: CQ Press, 2000), 337–50.

9. William H. Riker, "A Note on Ideology," in O'Toole, *American Intergovernmental Relations*, 97–98; and William H. Riker, *The Development of Modern Federalism* (Boston: Kluwer Academic, 1987).

10. Beer, *To Make a Nation*, 10–14.

11. David Walker, *The Rebirth of Federalism: Slouching Toward Washington*, 2nd ed. (New York: Chatham House Publishers, 2000), 22, 67. Also see Riker, *The Development of Modern Federalism*; and Riker, "Federalism," in *American Intergovernmental Relations*, ed. Laurence J. O'Toole (Washington, DC: CQ Press, 2000), 89–96; Morton Grodzins, "The Report of the President's Commission on National Goals: The American Assembly," *Goals for Americans* (Englewood Cliffs, NJ: Prentice-Hall), 1960, 265–92; and Harry N. Scheiber, "The Condition of American Federalism: An Historian's View," study submitted by the Subcommittee on Intergovernmental Relations Pursuant to S. Res. 205, 89th Congress, to the Committee on Government Operations, U.S. Senate, October 15 (Washington: U.S. Government Printing Office, 1966).

12. Affirmed by the Supreme Court in *Texas v. White* (7 Wallace 700 [1869]). Walker calls this a partly national, partly federal decision, 74.

13. Thomas Anton, *American Federalism and Public Policy* (Philadelphia, PA: Temple University Press, 1989), 182; O'Toole, *American Intergovernmental Relations*, 7; Stephen P. Erie, *Rainbow's End: Irish Americans and the Dilemmas of Urban Machine Politics, 1840–1985* (Berkeley: University of California Press, 1988), 75.

14. Robert H. Wiebe, *The Search for Order, 1877–1920* (New York: Hill and Wang, 1967), 190–99.

15. See William Riker, *Federalism: Origin, Operation, Significance* (Boston: Little, Brown, 1964); Scheiber, "The Condition of American Federalism"; Walker, 2000.

16. See John Kincaid, "From Cooperation to Coercion in American Federalism: Housing, Fragmentation, and Preemption, 1780–1992," *Journal of Law and Politics* 9 (Winter 1993): 333–433; Timothy Conlan, *New Federalism to Devolution: Twenty-five Years of Intergovernmental Reform* (Washington: Brookings Institution, 1998); Walker.

17. Walker; Jeffrey Pressman and Aaron Wildavsky, *Implementation: How Great Expectations in Washington Are Dashed in Oakland* (Berkeley: University of California Press, 1973).

18. Richard P. Nathan, Thomas L. Gais, and James W. Fossett, "Bush Federalism: Is There One, What Is It, and How Does It Differ?" paper presented at Annual Research Conference, Association for Public Policy Analysis and Management, Washington, DC, November 7, 2003.

19. Theresa A. Gullo and Janet M. Kelly, "Federal Unfunded Mandate Reform: A First-Year Retrospective," *Public Administration Review* 58, no. 5 (1998): 379–87.

20. Kincaid, "From Cooperation to Coercion in American Federalism," 28.

21. Conlan, *New Federalism to Devolution*, 1998, 391; Timothy Conlan and Robert L. Dudley, "Janus-Faced Federalism: State Sovereignty and Federal Preemption in the Rehnquist Court," *PS: Political Science and Politics* 38, no. 3 (2005): 363–66; John Dinan, "The State of American Federalism 2007–2008: Resurgent State Influence in the National Politicy Process and Continued State Policy Innovation," *Publius* 38 (Summer 2008): 381–415.

22. Richard A. Brisbin, "The Reconstitution of American Federalism? The Rehnquist Court and Federal-State Relations, 1991–1997," *Publius* 28 (Winter 1998): 189–215.

23. *National Labor Relations Board v. Jones and Laughlin Steel Corporation*, 301 U.S. 1 (1937); Susan Gluck Mezey, "The U.S. Supreme Court's Federalism Jurisprudence: *Alden v. Maine* and the Enhancement of State Sovereignty," *Publius* 30 (Winter 2000): 21–38.

24. *U.S. v. Lopez*, 115 S. Ct. 1624 (1995).

25. Brisbin, "The Reconstitution of American Federalism," 193, Mezey 23.

26. Kenneth T. Palmer and Edward B. Laverty, "The Impact of *United States v. Lopez* on Intergovernmental Relations: A Preliminary Assessment," in O'Toole, *American Intergovernmental Relations*, 160–74.

27. See *Seminole Tribes of Florida v. Florida*, 514 U.S. 549 (1995); *Florida Prepaid Postsecondary Education Expense Board v. College Savings Board*, 527 U.S. 666 (1999); *College Savings Bank v. Florida Prepaid Postsecondary Education Expense Board*, 527 U.S. 627 (1999); *New York v. United States* 505 U.S. 144 (1992).

28. *Alden v. Maine*, 527 U.S. 706 (1999); Carol S. Weissert and Sanford F. Schram, "The State of American Federalism, 1999–2000," *Publius* 30 (Winter 2000): 1–20.

29. Brisbin, "The Reconstitution of American Federalism," 197; *Shaw v. Reno*, 509 U.S. 630 (1993); *Daniel Kimel et al. v. Florida Board of Regents*, 528 U.S. 62; 120 S. Ct. 631 (2000); *City of Boerne v. Flores*, 521 U.S. 507 (1997).

30. John Dinan and Dale Krane, "The State of American Federalism," *Publius* 36 (2006): 327–74.

31. See *Cippollone v. Liggett Group*, 505 U.S. 504 (1992); *Geier v. American Honda Motor Company*, 529 U.S. 861(2000). See also Mary Davis, "Unmasking the Presumption in Favor of Preemption," *South Carolina Law Review* 53 (Summer 2002).

32. Timothy J. Conlan and François Vergniolle DeChantal, "The Rehnquist Court and Contemporary American Federalism," *Political Science Quarterly* 116, no. 2 (2001): 253–75; *U.S. v. Butler*; 297 U.S. 1 (1936); Charles R.Wise, "The Supreme Court's New Constitutional Federalism: Implications for Public Administration," *Public Administration Review* 61, no. 3 (2001): 343–58.

33. *Koslow v. Pennsylvania*, 302 F.3d 161 (3rd Cir. 2002), *cert. denied*, 123 S. Ct. 1353 (2003); *Robinson v. Kansas*, 295 F.3d 1183 (10th Cir. 2002), *cert denied*, 123 S. Ct. 2574 (2003); *Visnon v. Hawaii*, 288 F.3d 1145 (9th Cir. 2002), *cert. denied*, 123 S. Ct. 962 (2003). See also Dan Schweitzer, "The Year in Federalism," *National Environmental Enforcement Journal* 18, (2003): 11; and National Council of State Legislators, *Mandate Monitor*, 2008, www.ncsl.org.

34. Dinan and Krane, "The State of American Federalism." See also *Nevada v. Hibbs*, 538 U.S. 721 (2003); *Tennessee v. Lane*, 541 U.S. 509 (2004); and *Gonzales v. Raich*, 125 S. Ct. 2195 (2005).

35. "States Sue Federal Government," and "Texas, Arizona Sue Federal Government," *Migration News*, 1994, http://migration.ucdavis.edu; "Kempthorne to Sue Federal Government," *Idaho Observer*, January 2005, http://proliberty.com; *Davis v. United States EPA*, 336 F.3d 965 (2003 U.S. App).;

36. Dinan and Krane, "The State of American Federalism," 133–34.

37. Eric N. Waltenburg and Bruce Swinford, *Litigating Federalism: The States before the U.S. Supreme Court* (Westport, CT: Greenwood, 1999), 44–52, 58–64, 82–83.

38. Robert Barnes and Juliet Eilperin, "High Court Faults EPA Inaction on Emissions," *Washington Post*, April 3, 2007, www.washingtonpost.com; Alan Greenblatt, "Greenhouse Shift," *Governing*, May 2007, 17–18.

39. Christopher Banks and John Blakeman, "Chief Justice Roberts, Justice Alito, and New Federalism Jurisprudence," *Publius* 38 (Summer 2008): 576–600.

40. Lewis G. Irwin, *A Chill in the House: Actor Perspectives on Change and Continuity in the Pursuit of Legislative Success* (Albany: State University of New York Press, 2002), 56.

41. Loree Bykerk, "Organized Interests' Response to Unorthodox Lawmaking," paper presented at the Midwest Political Science Association Annual Meeting, Chicago, April 15–18, 2004; Elizabeth

O'Shaughnessy, "State Adjustments to Recent Congressional Lawmaking: Adapting to Stifling Federalism," doctoral dissertation, University of Illinois at Chicago, 2006.

42. Melvin Holli, *The American Mayor* (University Park: Pennsylvania State University Press, 1999), 89.

43. Richard Neustadt, *Presidential Power: The Politics of Leadership from FDR to Carter* (New York: John Wiley, 1989), 10.

44. David B. Magleby, David M. O'Brien, Paul C. Light, James MacGregor Burns, J. W. Peltason, and Thomas E. Cronin, *Government by the People, National, State, and Local Version*, 21st ed. (New York: Longman, 2006), 341.

45. Richard A. Musgrave, *Essays in Fiscal Federalism* (Washington, DC: Brookings Institution, 1965; Paul Peterson, *City Limits* (Chicago: University of Chicago Press, 1981).

46. Anton, *American Federalism and Public Policy*; and R. Douglas Arnold, *The Logic of Congressional Action* (New Haven, CT: Yale University Press, 1990). Also see Paul Peterson, *The Price of Federalism* (Washington, DC: Brookings Institution, 1995).

47. Herbert Wechsler, "The Political Safeguards of Federalism: The Role of the States in the Composition and Selection of the National Government, *Columbia Law Review* 54 (1955). Also see Grodzins, "The Report of the President's Commission on National Goals"; and Daniel Elazar, *American Federalism: A View from the States* (New York: Thomas Y. Cromwell, 1972).

48. Dale Krane, "The State of American Federalism, 2002–2003: Division Replaces Unity," *Publius* 33, no. 3 (2003): 1–45.

49. Paul Posner, *The Politics of Unfunded Mandates: Whither Federalism?* (Washington, DC: Georgetown University Press, 1998).

50. Sean Nicolson-Crotty, "National Election Cycles and the Intermittent Political Safeguards of Federalism," *Publius* 38, no. 2 (2008): 295–314.

51. *State of Connecticut et al. v. Margaret Spellings, Secretary of Education*, 549 F. Supp. 2d 16 (D. Conn., April 28, 2008); *School District of the City of Pontiac et al. v. Secretary of the United States Department of Education*, 512 F.3d 25 (6th Circuit, 2008); Mark Walsh, "Full Appeals Court to Reconsider Ruling That Revived NCLB Suit," *Education Week*, May 7, 2008, 8.

52. Bykerk, "Organized Interests' Response to Unorthodox Lawmaking," 12.

53. Frank R. Baumgartner and Bryan D. Jones, *Agendas and Instability in American Politics* (Chicago: University of Chicago Press, 1993), 233–34.

54. O'Bowman and Kearney, *The Resurgence of the States.*

55. Beverly A. Cigler, "Not Just Another Special Interest: Intergovernmental Representation," in *Interest Group Politics*, 4th ed., ed. Allan J. Cigler and Burdett A. Loomis (Washington, DC: CQ Press, 1995), 134.

56. Cigler; Alan Rosenthal, *The Third House: Lobbyists and Lobbying in the States*, 2nd ed. (Washington, DC: CQ Press, 2001).

57. David C. Nice and Patricia Frederickson, *The Politics of Intergovernmental Relations*, 2nd ed. (Chicago: Nelson-Hall, 1995), 38. See also R. Allen Hays, "Intergovernmental Lobbies: Toward an Understanding of Issue Priorities," *Western Political Quarterly* 44, no. 4 (1991): 1081–98; and Peterson, *City Limits.*

58. David Brady and Morris Fiorina, "Congress in the Era of the Permanent Campaign," in *The Permanent Campaign and Its Future*, ed. Norman J. Ornstein and Thomas E. Mann (Washington, DC: American Enterprise Institute and Brookings Institute, 2000), 134–61.

59. "Washington in Brief," *Washington Post*, June 30, 2006.

60. Eric Lipton, "Mayors Protest Cuts in Antiterrorism Funds," *New York Times*, June 22, 2006.

61. Jonathan Walters, "Uncivil Disunion," *Governing*, March 2004.

62. Josh Goodman, "Issues to Watch Legislatures 2006," *Governing*, January 2006.

63. David Hosansky, "The Other War over Mandates," *Governing*, April 1995.

64. Rosenthal, *The Third House*, 48–49. See also the Web sites of Florida League of Cities, www.flcities.com; New Jersey Association of Counties, www.njac.org; and Texas Municipal League, www. tml.org.

65. Donald F. Kettl, "PIGS without Pork," *Governing*, June 2007.

66. Russell L. Hanson, "The Interaction of Governments," in *Governing Partners: State-Local Relations in the United States*, ed. Russell L. Hanson, (Boulder, CO: Westview, 1998), 14.

67. Mavis Mann Reeves, "The States as Polities: Reformed, Reinvigorated, Resourceful," in *American Federalism, Third Century*, ed. John Kincaid, *Annals of the American Academy of Political and Social Science* 509 (May 1990): 83–93; see also Bowman and Kearney, *The Resurgence of the States*; Eric N. Waltenberg and Bruce Swinford, *Litigating Federalism: The States before the U.S. Supreme Court* (Westport, CT: Greenwood, 1999)

68. E. J. Dionne, "What Kind of Hater Are You?" *Washington Post*, March 15, 2006, A19.

69. Dinan and Krane, "The State of American Federalism," 348.

70. Barbara Rosewicz, "Bipartisanship No Pipe Dream for State Pols," January 30, 2007, www.stateline.org.

71. Rosewicz, "Bipartisanship No Pipe Dream for State Pols."

72. Elizabeth O'Shaughnessy, "State Adjustments to Recent Congressional Lawmaking: Adapting to Stifling Federalism," doctoral dissertation, University of Illinois at Chicago, 2006, 114–47.

Chapter 7

1. Eric Kelderman, "Mega-Donors Get Mixed Election Results," Stateline.org, Pew Research Center, February 5, 2007, http://pewresearch.org.

2. Sunshine Project, "Show Me the Money: Cash Clout in Illinois Politics," *Illinois Campaign for Political Reform*, www.ilcampaign.org.

3. Malia Zimmerman, "Off to the Races," *Hawaii Reporter*, July 26, 2006, www.hawaiireporter.com; Richard Borreca, "Democratic Majors Sit Out the Big Race," *Star Bulletin*, July 23, 2006, www.starbulletin.com; "2006 Gubernatorial General Election Results: Hawaii"; David Leip's Atlas of Presidential Elections, March 3, 2007, www.uselectionatlas.org.

4. Don Rose in *By the People*, documentary film available at www.uic.edu.

5. Associated Press, "Politics Often Family Business in West Virginia," *Parkersburg News*, July 11, 2004, Region News Section, 10A.

6. See Sidney Blumenthal, *Pledging Allegiance: The Last Campaign of the Cold War* (New York: HarperCollins, 1990), 300–1.

7. "Bush Campaign Theme Revealed," 2003, http://raena.net.

8. Jonathan Alter, "Memo to Kerry: Only Connect," *Newsweek*, May 31, 2004, www.newsweek.com.

9. Joel Bradshaw, "Who Will Vote for You and Why: Designing Strategy and Theme," in *Campaigns and Elections American Style*, ed. James Thurber and Candice Nelson (Boulder, CO: Westview, 1995), 42–43.

10. J. Cherie Strachan, *High-Tech Grass Roots: The Professionalization of Local Elections* (Lanham, MD: Rowman and Littlefield, 2003), 24.

11. "Midwest Local TV Newscasts Devote 2.5 Times As Much Air Time to Political Ads as Election Coverage, Study Says," Midwest News Index, 2006, www.joycefdn.org; Craig Varoga, "Preparing for the Future," *Campaigns and Elections*, January 2007, 72.

12. Dick Morris, "Bigger Isn't Always Better When It Comes to Media Buys," *Campaigns and Elections*, June 2007, 62.

13. Todd Blair and Garrett Biggs, "Cable Advertising: An Underrated Medium for Local Elections," *Campaigns and Elections*, September 12, 2005.

14. Dick Morris, "Direct Mail? Get a Horse!" *Campaigns and Elections*, May 2007, 54.

15. Hal Malchow, "Ten Principles of Effective Political Mail," *Campaigns and Elections*, May 2007, 64–65.

16. Personal interview with candidate by Betty O'Shaughnessy, March 7, 2007.

17. Tom Rivers, "Commentary: Memo to Candidates: Press Flesh," *Batavia (N.Y.) Daily News*, November 17, 2005, 4A.

18. Absentee and Early Voting, National Council of State Legislatures, October 9, 2008, www.ncsl.org.
19. Ian Urbina, "Casting Ballot from Abroad Is No Sure Bet," *New York Times*, June 13, 2007, www.nytimes.com.
20. Chris Taylor, "How Dean Is Winning the Web," *Time*, July 14, 2003.
21. Thomas Collins, "Web, Software Give Candidates Voter Lowdown," *Palm Beach (Fla.) Post*, February 27, 2004, Local Section, 1C.
22. O. K. Carter, "Smartest Candidates Know Where to Spend," *Fort Worth Star-Telegram*, May 2, 2004, Advance Metro Section, 1B.
23. Lorraine Swanson, "Candidates Enter the Blogosphere," *Edgewater News-Star*, January 31, 2006, 1.
24. "Poll: Bloggers, Citizen Reporters to Play Key Role in Journalism's Future," *Editor and Publisher*, February 15, 2007, www.editorandpublisher.com.
25. Chad Dotson, "The Newest Member of the Staff," *Campaigns and Elections*, February 2007, 55.
26. Amy Schatz, "Candidates Find a New Stump in the Blogosphere," *Wall Street Journal Online*, February 14, 2007, http://online.wsj.com.
27. Waldo Jaquith, "They've Got Your Back," *Campaigns and Elections*, February 2007, 54.
28. Richard Auxier and Alex Tyson, "Uploading Democracy: Candidates Field YouTube Questions," Pew Research Center, July 24, 2007, http://pewresearch.org.
29. The information in this section was provided by Mac Hansbrough, NTS, Inc., www.ntsdc.com.
30. Press release by David Orr, Cook County Clerk, "Incumbency Big Winner in Tuesday's Election," April 6, 2005, http://voterinfonet.com.

Chapter 8

1. Frank B. Feigert and M. Margaret Conway, *Parties and Politics in America* (Boston: Allyn & Bacon, 1976); see table 2.2, Electoral Party Systems in the United States, 28–29.
2. The story was originally told by Alan Rosenthal in his book *The Third House: Lobbyists and Lobbying in the States* (Washington, DC: CQ Press, 1993); reprinted in Christopher Mooney and Barbara Van Dyke-Brown, *Lobbying Illinois* (Springfield: University of Illinois at Springfield, 2003), 3.
3. "President's Report," State University Annuitants Association, *UIC Chapter Newsletter*, July 2000, 1.
4. The first definition comes from Mooney and Van Dyke-Brown, *Lobbying Illinois*, 1; the second comes from Thomas Dye and Susan MacManus, *Politics in States and Communities*, 11th ed. (New York: Prentice Hall, 2003), 125.
5. Clive S. Thomas and Robert Hrebenar, "Interest Groups in the Fifty States," *Comparative State Politics* 20, no. 4 (1999): 7, quoted in Dye and MacManus, 123; Anthony Nownes and Patricia Freeman, "Interest Group Activity in the States, *Journal of Politics* 60 (February 1998): 92, quoted in Dye and MacManus, 125.
6. For state data, see *The Book of the States* (Lexington, KY: Council of State Governments, various years); for cities and counties, see the city or county clerk Web site.
7. Rosenthal, *The Third House*, 17.
8. See Rosenthal, *The Third House*, 18–19; and Clive S. Thomas and Ronald J. Hrebenar, "Interest Groups in the States," in *Politics in the American States: A Comparative Analysis*, 7th ed., ed. Virginia Gray, Russell L. Hanson, and Herbert Jacob (Washington, DC: CQ Press, 1999), 127–28.
9. Tom Abate, "Silicon Valley Leaders to Lobby Lawmakers," *San Francisco Chronicle*, April 22, 2005, C3.
10. Elizabeth A. O'Shaughnessy, *State Adjustments to Recent Congressional Lawmaking*, doctoral dissertation, University of Illinois at Chicago, 2006, 97–99; Beverly Cigler, "Not Just Another Special Interest: Intergovernmental Representation," in *Interest Group Politics*, 4th ed., ed. Allen J. Cigler and Burdett A. Loomis (Washington, DC: CQ Press, 1995), 134.

11. Jerry Briscoe and Charles G. Bell, "California Lobbyists—Preliminary Report" Institute of Governmental Affairs, University of California, Davis, cited in Rosenthal, *The Third House*, 26; Jeffrey H. Birnbaum, "Women, Minorities Make Up New Generation of Lobbyists," *Washington Post*, May 1, 2006, D1.

12. Thomas and Hrebenar, "Interest Groups in the Fifty States," 7.

13. Ray Long, "Top GOP Official's Fees Get Scrutiny; Pension Board to Quiz Investment Firm," *Chicago Tribune*, August 9, 2005, 1 and 4.

14. Marc Perrusquia, "Stung—John Ford, Other Lawmakers Charged with Taking Bribes in FBI's Tennessee Waltz," *Commercial Appeal* (Memphis, Tenn.), May 27, 2005, A1.

15. Anthony J. Nownes and Patricia Freeman, "Interest Group Activity in the States," *Journal of Politics* 60 (February 1998): 92.

16. Mooney and Van Dyke-Brown, *Lobbying Illinois*, 29–30.

17. Mooney and Van Dyke-Brown, *Lobbying Illinois*, 35.

18. The American League of Lobbyists, "Code of Ethics," as cited in Mooney and Van Dyke-Brown, *Lobbying Illinois*, 114.

19. American Society of Association Executives, "Standards of Conduct," as cited in Mooney and Van Dyke-Brown, *Lobbying Illinois*, 116.

20. Michael Cooper, "Lobbying Times Three: Gaining from Gridlock: Lobbyists Making Big Money in Albany's Chronic Logjams," *New York Times*, March 15, 2005, www.nytimes.

21. For more information on these investigations, see Bob Woodward and Carl Bernstein, *All the President's Men* (New York: Simon and Schuster, 1974), and *The Final Days* (New York: Simon and Schuster, 1976).

22. Interview with Harris Meyer by Dick Simpson, Chicago, May 28, 2005.

23. Edward Smith, "Disappearing Act: A Declining Statehouse Press Corps Leaves Readers Less Informed about Lawmakers' Efforts," *State Legislatures*, May 2008, 28.

24. Mark J. Magyar, *New Jersey Reporter*, January 2004, 7–9; reprinted in Thad L. Beyer, *State and Local Government, 2004–2005* (Washington, DC: CQ Press, 2004), 79.

25. Christopher Swope, "Blogrolling," *Governing*, July 2005, 22.

26. Christopher Swope, "Instant Influence: A New Generation of Web Scribes Is Shaking Up State Capitol Politics," *Governing*, July 2005, 20–25.

27. See "Prologue: Florence Scala," in Studs Terkel, *Division Street: America* (New York: Random House, 1967), 1–10.

28. Greg Lucas, "Beale Community Stands Down," *San Francisco Chronicle*, May 14, 2005, B2.

29. Barbara Rose, "Local 880: Labor's New Up-and-Comer: Low-Income Workers Organize, Get Respect," *Chicago Tribune*, June 5, 2005, 1 and 12.

30. Janice Fine, "Building Community Unions," *Nation*, January 1, 2001, 18–22.

31. Rosenthal, *The Third House*, 225–227.

32. Martin Espinoza, "Political Protest Makes History," *Press Democrat*, May 2, 2006, www.pressdemocrat.com; Peter Prengaman, "Minuteman Kick Off 12-city Tour for Immigration Reform," *Houston Chronicle*, May 3, 2006, www.houstonchronicle.com.

33. Alinsky's recommendations for community organizing can be found in his book *Rules for Radicals* (New York: Random House, 1969). An analysis of the Alinsky methods can be found in Dick Simpson and George Beam, *Strategies for Change* (Chicago: Swallow, 1976) and *Political Action* (Chicago: Swallow, 1984).

Chapter 9

1. The figures in this paragraph are from *Governing* magazine's *Sourcebook 2005* (Washington, DC: CQ Press, 2005).

2. This chapter draws extensively on "Modest Tax Effort from a Wealthy State," in Samuel K. Gove and James D. Nowlan, *Illinois Politics and Government: The Expanding Metropolitan Frontier* (University of Nebraska Press, 1995).

3. U.S. Bureau of the Census, Governments Division, "State and Local Government Finances: 2004–2005," and other years, www.census.gov; Kathleen O'Leary Morgan and Scott Morgan, eds., *State Rankings 2008: A Statistical View of America* (Washington, DC: CQ Press 2008), 298.

4. Department of Health and Human Services, Federal Financial Participation in State Assistance Expenditures, FY 2008, http://aspe.hhs.gov.

5. Kaiser Commission on Medicaid and the Uninsured, "Medicaid Facts: The Medicaid Program at a Glance," #2735-02, Henry J. Kaiser Foundation, March 2007, www.kff.org.

6. *Illinois State Budget Fiscal Year 2008*, Office of the Governor, Springfield, Illinois, 2007.

7. Peter Nardulli, "Symposium: Prison Crowding: The Misalignment of Penal Responsibilities and State Prison Crises: Costs, Consequences, and Corrective Actions," *University of Illinois Law Review* 365 (1984).

8. "Tight State Budget Hurting Real People," *Peoria Journal-Star*, April 10, 2005, A4.

9. *Sourcebook 2008*, www.sourcebook.governing.com.

10. Morgan and Morgan, *State Rankings 2008*, 311.

11. All figures are from Morgan and Morgan, *State Rankings 2008*.

12. See comScore Networks, "comScore Forecasts Total E-Commerce Spending by Consumers Will Reach Approximately $170 Billion in 2006," August 2, 2006, www.comscore.com.

13. "Public Officials of the Year," *Governing*, November 2003, www.governing.com.

14. Richard Wolf, "States Consider Tobacco Tax Hikes," *USA Today*, 1.

15. Ohio Office of Management and Budget, www.obm.ohio.gov.

16. Morgan and Morgan, *State Rankings 2008*, 330.

17. State of Indiana budget forecast, www.in.gov.

18. Jack R. Van Der Slik, "Legalized Gamblers: Predatory Policy," *Illinois Issues*, March 1990, 30.

19. For a delightful, highly instructive essay on budgeting, see Robert Mandeville, "It's the Same Old Song," *Illinois State Budget, Fiscal Year 1991*, Bureau of the Budget, Springfield, March 1990.

20. Of its own funds, the City of Chicago spends less than 7 percent of its operating budget on health, education, welfare, and the arts combined.

21. Mandeville, "It's the Same Old Song."

22. This section draws on Craig Bazzani, "The Executive Budget Process," in *Inside State Government*, ed. James Nowlan (Urbana, IL: Institute of Government and Public Affairs, 1982).

23. Morgan and Morgan, *State Rankings 2008*, 319–20.

24. Press release, "Nearly $43 Million Diverted from General Revenue Fund in FY89," Office of the Comptroller, Springfield, Illinois, May 24, 1990.

25. Charles Mahtesian, "The Stadium Trap," *Governing*, May 1998, 22–26.

26. See Elizabeth O'Shaughnessy, "State Adjustments to Recent Congressional Lawmaking: Adapting to Stifling Federalism," doctoral dissertation, University of Illinois at Chicago, 2006, 172–93, for a comprehensive discussion of the politics of the national government's role in both the Individuals with Disabilities Education Act and the No Child Left Behind Act.

27. Jack McHugh and Steve Stanek, " 'Shift-and-Shaft' Tax Proposal Ekes Out Win in Michigan," *Budget and Tax News*, November 1, 2004.

28. Thomas J. Anton, *The Politics of State Expenditure in Illinois* (Urbana: University of Illinois Press, 1966).

29. Press release, Illinois State Board of Education, Springfield, January 19, 1989.

Chapter 10

1. Robin Hambelton and Jill Gross, eds., *Governing in a Global Era: Urban Innovation, Competition, and Democratic Reform* (New York: Palgrave, 2007).

2. Hambelton and Gross, *Governing in a Global Era.*

3. Bruce Katz and Robert Lang, eds. *Redefining Urban and Suburban America: Evidence from Census 2000*, vol. 1 (Washington, DC: Brookings Institution, 2003), 1–11.

4. Beatriz Ponce de León, *A Shared Future: Economic Engagement of Greater Chicago and Its Mexican Community* (Chicago: Chicago Commission on Global Affairs, 2006).

5. Pierre Devise, *Chicago's Widening Color Gap* (Chicago: Interuniversity Social Research Committee, 1967).

6. Alan Ehrenhalt, "The Morphing Megalopolis," *Governing*, September 2001, 11–12.

7. Dennis Judd and Todd Swanson, *City Politics* (New York: Longman, 2002), 332–33; Peter Dreier, John Mollenkopf, and Todd Swanstrom, "Regionalisms Old and New," in *The Urban Politics Reader*, ed. Elizabeth A. Strom and John H. Mollenkopf (London: Routledge, 2007), 303–13.

8. E. Terrence Jones and Elaine Hays, "Metropolitan Citizens in St. Louis," in *Inside Urban Politics*, ed. Dick Simpson (New York: Longman, 2003), 286–92.

9. Peter Dreier, John Mollenkopf, and Todd Swanstrom, *Place Matters: Metropolitics for the Twenty-first Century* (Lawrence: University Press of Kansas, 2001), 306–7.

10. See Myron Orfield, *Metropolitics: A Regional Agenda for Community and Stability* (Washington, DC: Brookings Institution, 1997), and *American Metropolitics* (Washington, DC: Brookings Institution, 2002).

11. Josh Goodman, "The Tax Grab Game," *Governing*, April 2007, www.governing.com.

12. Goodman, "The Tax Grab Game."

13. "The Challenge of Governance in Houston," *Houston Chronicle*, in Simpson, *Inside Urban Politics*, 280–85.

14. J. Edwin Benton, "County Governments: 'Forgotten' Subjects in Local Government Courses?" *Journal of Political Science Education* 3, no. 2 (2007): 112.

15. David B. Magleby, David M. O'Brien, Paul G. Light, J. W. Peltason, and Thomas E. Cronin, *State and Local Politics: Government by the People* (Upper Saddle River, NJ: Prentice Hall, 2008), 154–55.

16. David Y. Miller, "Exploring the Structure of Regional Governance in the United States, in *Urban and Regional Policies for Metropolitan Livability*, ed. David K. Hamilton and Patricia S. Atkinds (Armonk, NY: M. E. Sharpe, 2008), 14.

17. Fred Siegel, "Is Regional Government the Answer?" in *The Politics of Urban America: A Reader*, 3rd ed., ed. Dennis Judd and Paul Kantor (New York: Longman, 2002), 404–73.

18. See Evan McKenzie, *Privatopia: Homeowner Associations and the Rise of Residential Private Government* (New Haven, CT: Yale University Press, 1994).

19. McKenzie, *Privatopia*; Pamela Dittmer McKuen, "Ombudsman, Mediator, Arbitrator, Educator: Does Illinois Need Someone to Deal with Condo Disputes?" *Chicago Tribune*, June 17, 2007, www.chicagotribune.com.

20. Miller, "Exploring the Structure of Regional Governance in the United States," 13.

21. Kenneth K. Wong, "Does Mayoral Control Improve Performance in Urban Districts?" in *When Mayors Take Charge: School Governance in the City*, ed. Joseph P. Viteritti (Washington, DC: Brookings Institution, 2009), 79.

22. Rob Gurwitt, "Battered School Board," *Governing*, May 2006, 38–45.

23. Mark Muro et al., "Reconnecting Massachusetts Gateway Cities: Lessons Learned and an Agenda for Renewal," Joint project of MassInc and the Metropolitan Policy Program, the Brookings Institute, February 2007, http://media.brookings.edu.

24. Neal Peirce, "What's the Survival Formula for Our Second-Tier Cities?" Washington Post Writers Group, March 4, 2007, npeirce@citistates.com.

25. "Chicago Opening Development Office in Shanghai, China," Newsbriefs, *Chicago Flame*, University of Illinois at Chicago, February 19, 2007, 3.

26. Hambleton and Gross, *Governing Cities in a Global Era.*

27. Hank Savitch and Paul Kantor, *Cities in the International Marketplace: The Political Economy of Urban Development in North America and Western Europe* (Princeton, NJ: Princeton University Press, 2002).

28. David Ranney, *Global Decisions, Local Collisions: Urban Life in the New World Order* (Philadelphia: Temple University Press, 2003), 11.

29. Paul Peterson, *City Limits* (Chicago: University of Chicago Press, 1981).

30. Executive Order: Establishment of the White House Office of Urban Affairs, February 19, 2009, www.whitehouse.gov.

31. Alan Greenblatt, "Obama and the Cities: The President Has a Whole New Notion of Urban Policy," *Governing*, April 2009, 27.

32. Mark Muro and Jennifer Bradley, "ARRA on the Ground," Citiwire.net, April 2, 2009.

33. Christy Levy, "Biden: Successful Cities Key to Economic Recovery," *UIC News*, April 29, 2009, 1 and 9.

34. "Thirty-one States Join to Create Climate Registry," May 9, 2007, www.greenbiz.com.

35. Neal Peirce, "Global Warming: A State and City Rescue," *Washington Post*, May 27, 2007, www.washingtonpost.com. The quotes are from the article.

36. "Global Views 2004: Comparing Mexican and American Opinion and Foreign Policy," Chicago Council on Foreign Relations, www.ccfr.org.

37. "Perceptions and Misconceptions: How We See Each Other in Mexico-U.S. Relations," Woodrow Wilson International Center for Scholars, Mexican Institute, February 27, 2004, www.wilsoninstitute.org.

38. Jack Downey, "Canada's Culture: Definition of the Indefinable," www.canadianculture.com.

39. Steven Kull and Doug Miller. "Global Public Opinion on the US Presidential Election and US Foreign Policy," Principal Investigator Poll Organization, www.pipa.org.

40. Wayne A. Cornelius, Todd A. Eisenstadt, and Jane Hindley, eds., *Subnational Politics and Democratization in Mexico* (San Diego: University of California, 1999), 489.

41. Merilee S. Grindle, "Mexico," in *Introduction to Comparative Politics: Political Challenges and Changing Agendas*, ed. Mark Kesselman, Joel Krieger, and William A. Joseph (Boston: Houghton-Mifflin, 2000), 406.

42. Ginger Thompson, "Corruption Hampers Mexican Police in Border Drug War," *New York Times*, July 5, 2005, www.nytimes.com.

43. Cornelius, Eisenstadt, and Hindley, *Subnational Politics and Democratization in Mexico*.

44. "Transparency International Corruption Perceptions Index 2004," www.transparency.org.

45. Enrique Krauze, "You and Me," speech to the Woodrow Wilson International Center for Scholars, Mexican Institute, February 27, 2004, www.wilsoninstitute.org.

46. "Between Here and There," *Economist*, July 7, 2001.

47. Harold Kaplan, *Urban Political Systems: A Functional Analysis of Metro Toronto* (New York: Columbia University Press, 1967).

48. Government of Quebec, www.stat.gouv.qc.ca.

49. Scott Foster, "Study Finds Surgery Waiting Periods for Ontario Cancer Patients Have 'Increased Substantially' over the Last Decade." Canadian Medical Associations, August 2, 2005, www.cma.com.

Chapter 11

1. These figures are from the National Education Association and the U.S. Department of Education as reported in *Sourcebook 2005*, a publication of *Governing* magazine (Washington, DC: CQ Press, 2005).

2. U.S. Census Bureau, 2002; U.S. Department of Commerce, Bureau of Economic Analysis, 2003; and the U.S. Department of Education, National Center for Education Statistics, 2003.

3. From the Organization for Economic Cooperation and Development, 2003, as reported in Dennis Patrick Leyden, *Adequacy, Accountability, and the Future of Public Education Funding* New York: Springer, 2005), chap. 1.

4. Kern Alexander and Richard G. Salmon, *Public School Finance* (Boston: Allyn and Bacon, 1995), 7–8.

5. David Nasaw, *Schooled to Order: A Social History of Public Schooling in the United States* (New York: Oxford University Press, 1979), 30.

6. Higher Education Resource Hub, "Land Grant Act: Histories and Institutions," June 18, 2007, www.higher-ed.org.

7. Alexander and Salmon, *Public School Finance*, 10.

8. Nasaw, *Schooled to Order*, 109, 120.

9. Leyden, *Adequacy, Accountability, and the Future of Public Education Funding*, 4.

10. Figures are from U.S. government sources as reported in Leyden, *Adequacy, Accountability, and the Future of Public Education Funding*, 6; and National Center for Education Statistics as reported in Kathleen O'Leary Morgan and Scott Morgan, eds., *State Rankings 2006* (Washington, DC: CQ Press, 2006).

11. *The Book of the States 2006* (Lexington, KY: Council of State Governments, 2006).

12. Diane Ravitch, ed., *Debating the Future of American Education* (Washington, DC: Brookings Institution, 1995), 3.

13. Frederick Wirt and Michael Kirst, *Schools in Conflict: The Politics of Education*, 2nd ed. (Berkeley, CA: McCutchan, 1989).

14. David T. Conley, *Who Governs Our Schools: Changing Roles and Responsibilities* (New York: Teachers College Press, 2003), 7.

15. Conley, *Who Governs Our Schools*, 37.

16. *Education State Rankings, 2005–2006* (Lawrence, KS: Morgan Quitno, 2005).

17. Conley, *Who Governs Our Schools*, 13, 57.

18. For more extensive discussion of the Kentucky reforms, see Samuel Mitchell, *Tidal Waves of School Reform* (Westport, CT: Praeger, 1996), chap. 4.

19. Karen W. Arenson, "Educators React to Shift in Leadership at Gates Foundation," *New York Times*, November 4, 2006.

20. Arenson, "Educators React to Shift in Leadership."

21. National Education Association, www.nea.org; American Federation of Teachers, www.aft.org.

22. "Kansas School Spending below Average, but Taxpayers Pay More," Associated Press, March 18, 2005, Lexis-Nexis.

23. National Education Association, www.nea.org.

24. Michael Martinez, "School Plan Sparks L.A. Tiff," *Chicago Tribune*, October 23, 2006.

25. Alan Greenblatt, "The Impatience of Paul Vallas," *Governing*, September 2005.

26. Susan Snyder, "Vallas in with Roar, out with Rancor," *Philadelphia Inquirer*, June 17, 2007.

27. Thomas Payzant, "A for Improvement," *Governing*, November 2005, www.governing.com.

28. Greenblatt, "The Impatience of Paul Vallas."

29. Conley, *Who Governs Our Schools*, 3–4.

30. See Elizabeth O'Shaughnessy, "State Adjustments to Recent Congressional Lawmaking: Adapting to Stifling Federalism," doctoral dissertation, University of Illinois at Chicago, 2006, for a comprehensive discussion of the politics of the national government's role in the Individuals with Disabilities Education Act and the No Child Left Behind Act.

31. O'Shaughnessy, "State Adjustments to Recent Congressional Lawmaking," 179; New American Foundation, "Individuals with Disabilities Education Act: Funding Distribution," 2009, Federal Education Budget Project, febp.newamerica.net.

32. Frederick M. Hess, "No Child Left Behind: Trends and Issues," *The Book of the States 2006* (Lexington, KY: Council of State Governments, 2006), 474.

33. Paul L. Kimmelman, *Implementing NCLB: Creating a Knowledge Framework to Support School Improvement* (Thousand Oaks, CA: Corwin, 2006), 22.

34. Clint Bolick, "Four Million Children Left Behind," *Wall Street Journal*, September 7, 2006.

35. "Fiscal Year 2010 Budget Request Fact Sheet," U.S. Department of Education; Education Policy Program, "Summary and Analysis of President Obama's Education Budget Request, Fiscal Year 2010," New America Foundation, www.newamerica.net; Alyson Klein, "Obama's Choices Scrutinized, Programs' Funding Is Debated in Light of Stimulus Package," *Education Week*, May 18, 2009, www.edweek.org.

36. Alexander and Salmon, *Public School Finance*, 97–99.

37. Alexander and Salmon, *Public School Finance*, 153.

38. Alan Ehrenhalt, "Schools+Taxes+Politics=Chaos," *Governing*, January 1999, 30.

39. *Brown v. The Board of Education of Topeka, Kansas*, 347 U.S. 483 (1954); *Plessy v. Ferguson*, 163 U.S. 537 (1896).

40. Arthur Wise, *Rich Schools, Poor Schools: The Promise of Equal Educational Opportunity* (Chicago: University of Chicago Press, 1968), as discussed in Leyden, *Adequacy, Accountability, and the Future of Public Education Funding*, 124.

41. *Serrano v. Priest*, 5 Cal. 3d 584 (1971) (Serrano I).

42. *Rodriguez v. San Antonio Independent School District*, 411 U.S. 1 (1973).

43. Leyden, *Adequacy, Accountability, and the Future of Public Education Funding*, chap. 5.

44. Alexander and Salmon, *Public School Finance*, 107.

45. Leyden, *Adequacy, Accountability, and the Future of Public Education Funding*, 155ff.

46. Leyden, *Adequacy, Accountability, and the Future of Public Education Funding*, 139.

47. As reported in Alexander and Salmon, *Public School Finance*, 349.

48. Leyden, *Adequacy, Accountability, and the Future of Public Education Funding*, 139–41.

49. Paul Tough, "What It Will Really Take to Close the Education Gap," *New York Times Magazine*, November 26, 2006, 44–77.

50. *A Brief Profile of America's Private Schools*, National Center for Education Statistics, U.S. Department of Education, NCES 2003-417, 2003.

51. Hubert Morken and Jo Renee Formicola, *The Politics of School Choice* (Lanham, MD: Rowman and Littlefield, 1999), 2.

52. Morken and Formicola, *The Politics of School Choice*; see also John E. Chubb and Terry M. Moe, *Politics, Markets, and America's Schools* (Washington, DC: Brookings Institution, 1990); Reason Foundation, www.reason.org; Heritage Foundation, www.heritage.org; Separation of School and State Alliance, www.schoolandstate.org.

53. Wisconsin Department of Public Instruction, "Milwaukee Parental Choice Program: Facts and Figures for 2008–2009, www.dpi.wi.gov; Alan J. Borsuk, "Vouchers to Pass 4100 Million Mark," *Milwaukee Journal-Sentinel*, November 21, 2006.

54. U.S. Department of Education, National Center for Education Statistics, "1.1 Million Home-schooled Students in the United States in 2003," nces.ed.gov; "No School, and the Child Chooses What to Learn," *New York Times*, November 26, 2006.

55. "No School."

56. Data from *Digest of Education Statistics 2006*, National Center for Education Statistics, Washington, DC; in *State Rankings 2008*, 152.

57. *State Rankings 2008*, 146, 150, 151; "Fiscal Year 2010 Budget Request Fact Sheet," U.S. Department of Education; Education Policy Program, "Summary and Analysis of President Obama's Education Budget Request, Fiscal Year 2010," New America Foundation.

58. *The Control of the Campus: A Report on the Governance of Higher Education*, Carnegie Foundation for the Advancement of Teaching, Washington, DC, 1982, 80, 71.

59. Richard Hofstadter and Wilson Smith, eds., *American Higher Education: A Documentary History*, vol. 1 (Chicago: University of Chicago Press, 1961), 11.

60. Lyman A. Glenny and Thomas K. Dalglish, *Public Universities, State Agencies, and the Law* (Berkeley: Center for Research and Development in Higher Education, University of California, Berkeley, 1973), 17–18.

61. *Control of the Campus*, 10–11.
62. *Control of the Campus*, 17.
63. Florida Department of Education, "Florida's Community Colleges," www.fldoe.org; State University System of Florida, www.flbog.org; Richard C. Richardson, Kathy Reeves Bracco, Patrick Callan, and Joni E. Finney, *Designing State Higher Education Systems for a New Century* (Washington, DC: American Council on Education and Oryx Press, 1999), chap. 7.
64. *Designing State Higher Education Systems*, 92.
65. California Postsecondary Education Commission, www.cpec.ca.gov; *Designing State Higher Education Systems*, 50.
66. *Bakke v. Regents of University of California*, 438 U.S. 265 (1978).
67. Philip G. Altbach, Robert O. Berdahl, and Patricia J. Gumport, eds., *American Higher Education in the Twenty-First Century: Social, Political, and Economic Challenges*, 2nd ed. (Baltimore, MD: Johns Hopkins University Press, 2005), 531.
68. *American Higher Education*, 531; *Gratz v. Bollinger*, 123 S.Ct. 2411 (2003); *Grutter v. Bollinger*, 123 S.Ct. 2325 (2003).
69. *American Higher Education*, 198–99.

Chapter 12

1. Mark Hugo Lopez and Paul Taylor, "Dissecting the 2008 Electorate, Most Diverse in U.S. History," Pew Research Center, April 30, 2009, www.pewresearch.org.
2. Thomas J. Gradel, Dick Simpson, and Andris Zimelis with Kirsten Byers, David Michelberger, Chris Olson, and Nirav Sanghani, "The Depth of Corruption in Illinois: Anti-Corruption Report Number 2," May 13, 2009, www.uic.edu.

GLOSSARY

A

Absentee voting Voting before Election Day, generally by mailing an absentee ballot. In some states, a voter must provide an acceptable reason for voting absentee.

Accreditation The approval from a national or regional agency that an institution is providing a professionally acceptable level of education.

Acculturated Socialized into another country's norms, culture, and ideals.

American Federation of Teachers (AFT) The second largest teachers union in the United States, with over 1.3 million members.

Anglophones People whose primary language is English.

Antifederalists The 18th-century faction opposed to ratification of the U.S. Constitution.

Appellate court A court in which plaintiffs and defendants appeal the outcome from a trial; if legal or technical errors occurred in the trial, the appellate court may order that the case be reheard in trial court with a new jury.

Appropriation The amount that the legislature has authorized the government to spend on a specific budget category.

Arraignment A pre-trial appearance where accused persons are brought before a judge to hear the charges filed against them and to file a plea of either guilty, not guilty, or no contest.

Attorney general The primary legal advisor and chief law enforcement officer of a state.

B

Balanced budget A budget in which revenues equal expenditures.

Bicameral legislature A legislative body composed of two chambers, as found in most states.

Biennial Continuing or lasting for two years. A two-year period is a biennium.

Block grant A federal grant given to a state for a general purpose, such as community development.

Blog A Web log; an online journal or diary.

Broker Legislator who makes deals among varied interests and across party lines on controversial issues, getting them enacted into laws.

Budget Plan for spending financial resources, which include tax revenues, fees, lottery profits, interest income, federal transfers to the states, borrowed money, and monies from other sources.

Burden of proof The burden of the prosecutor to prove the defendant's guilt beyond a reasonable doubt.

Bureaucracy The government agencies and offices and their staffs that administer government services and implement policy.

C

Categorical grant A federal grant given to a state for a specific purpose, such as school lunch programs.

Caucus A group of legislators of the same political party or faction, or an informal, closed meeting of the group.

Charter school A public school that has legal authority to provide a different type of education within a public school district, often without the restrictions under which the regular schools operate.

Chief executive The highest elected or appointed official of a state, county, city, or separate local unit of government such as the local school system.

Chief judge A judge who manages the court building and employees, assigns cases to fellow judges, and handles administrative matters.

Chief of staff The principal assistant to the governor who generally oversees policy formulation and implementation, negotiates for the governor, and manages the day-to-day operations of the governor's office.

City charter A document that provides the framework for the laws of a given town or city.

Civic group An organized local group whose members promote the public interest.

Civility Civil conduct: acting with politeness, thoughtfulness, and courtesy.

Clear statement rule Supreme Court ruling that holds that a federal law cannot be interpreted to interfere with state law unless Congress made a plain statement in its legislation to that effect.

Clearance statistics The number of crimes committed compared with the number of crimes that resulted in the arrest of the responsible parties.

Commission form of government Local government in which three to five elected commissioners serve as executives and as the legislative body of city.

Committee-dominated legislature A legislature in which the committees do the major work on legislation, and committee recommendations are guides for voting by individual legislators of both parties.

Common-interest development (CID) A private community whose land and common facilities are owned by the members of the housing association and governed by a separate condominium or homeowners association board of directors.

Community organization A group formed by the people of a community to promote the community's general well-being or provide a social service.

Commuter tax A tax levied on the people who work in a city or other taxing jurisdiction but do not reside there.

Contract cases Civil cases in which one party sues another for violating an agreement or making an agreement under false pretenses.

Contract with America A 1994 plan promoted by Speaker of the House Newt Gingrich to eliminate many entitlement programs and transfer governmental powers to the states.

Cooperative federalism A system of government in which powers and duties are shared between state, local, and national governments.

Correctional population All individuals under some form of court supervision, including probation, prison, jail, and parole.

Council-manager form of government Form of city government in which the mayor and city council appoint a professional city manager, who reports to them.

Court of last resort The highest court to which a case can be appealed.

D

Delegate A legislator or other elected official who will vote and act based on the opinions of the majority of constituents.

Descriptive representation theory The theory that constituents are best represented by legislators who are similar to them in characteristics such as race, ethnicity, and socioeconomic level.

Deterrence Any crime-fighting method that seeks to prevent crime, such as by increasing the potential costs of committing a crime or the probability of getting caught.

Devolution Returning responsibility for many functions of government to the states.

Dillon's rule All local governments are creature of the state and have only the powers explicitly granted by the state government.

Distributive services Ordinary city services such as police protection, parks, garbage pickup, and street repair which can be distributed to all citizens.

District attorney Local public official who represents the government in the prosecution of accused criminals.

Dual federalism A system of government in which the federal and state governments are supreme in their own separate spheres of government.

E

Early voting Voting before Election Day in person at specified locations.

Earmark A provision in legislation that specifies certain spending priorities to a limited number of projects, programs, or grants or funding to be set aside for a special purpose or recipient.

Enumerated powers Those powers specifically granted to the federal government by the Constitution.

Expert Legislator who becomes a respected authority to whom other lawmakers turn for cues in voting.

Extradition The procedure by which one government delivers a fugitive to the legal jurisdiction of another government.

Exurbs Developed areas beyond a central city and its suburbs.

F

Federal preemption A federal law that has priority over an existing state law or regulation.

Federalism The constitutional relationship between the states and the federal government.

Federalists The 18th-century faction supporting ratification of the U.S. Constitution.

Fiscal federalism The theory that the costs and performance of particular government tasks are best found at certain levels of government.

Floor leader The member of the legislature who determines the agenda and sometimes the procedural rules.

Foundation level The minimum amount of combined local and state financial support that will be provided each pupil in all school districts in a state.

Francophones People whose primary language is French.

Free media Media coverage not paid for by the campaign, such as news reports of a press conferences or other actions or words of the candidate.

Full faith and credit clause Article IV, section 1, of the Constitution, which says that states are to recognize each other's public acts, records, and judicial proceedings.

G

Gated community A private community with limited access points and security guards to protect the property.

General fund The fund that supports general state government operations and major functions such as education.

Gerrymandering Drawing district lines to benefit a particular political party or ethnic, racial, or other group.

Globalization The growth of connections such as international trade, immigration, and communications throughout the world.

Grassroots lobbying Lobbying that arises (or appears to arise) from the concerns of many individuals rather than being undertaken by representatives of an organization.

H

Home schooling Schooling that is carried out in the home, usually by a parent, rather than in an established school.

Horizontal federalism The relationships between and among states.

I

Incarceration rate The number of people being detained in jail or prison in comparison to the population.

Initiative A process in which voters petition to place a proposed law or constitutional amendment on the ballot for a direct vote by the electorate.

Initiative process Direct democratic action by voters through a petition to place a measure on the ballot for the next election.

Interest group An organized group that tries to influence policy affecting the group's particular area of interest.

Intergovernmental lobby A coalition of state or local government officials or professional associations of government officials who lobby in Washington.

Intergovernmental relations Within the United States, the relationships between and among the federal, state, and local governments.

J

Jury nullification A jury decision acknowledging the accused has violated a law but declaring the defendant "not guilty," sometimes to protest a law that seems unfair.

L

Lawmaker Legislator who enjoys putting proposals into the best shape possible and advocating their adoption.

Legal staff Executive office staff who provide advice on the extent of the governor's statutory and constitutional authority and the legal implications of individual decisions.

Legislative staff Executive office staff who work with legislative leaders and individual lawmakers, pushing the chief executive's agenda, counting heads, and handling lawmaker requests for executive support of their legislation.

Line-item veto The power to strike individual items from a budget appropriation without negating the entire budget.

Lobbying An organized effort to influence public policy on behalf of a particular interest area.

Local control The concept that school affairs should be governed primarily by the local school district rather than by state and federal government laws and regulations.

Local property tax A tax assessed on residential, commercial, industrial, and farm property according to the value of the property.

M

Magnet school A specialized public school that focuses on a particular field in education, such as the arts, science, or mathematics.

Majority vote rule A candidate must receive at least 50 percent of the vote to win the election.

Mayor-council form of government Form of city government in which an elected mayor serves as the chief executive, and an elected city council serves as the legislature.

Medicaid The federal program that supports health care for low-income people and nursing home care for low-income and even some middle-income elderly; the federal government shares the cost with the states.

Metropolitan region The large area of suburbs, smaller towns, and counties around a central city.

Mordida A Mexican colloquial term for bribe, from the word meaning "bite."

N

National Education Association (NEA) The largest teachers union in the United States, with 3.2 million members.

New Deal The federal programs of President Franklin Roosevelt put in place during the 1930s to overcome the effects of the Great Depression.

Nixon's New Federalism A plan to make government more efficient by shifting power and authority to state and local officials.

Nonpartisan election An election in which candidates run for office without any party label.

Norm A standard developed to guide the behavior of members of a group.

O

Office of Management and Budget (OMB) The government office responsible for developing the budget for the chief executive or sometimes for the legislature.

Ombudsman An official who is designated to look into individual concerns or complaints and seek a solution.

P

Paid advertising Advertisements paid for by the campaign in one or more media, including press, radio, television ads, and direct mail.

Paid lobbyist A full-time professional lobbyist hired by an organization to influence government legislation, regulations, or decisions.

Participatory democracy or pure democracy A system of government in which citizens make the governmental decisions directly in assembly together.

Party-dominated legislature A legislature in which votes are determined by party affiliation.

Patronage The ability of a chief executive to appoint supporters to government jobs based on their campaign work and party loyalty.

Peer review Review by a panel of professionals in the same field.

Personnel office Executive office that manages and makes recommendations to the chief of staff and governor on staff hiring and appointments to independent state boards and commissions.

Plea bargain An agreement in which the prosecutor reduces the charge or recommends a minimum or suspended sentence if the accused pleads guilty.

Plurality vote rule The candidate receiving the most votes of all candidates running wins the election.

Pocket veto The power to veto legislation by not signing it within a specific time.

Political machine A political organization, usually a political party, characterized by patronage, precinct work, favoritism, and party loyalty.

Pork barrel legislation Legislation that specifically benefits a member of Congress's district or state by providing funding for jobs or projects.

Precinct The smallest election unit in each state, comprising about 500 registered voters.

President of the senate The official leader of the state senate, with the power to appoint committees and guide legislation through the legislative process as determined by the specific rules of the body.

Press office Executive office that responds to inquiries from the press corps, issues press releases about accomplishments, writes speeches, and organizes press conferences.

Privileges and immunities clause The clause in Article IV, section 2, of the Constitution that states: "the citizens of each state shall be entitled to all privileges and immunities of citizens in the several states."

Progressive movement A late 19th- and early 20th-century movement that sought to end corruption in corporations and government by regulating businesses and weakening party machines by giving citizens more direct participation in the electoral process and government.

Progressive tax A tax set at a higher percentage for higher-level incomes than for lower-level incomes.

Proportional tax A tax set at the same percentage for all income levels.

Public defender Attorney funded by taxpayer dollars to people without financial means so they may properly defend themselves in legal proceedings.

R

Rainy day fund Monies set aside in good times for use during weak economic times when revenues might not be adequate to meet demands for services.

Reagan's New Federalism A plan to reduce the role of the federal government, especially through shifting responsibility for many domestic programs to the states.

Recall A process in which voters petition to hold a special election that will decide whether to remove a public official from office.

Recidivism The tendency to commit another crime and be returned to prison.

Reciprocity A return of a favor or a vote in kind.

Redistributive services Services like public housing or welfare payments that wealthier taxpayers fund and poorer taxpayers receive, which provide a limited redistribution of wealth.

Reduction veto A veto that reduces a particular appropriation.

Referendum A process in which voters petition to place a referendum question on the ballot; a referendum is advisory to the government, not binding law.

Regressive tax A tax that extracts a higher percentage of the income of a poor person than of a wealthy person.

Representative democracy A system of government in which citizens elect government officials who make the actual governmental decisions.

Retention election Election held to decide whether a sitting judge will remain on the bench.

Revenue sharing The distribution of federal moneys to state and local governments.

Ritualist Legislator who enjoys the status of holding office but does not work hard to develop legislation.

Roll call vote Each legislator votes when called on by name.

S

Scheduler Executive assistant who allocates the governor's time for meetings with staff, legislators, interest groups, and community delegations, as well as private and family time.

School choice An alternative to conventional public schooling that lets parents choose among the schools in a district, including magnet and charter schools.

School code That part of a state's statutes devoted to education.

School mandate A law or regulation that local school districts must follow.

School voucher A grant of money to a family to assist in paying tuition for a child to attend a private school.

Separate funds or dedicated funds Funds dedicated to specific purposes and often supported by special fees.

Sin tax A tax levied on goods or services that many consider vices, such as tobacco, alcoholic beverages, or gambling.

Social movement Activism by people who share a common concern about a social issue and come together in often ad-hoc organizations to pressure the government or private corporations for change.

Sovereign Immunity States cannot be sued by citizens of another state.

Speaker of the house of representatives The official leader of the state house of representatives, with the power to appoint committees and guide legislation through the legislative process as determined by the specific rules of the body.

Special committees A legislative committee appointed for a short period of time to study or address a particular issue.

Standing committee A permanent legislative committee that writes laws or statutes for a specific issue area, such as transportation or education.

State budget bureau Executive office that prepares the annual or biennial budget for the governor and monitors spending by the state agencies; also called the office of management and budget.

State constitution A document that provides the framework for the laws of a given state.

Strong mayor government Mayor-council government in which the mayor serves as the sole executive, prepares the budget, and has veto power over the city council.

T

Target number The number of votes needed to win a particular election.

Tenure A status granted after a trial period that gives protection from arbitrary dismissal.

Term limits Laws that restrict how many times an elected official can run for reelection.

Tort A legal case in which an individual, a business or organization, or the government is accused of committing a civil wrong by violating others' bodily, property, or legal rights, or for breaching a duty owed under statutory law.

Transnational corporation A firm whose facilities and employees span more than one nation.

Trial court The court where all cases are initially heard or otherwise settled.

Trustee A legislator or other elected official who will vote and act based on personal judgment about what is in the best interests of constituents.

Truth in sentencing A policy requiring that convicted criminals serve a certain percentage of their sentence before they are eligible for release.

Tyranny Oppressive governmental power in the hands of one, a few, or many who use their power to deny the civil rights of citizens or thwart the public good.

U

Unfunded mandate Any federal legislation that would require state and local governments to spend their own funds to carry out nationally mandated programs.

Unicameral legislature A legislative body composed of one chamber, as found in Nebraska.

Unschooling Unstructured schooling, usually home based, in which children are allowed to explore their worlds unencumbered by the usual teacher-student relationship.

Urban sprawl Unplanned, uncontrolled development of land at the boundary of an urban area.

Urbanization The growth of cities and urban areas.

V

Veto power The power of a chief executive to negate the legislature's enactment of a law. A veto may be overridden by a special majority vote of the legislature, such as two-thirds or three-fifths of each legislative branch.

Voice vote Legislators vote by saying "Aye" as a group or "Nay" as a group when asked whether they approve or disapprove of the proposed legislation.

W

Weak mayor government Mayor-council government in which the mayor shares executive power with other government officials and lacks veto power.

CREDITS

Chapter 1

Chapter 2

Chapter 3

Chapter 4

Chapter 5

Chapter 6

INDEX